Springer Series in Information Sciences 22

Editor: M. R. Schroeder

Springer
Berlin
Heidelberg
New York
Barcelona
Hong Kong
London
Milan
Paris
Singapore
Tokyo

D1422242

Springer Series in Information Sciences

Editors: Thomas S. Huang   Teuvo Kohonen   Manfred R. Schroeder
Managing Editor: H. K. V. Lotsch

Volumes 1–29 are listed at the end of the book.

E. Zwicker   H. Fastl

Psychoacoustics

Facts and Models

Second Updated Edition
With 289 Figures

 Springer

Professor Dr.-Ing. Eberhard Zwicker †

Institut für Elektroakustik,
Technische Universität München

Professor Dr.-Ing. Hugo Fastl

am Lehrstuhl für Mensch-Maschine-Kommunikation,
Technische Universität München,
Arcisstrasse 21, D-80333 München, Germany
e-mail: fastl@mmk.ei.tum.de

Series Editors:

Professor Thomas S. Huang

Department of Electrical Engineering and Coordinated Science Laboratory,
University of Illinois, Urbana, IL 61801, USA

Professor Teuvo Kohonen

Helsinki University of Technology, Neural Networks Research Centre, Rakentajanaukio 2 C,
FIN-02150 Espoo, Finland

Professor Dr. Manfred R. Schroeder

Drittes Physikalisches Institut, Universität Göttingen, Bürgerstrasse 42-44,
D-37073 Göttingen, Germany

Managing Editor:

Dr.-Ing. Helmut K. V. Lotsch

Springer-Verlag, Tiergartenstrasse 17,
D-69121 Heidelberg, Germany

ISSN 0720-678X
ISBN 3-540-65063-6 2nd Edition Springer-Verlag Berlin Heidelberg New York
ISBN 3-540-52600-5 1st Edition Springer-Verlag Berlin Heidelberg New York

Library of Congress Cataloging-in-Publication Data. Zwicker, Eberhard. Psychoacoustics: facts and models /
E. Zwicker, H. Fastl. – 2nd ed. p. cm. – (Springer series in information sciences; 22) Includes bibliographical
references and index. ISBN 3-540-65063-6 (softcover; alk. paper) 1. Psychoacoustics. I. Fastl, H. (Hugo),
1944– . II. Title. III. Series. QP461.Z92 1998 612.8'5–dc21 98-46050

Springer- Verlag is a company in the BertelsmannSpringer publishing group.
© Springer- Verlag Berlin Heidelberg 1990, 1999
Printed in Germany

The use of general descriptive names, registered names, trademarks, etc. in the publication does not imply, even
in the absence of a specific statement, the such names are exempt from the relevant protective laws and
regulations and therefore free for general use.

Typesetting: Data conversion by Steingraeber Satztechnik GmbH, Heidelberg
Computer to plate: Mercedes Druck GmbH, Berlin
Cover design: *design & production* GmbH, Heidelberg
SPIN: 10771514   56/3111/mf - 5 4 3 2 1 - Printed on acid-free paper

Preface to the Second Edition

Shortly after the appearance of the first edition of this book, the scientific community was shocked by the unexpected and untimely death of the great psychoacoustician Professor Eberhard Zwicker. The present second edition of Psychoacoustics – Facts and Models is meant as a tribute to my mentor Eberhard Zwicker, who was both an outstanding scientist and a dedicated teacher.

Therefore, the basic concept of the book has remained untouched. However, new results and references have been added in most chapters, in particular in Chap. 5 on pitch and pitch strength, Chap. 10 on fluctuation strength, Chap. 11 on roughness, and in Chap. 16 concerning examples of practical applications. In addition, occasional typographical errors have been corrected and some older material re-arranged. In essence, however, care was taken to keep the style of the original work.

The encouragement as well as the helpful and patient cooperation of Springer Verlag, especially of Dr. Helmut Lotsch, is gratefully acknowledged. My thanks go to the many students and co-workers who assisted in the preparation of the second edition, in particular Dipl.-Ing. Wolfgang Schmid and Dipl.-Ing. Thomas Filippou.

Munich, January 1999   *H. Fastl*

Preface to the First Edition

Acoustical communication is one of the fundamental prerequisites for the existence of human society. In this respect the characteristics of our receiver for acoustical signals, i.e. of the human hearing system, play a dominant role. The ability of our hearing system to receive information is determined not only by the qualitative relation between sound and impression, but also by the quantitative relation between acoustical stimuli and hearing sensations. With the advent of new digital audio techniques, the science of the hearing system as a receiver of acoustical information, i.e. the science of psychoacoustics, has gained additional importance. The features of the human hearing system will have to be taken into account in planning and realizing future acoustical communication systems in economically feasible projects: Each technical improvement in this area will be judged by listening and relating the result of listening to the cost.

In the years from 1952 to 1967, the research group on hearing phenomena at the Institute of Telecommunications in Stuttgart made important contributions to the quantitative correlation of acoustical stimuli and hearing sensations, i.e. to psychoacoustics. Since 1967, research groups at the Institute of Electroacoustics in Munich have continued to make progress in this field. The correlation between acoustical stimuli and hearing sensations is investigated both by acquiring sets of experimental data and by models which simulate the measured facts in an understandable way. This book summarizes the results of the above-mentioned research groups in two ways. First, the content of many papers originally written in German is made available in English. Second, the known psychoacoustical facts and the data produced from models are united to give an integrated picture and a deeper understanding. The references are confined to papers published by the two research groups mentioned, although there are naturally many more relevant papers in the literature.

The book is aimed primarily at research scientists, development engineers, and research students in the fields of psychoacoustics, audiology, auditory physiology, biophysics, audio engineering, musical acoustics, noise control, acoustical engineering, ENT medicine, communication and speech science. It may also be useful for advanced undergraduates in these disciplines. A special feature of the book is that it combines psychoacoustical facts, descriptive models, and applications presented in the form of examples with hints for the solution of readers' problems.

The first three chapters give an introduction to the stimuli and procedures used in the experiments, to the basic facts of hearing, and to information processing in the auditory system. The important role played by the active processing within the inner ear is stressed in order to understand frequency selectivity and nonlinear behaviour of our hearing system. The next four chapters deal with frequency resolution and temporal resolution expressed in masking, pitch, critical bands and excitation, as well as just-noticeable changes in the sound parameters. The different kinds of pitch are described in Chap. 5, and the following six chapters deal with the basic sensations of loudness, sharpness, fluctuation strength, roughness, subjective duration, and rhythm. The next two chapters concern the ear's own nonlinear distortion and binaural hearing, with emphasis given to the topics that have been covered by the two research groups. The last chapter provides examples of applications, which will be of special interest to those engaged in finding practical solutions.

For didactical reasons, the text is not interrupted by the inclusion of references. However, at the end of the volume, the relevant literature published by the Stuttgart and Munich groups is cited, as is the literature dealing with the various applications given in the final chapter. The equations appearing in the book are given as "magnitude equations", containing not only symbols but also the units in which the variables are to be expressed. This should help to avoid mistakes since one can check the units of the calculated quantity.

Some of the figures contain more information than is needed for the immediate discussion. This is simply a device to save space and the additional information is invariably discussed at a later point in the text.

We would like to acknowledge the helpful and patient cooperation of Springer-Verlag. We thank the many individuals who contributed to the realization of this book, notably, Mrs. Angelika Kabierske for drawing the figures, Mrs. Barbi Ertel for typing the text, Dr. Frances Harris, Dr.-Ing. Tilmann Zwicker, and Dipl.-Ing. Gerhard Krump for reading drafts, and Dr. Bruce Henning for many very fruitful discussions and suggestions.

Munich, June 1990   *E. Zwicker*
                                                                     H. Fastl

Contents

1. Stimuli and Procedures

In this chapter, some fundamental correlations between the temporal and the spectral characteristics of sounds are briefly reviewed. The transformations of electric signals into sound by loudspeakers and headphones are described. Moreover, some psychophysical methods and procedures are mentioned. Finally, the relationship between stimuli and hearing sensations in general and the processing of raw data in psychoacoustics are discussed.

1.1 Temporal and Spectral Characteristics of Sound

Some temporal and spectral characteristics of sounds frequently used in psychoacoustics are outlined in Fig. 1.1. Sounds are easily described by means of the time-varying sound pressure, $p(t)$. Compared to the magnitude of the atmospheric pressure, the temporal variations in sound pressure, caused by sound sources are extremely small. The unit of sound pressure is the PASCAL (Pa). In psychoacoustics, values of the sound pressure between 10^{-5} Pa (absolute threshold) and 10^2 Pa (threshold of pain) are relevant. In order to cope with such a broad range of sound pressures, the sound pressure level, L, is normally used. Sound pressure and sound pressure level are related by the equation

$$L = 20 \log (p/p_0) \, \text{dB} \, . \tag{1.1}$$

The reference value of the sound pressure p_0 is standardized to $p_0 = 20\mu$ Pa.

Besides sound pressure and sound pressure level, the sound intensity, I, and sound intensity level are also relevant in psychoacoustics. In plane travelling waves, sound pressure level and sound intensity level are related by the equation

$$L = 20 \log (p/p_0) \, \text{dB} = 10 \log (I/I_0) \, \text{dB} \, . \tag{1.2}$$

The reference value I_0 is defined as $10^{-12} \, \text{W/m}^2$.

In particular, when dealing with noises, it is advantageous to use, instead of sound intensity directly, its density, i.e., the sound intensity within a bandwidth of 1 Hz. The expression "noise power density" – although not quite correct – is also used. The logarithmic correlate of the density of sound intensity is called sound intensity density level, usually shortened to density

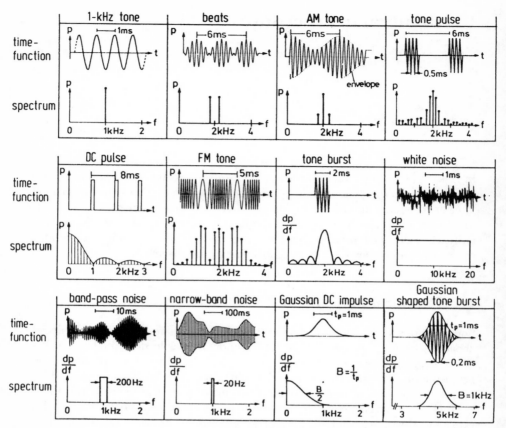

Fig. 1.1. Time functions and associated frequency spectra of stimuli commonly used in psychoacoustics

level, l. For white noise, which shows a density level independent of frequency, l and L are related by the equation

$$L = [l + 10 \log (\Delta f/\text{Hz})] \text{ dB} , \qquad (1.3)$$

where Δf represents the bandwidth of the sound in question measured in Hertz (Hz).

The panel "1-kHz tone" in Fig. 1.1 shows that a continuous sinusoidal oscillation of the time function of sound pressure p, with a temporal distance of 1 ms between the maxima, corresponds to a spectrum with just one spectral line at 1 kHz.

The panel "beats" is most easily explained in the spectral domain, where a combination of two pure tones of the same amplitude is displayed. The corresponding time function clearly shows a strong variation of the temporal envelope.

The panel "AM tone" depicts both the time function and the spectrum of a sinusoidally amplitude-modulated 2-kHz tone. The time function shows the sinusoidal oscillation, the envelope of which varies with the modulation frequency. The corresponding spectrum illustrates that an AM tone is described by three lines. The level differences, ΔL, between the centre line at 2 kHz on the one hand and either the lower or the upper side line on the other, are related to the degree of modulation, m, by the equation

$$\Delta L = 20 \log (m/2) \, \text{dB} \; . \tag{1.4}$$

The period of the envelope fluctuation shown, 6 ms, corresponds to a modulation frequency of 167 Hz, which in the spectral domain represents the frequency difference between the centre line, called the carrier, and the upper and lower side lines, respectively.

The panel "tone pulse" shows both the time function and the spectrum of a pure tone that is rectangularly gated at regular intervals. The tone frequency is 2000 Hz and the gating interval is 6 ms. In the spectral domain, the spacing between the lines corresponds to the gating frequency of 167 Hz.

The panel "DC pulse" shows a similar situation. In this case however, a DC-voltage rather than a pure tone is gated at regular intervals. The duration of the DC pulses is 1 ms, the spacing of the DC pulses 8 ms. The corresponding spectrum shows lines with a separation of the reciprocal of 8 ms, i.e. 125 Hz. At frequencies corresponding to 1/1 ms, 2/1 ms, 3/1 ms and so on, the amplitude of the spectral lines shows distinct minima.

The last of the given examples which produces discrete or line spectra is an "FM tone". The frequency of a tone at 2 kHz is sinusoidally frequency-modulated between 1 and 3 kHz with a modulation frequency of 200 Hz. The related amplitude spectrum is symmetrical with respect to 2 kHz, and follows in its envelope a Bessel-function. If the modulation index, i.e. the ratio between frequency deviation and modulation frequency, is small then most lines of the Bessel spectrum disappear, and the resulting spectrum is similar to the spectrum of an AM tone with one centre line and two side lines. However, compared to the AM tone, the side lines of the FM tone with small modulation index are shifted in phase by 90 degrees.

The panel "tone burst" in Fig. 1.1 is the first example of a series of sounds which produce continuous rather than line spectra. The function illustrates a single 2-kHz tone burst of 2-ms duration. The corresponding spectrum shows a maximum at 2 kHz and minima with spacings of 500 Hz. Thus the spectrum of the single tone burst is comparable to the spectra of tone pulses or DC pulses. Although tone pulses and DC pulses produce line spectra, a single tone burst produces a continuous spectrum.

White noise is an important example of a sound producing a continuous spectrum. In psychoacoustics, for practical reasons, the bandwidth of the white noise is normally limited to 20 Hz–20 kHz. As can be seen in the panel "white noise" in Fig. 1.1, the spectral density is independent of frequency in the whole range of 20 kHz. It should be mentioned that this holds for the long term spectrum, whereas the short term spectrum of white noise may show some frequency dependence. The time function of white noise exhibits a Gaussian distribution of amplitudes.

If the bandwidth of white noise is restricted by a filter, we get band-pass noise. The panel "band-pass noise" in Fig. 1.1 shows a typical example of the time function for a band-pass noise at 1 kHz with 200-Hz bandwidth, Δf. The time function displayed is a single occurrence, which does not recur periodically. As with white noise, the rule for band-pass noise is that at a specific instant, the amplitude can only be given with a certain probability; the probability function shows a Gaussian distribution. The speed of the envelope fluctuation is limited by the bandwidth of the filter. To a first approximation, the time function of the band-pass noise shown can be considered as a 1-kHz tone, which is randomly amplitude (and phase) modulated. On average, the number n of the envelope maxima per second can be approximated by the formula

$$n = 0.64 \, \Delta f \, . \tag{1.5}$$

Therefore, the "effective" modulation frequency, f^*_{mod}, of a bandpass noise with bandwidth Δf can be approximated by the formula

$$f^*_{\mathrm{mod}} = 0.64 \, \Delta f \, . \tag{1.6}$$

In the case of a band-pass noise with 200-Hz bandwidth, this means that envelope maxima should occur with an average distance of about 8 ms. The time function in the panel "band-pass noise" in Fig. 1.1 indicates that this approximation is valid.

The panel "narrow-band noise" shows the same features as discussed for band-pass noise. However, in this case, the bandwidth is only 20 Hz, the envelope fluctuates very slowly, and the temporal distance of envelope maxima increases on average to about 80 ms. The variation of the time function indicates that the narrow-band noise can be considered in a first approximation as a pure tone at 1 kHz, which is randomly amplitude modulated.

The panel "Gaussian-DC impulse" shows both the time function and spectrum of a DC impulse with a Gaussian-shaped envelope. The Gaussian shape represents an optimum trade between the speed of variation in the temporal envelope and the bandwidth of the associated spectrum, in that the product of the bandwidth and duration is a minimum for the Gaussian shape. In the example, the time $t_{\mathrm{p}} = 1 \, \mathrm{ms}$ is chosen in such a way that a rectangularly shaped time function with the same maximum sound pressure encompasses the same area under the curve, as does the Gaussian-DC impulse. In this case, the duration t_{p} is measured just below half the maximum value of the sound pressure, at exactly 0.456 times the maximum sound pressure. The corresponding bandwidth in the spectral domain amounts to about 500 Hz in this example.

The panel "Gaussian-shaped tone burst" shows the time function and the spectrum of a type of gated tones, which is preferred in psychoacoustics

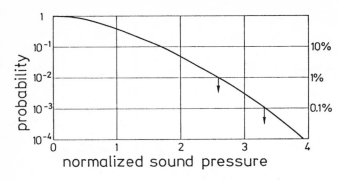

Fig. 1.2. Probability with which the sound pressure of Gaussian noise exceeds a given sound pressure, normalized with respect to its RMS value

because of the relatively steep slope of its temporal envelope, and, at the same time, its relatively narrow spectral distribution. The example given in Fig. 1.1 shows the situation for a single Gaussian-shaped tone impulse. If the impulse is repeated at a rate of 1 Hz, then the spectral envelope remains unchanged, however a line spectrum with a line spacing of 1 Hz occurs.

As mentioned above, for noise signals their maximum amplitude cannot be given, because in Gaussian noise the amplitude varies according to a Gaussian distribution. This means that only the probability can be indicated for which the sound pressure exceeds a given value. In Fig. 1.2, this probability is given as a function of the actual sound pressure, normalized with respect to its long-term root mean square (RMS) value. The probability that the actual sound pressure lies above the RMS value decreases with the ratio of the actual sound pressure to RMS value. If clipping of a noise signal can be tolerated 1% of the time, this means that a sound pressure with an amplitude 2.6 times that of the RMS value has to be transmitted undistorted. For psychoacoustic experiments, a more strict limit is necessary, because peak clipping can be tolerated only 0.1% of the time. Therefore, a sound pressure that exceeds the RMS value by a factor of 3.4 has to be transmitted without distortion. For practical purposes, this means that with noise signals the reading on the level meter has to be reduced by 10 dB, in comparison with the reading for pure tones, in order to avoid severe distortion of the noise signals.

1.2 Presentation of Sounds by Loudspeakers and Earphones

In psychoacoustic experiments, the transformation of electric oscillations into sound waves is usually achieved by loudspeakers or earphones. In both cases, the frequency response and the nonlinear distortions produced by the transducer are of great importance. Figure 1.3 shows the frequency response of a cabinet which contains three loudspeakers: electrodynamic speakers for low and midfrequencies and a piezo-electric-horn for high frequencies. The frequency response (given as L_1 over f) of this assembly when measured in the

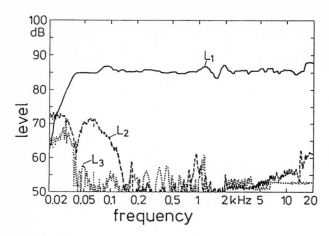

Fig. 1.3. Frequency response L_1 of a loudspeaker cabinet in an anechoic chamber together with the responses of quadratic (L_2) and cubic (L_3) distortion products, which have been shifted upwards by 20 dB

anechoic chamber is flat, within ± 2 dB in a frequency range between 35 Hz and 16 kHz. The frequency response of the distortion products L_2 (at $2f$) for quadratic distortions and L_3 (at $3f$) for cubic distortions is also given in Fig. 1.3, however with the zero level shifted up by 20 dB. For psychoacoustic applications, distortion factors of only 0.1% or less can be permitted, corresponding to a level difference of 60 dB. Taking into account the average level L_1 of 85 dB and the shifted zero level for L_2 and L_3, this would mean that the level of the corresponding distortion components should not exceed 45 dB on the scale used. The results plotted in Fig. 1.3 clearly show that in the whole frequency range, a distortion factor as low as 0.1% is rarely reached. However, in a frequency range above about 150 Hz, the distortion factor amounts on average to about 0.3%, which is a relatively good figure for loudspeaker representation.

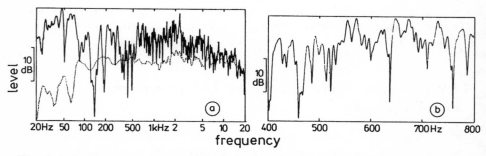

Fig. 1.4a,b. Frequency response (**a**) of a loudspeaker in a normal living room (*solid*) and in an anechoic chamber (*dotted*). Panel (**b**) shows, on an enlarged frequency scale, the response in a living room

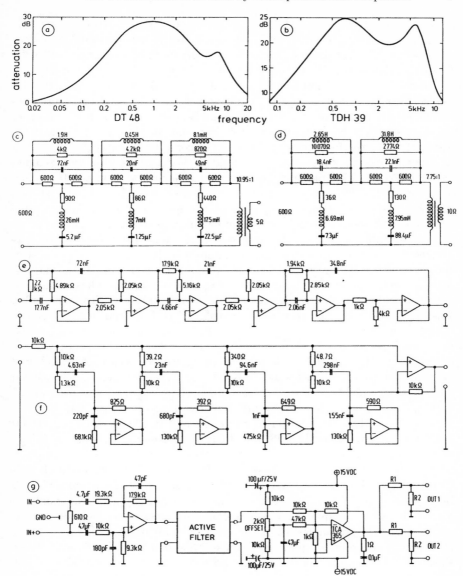

Fig. 1.5a–g. Frequency dependence of the attenuation of free field equalizers developed for the earphones DT 48 (**a**) and TDH 39 (**b**). The corresponding passive LCR-networks are indicated in (**c**) and (**d**), active circuits are given in (**e**) and (**f**), respectively. The network shown in (**g**) determines that 1 V at the input corresponds to 80 dB SPL also for the active equalizer

If sounds are represented via a loudspeaker that is not in an anechoic chamber but in a "normal" room, such as a living room, additional complications occur. The frequency characteristic of the room is superimposed on the frequency characteristic of the loudspeaker. An example is given in Fig. 1.4. The dotted line in the left panel represents the frequency characteristic of a loudspeaker measured in an anechoic chamber, and the solid line represents the frequency characteristic of the same loudspeaker in a living room. From the data shown in Fig. 1.4a, it becomes clear that the resonances of the room distinctly alter the frequency response of the combination. Figure 1.4b shows part of the frequency response of the loudspeaker plus room on an expanded frequency scale. This plot reveals very sharp, narrow dips in the frequency response. If the frequency of a pure tone varies only slightly near such a dip, then the small frequency variation is transformed into a large amplitude variation, and this leads to clearly audible loudness differences.

These problems can mostly be overcome if sounds are presented via earphones. One advantage is that the earphones usually used in psychoacoustics show very little nonlinear distortion (less than 0.1% or $-60\,$dB) in the frequency range of interest. The frequency response of earphones has to be measured on real ears, because current couplers can produce misleading results. Therefore, the frequency response of earphones is measured in the anechoic chamber by subjectively performed loudness comparisons of tones, presented via a loudspeaker or via earphones. The exact details of this procedure are described in DIN 45619 T.1. Because the frequency responses of headphones often used in psychoacoustics show a band-pass characteristic when measured on real ears, equalizers have been developed. The combination of earphone and equalizer gives a free field equivalent frequency response, which is flat within $\pm 2\,$dB. The attenuation characteristics of free field equalizers developed for the earphones DT 48 and TDH 39 are shown in Fig. 1.5. These attenuation characteristics also illustrate the free field equivalent frequency response of the respective earphones (DT 48 in a, TDH 39 in b). In addition, circuit diagrams are given for the realization of the equalizers with both passive and active components. For $1\,$V at the input of the equalizer, the combination of equalizer plus earphone produces a free field equivalent sound pressure level of $80\,$dB. If the earphones are used without equalizer then it must be kept in mind that they act like bandpass filters and change the sounds presented. This means that both the tone colour and the loudness are dramatically affected, particularly with broad-band sounds.

1.3 Methods and Procedures

In the following section, several methods that are frequently used in psychoacoustics will be discussed. The main differences between the methods are that they are designed for different types of psychoacoustical tasks, and that it takes different amounts of time to arrive at a relevant result.

Method of Adjustment. In this method, the subject has control over the stimulus. For example, the subject can vary the level of a pure tone until it is just audible. In another experiment, the subject might vary the frequency of a tone until its pitch is equal to the pitch of a reference tone, or, in another case, until its pitch forms the musical interval of an octave with respect to the pitch of the reference tone.

Method of Tracking. In the method of tracking, the subject also controls the stimulus; however, in contrast to the method of adjustment, the subject controls only the direction in which a stimulus varies. In the measurements of absolute thresholds, for example, the subject increases and decreases the level of a pure tone in such a way that the situations "tone audible" and "tone inaudible" follow each other in sequence. If the excursions of the level are plotted as a function of frequency, this method is called Békésy tracking. The mean value of the zigzag curves is taken as an indication of the value in question. Although traditionally this averaging procedure is performed by eye, the method of tracking can also be implemented by a computer, which stores the reversals and automatically calculates the appropriate average.

Magnitude Estimation. In this method, stimuli are assigned numbers corresponding to the perceived magnitude in some dimension. For example, a sequence of stimuli can be assigned numbers corresponding to their perceived loudness. Form the ratio of the numbers, the ratio of loudness can be deduced. In addition it is sometimes useful to present a standard, which is called the anchor sound. In this case, pairs of stimuli are presented, and the first stimulus of each pair is kept constant. This standard, or anchor, is assigned a numerical value, say 100, which might represent its loudness. Relative to this value, the loudness of the second sound has to be scaled. If, for example, the second sound is three times as loud as the first sound, then the subject has to respond with the number 300. Instead of magnitude estimation, magnitude production can also be used. In this case, the subject is given a ratio of numbers, and has to adjust a second stimulus in such a way that the ratio of psychoacoustical magnitudes (e.g. loudness) corresponds to the ratio of numbers given by the experimenter.

All the psychophysical methods discussed so far have the common feature that a final value of threshold or of ratio can be deduced from a single trial. In the first two methods described, the subject is actively involved in the task by controlling the stimulus. Sometimes such an activity may produce a bias, as for example in loudness comparisons. In such cases, the average of the results of the two measurements, one with varying sound "A" and another with varying sound "B" leads to the value of interest.

In the following methods, the value of interest is usually deduced from the responses of the subject via psychometric functions.

Yes–No Procedure. In this method, the subject has to decide whether a signal was present or not. There is only one interval in which either the signal occurs or doesn't occur. This means that this procedure is a "one interval-two alternative forced-choice procedure", because the subject is not allowed to answer "I don't know whether or not a signal was present", but has to decide "yes" or "no".

Two-Interval Forced Choice. In this procedure, the subject is presented with two intervals, and has to decide whether the signal occurs in the first or second interval. Sometimes three or four intervals may be used and the task of the subject is to decide in which interval the sound is different with respect to some quality, say loudness or pitch. With these procedures, feedback is frequently given. This means that after each trial the subject is informed of the right answer, usually by a light indicating the interval that contained the signal.

Adaptive Procedures. Whereas in classic forced-choice procedures the stimuli to be presented are chosen by the experimenter, in adaptive procedures the stimulus presented in a trial depends on the answers given by the subject in preceeding trials. These procedures are also called "up-down" procedures. If, for instance, the absolute threshold is measured in an adaptive procedure, then the sound pressure is lowered until it can be taken for granted, that the subject doesn't hear the stimulus. The sound pressure is then increased until the subject clearly hears the stimulus, and after that it is again decreased. The step size is reduced with the number of reversals. When a predetermined small step size is reached, a value can be calculated with an accuracy of half the final step size by averaging the last few reversals. This means that adaptive procedures show some similarity to the method of tracking, because they yield a final value without the explicit use of psychometric functions.

Comparison of Stimulus Pairs. If the effects of variations along different stimulus dimensions are to be evaluated, the method of comparison of stimulus pairs has to be used. In this method, a pair of stimuli AB is different along one dimension, say loudness, whereas the subsequent pair CD is different along another dimension, say pitch. The task of the subject is to decide whether the perceived difference between stimuli AB of the first pair is larger than the perceived difference between the stimuli CD of the second pair. From this type of experiment, the equality of variations along different stimulus dimensions can be deduced.

Results obtained in psychoacoustical experiments generally depend on the procedure used. As a rule the sensitivity of a subject is enhanced if a comparison among several alternatives is possible. Concerning the measurement time and efficiency of the different procedures, methods which *directly* yield an estimate, such as adjustment, tracking or magnitude estimation, are very time efficient. However, with procedures which require a psychometric function like "yes–no" and "multiple alternative forced-choice" procedures, it takes many trials, and hence much time to arrive at stable results. With respect to adaptive procedures, the time necessary to arrive at a psychoacoustically meanigful value depends significantly on the details of the algorithm implemented. A compromise has to be found between the final step size and the number of trials necessary, because higher accuracy, i.e. a smaller step size, goes with a larger number of trials.

1.4 Stimuli, Sensations, and Data Averaging

In this section, the relation between stimuli described in physical terms, and the hearing sensations elicited by these stimuli, is assessed. The step size of stimuli is compared to steps of the sensation, and the notions of thresholds, ratios, and equality of sensation are addressed. A procedure for data averaging which can cope with the nonlinear relation between a sensation scale and different transformations of one and the same stimulus scale is proposed.

The most important physical magnitude for psychoacoustics is the time function of sound pressure. The stimulus can be described by physical means in terms of sound pressure level, frequency, duration and so on. The physical magnitudes mentioned are correlated with the psychophysical magnitudes loudness, pitch, and subjective duration, which are called hearing sensations. However, it should be mentioned that the pitch of a pure tone depends not only on its frequency, but also to some extent on its level. Nonetheless, the main correlate of the hearing sensation pitch is the stimulus quantity frequency. Physical stimuli only lead to hearing sensations if their physical magnitudes lie within the range relevant for the hearing organ. For example, frequencys below 20 Hz and above about 20 kHz dot not lead to a hearing sensation whatever their stimulus magnitude. Just as we can describe a stimulus by separate physical characteristics, so we can also consider several hearing sensations separately. For instance, we can state "the tone with the higher pitch was louder than the tone with the lower pitch". This means that we can attend separately to the hearing sensation "loudness" on one hand and "pitch" on the other. A major goal of psychoacoustics is to arrive at sensation magnitudes analogous to stimulus magnitudes. For example, we can state that a 1-kHz tone with 20 mPa sound pressure produces a loudness of 4 sone in terms of hearing sensation. The unit "sone" is used for the hearing sensation loudness, in just the same way as the unit "Pa" is used for the sound pressure. It is most important not to mix up stimulus magnitudes such as "Pa" or "dB" and sensation magnitudes such as "sone".

The relationship between physical magnitudes of the stimulus and magnitudes of the correlated hearing sensations can be given either by equations or graphs. Figure 1.6 shows an example where the stimulus magnitude is plotted along the abscissa and the sensation magnitude along the ordinate. Although the correlation between stimulus and sensation is displayed by a continuous curve, it should be realized that tiny variations in stimulus magnitude (from A_0 to A_1, say) may *not* lead to a variation in sensation magnitude.

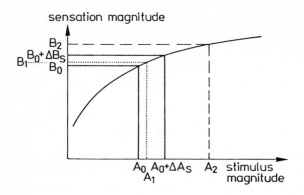

Fig. 1.6. Example of sensation magnitude versus stimulus magnitude

This is because the variation from B_0 to B_1 may be within the step size ΔB_s and only steps greater than ΔB_s produce sensations different enough to hear. Thus in Fig. 1.6, ΔB_s might represent the smallest variation in stimulus magnitude leading to a difference in sensation magnitude. If the stimulus magnitude is increased from A_0 to A_2, the variation is clearly reflected in sensation magnitude, because the corresponding change in sensation magnitude is well above the minimal step size of sensation ΔB_s. The step size of the stimulus, ΔA_s, that leads to a difference in the hearing sensation, ΔB_s, is typical for psychoacoustic tasks, which are called "difference thresholds" or just "thresholds".

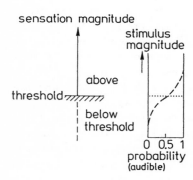

Fig. 1.7. Determination of threshold, i.e. that stimulus magnitude (or stimulus increment) for which the corresponding sensation or sensation increment is audible with 50% probability

An extreme example of a threshold is absolute threshold, i.e. the level of a pure tone necessary to be just audible. The threshold is not fixed for all time, but depends somewhat on the circumstances. Therefore, only a probability can be given that a certain stimulus level will lead to just-audible hearing sensations. This reasoning is illustrated in Fig. 1.7. Both the stimulus magnitude and the sensation magnitude increase in the vertical direction. However, below threshold the stimulus does not lead to a sensation. In the right panel of Fig. 1.7, the probability that a sensation is produced for different stimulus magnitudes is given. The conventional threshold is chosen to correspond to the probability 0.5. This means that in 50% of the trials, the "threshold" stimulus leads to a sensation, whereas in the other 50% of the trials, no sensation is produced. As a rule, it is rather easy for the subject to determine a threshold.

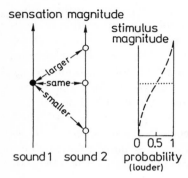

Fig. 1.8. Determination of equality, i.e. that stimulus magnitude for which the corresponding sensation sounds louder than a comparison sensation with 50% probability

A somewhat more complicated task for the subject is to assign *equality* to sounds. This task is explained in Fig. 1.8 by an example. Sound 2 is compared to sound 1 with respect to loudness. Considering the sensation magnitude, it is clear that the same loudness is only reached if the marks at sound 1 and 2 are positioned at the same height. The right panel shows the correlated stimulus magnitude of sound 2 with respect to the probability that the subject responds "sound 2 is louder". Although the distribution for the task equality is shallower than that for "threshold", subjects as a rule report no difficulty in producing points of equality.

A more complicated task for the subject is to produce ratios of sensations. In Fig. 1.9, an example for the case in which the perceived magnitude of a sensation should be halved is given. The sensation magnitudes of sound 1 and 2 are shown as vertical arrows. The sensation magnitude of sound 1 is taken

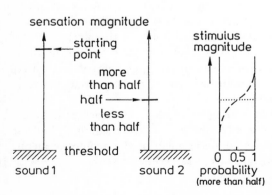

Fig. 1.9. Determination of the ratio "half" sensation, i.e. that stimulus magnitude for which, with 50% probability, the corresponding sensation sounds half (as loud, for example) as a comparison sensation

as a starting point, and relative to this value sound 2 has to be changed in
such a way that it produces half the sensation magnitude elicited by sound 1.
The right panel of Fig. 1.9 shows the stimulus magnitude for the probability
that the subject perceives sound 2 as producing more than half the sensation
magnitude produced by sound 1. Again, the probability 0.5 is defined as
representing the required ratio of one half.

Because the results of the same person in different trials (intraindividual
differences) as well as the results from different subjects (interindividual differences) can vary substantially, it is advisable to perform several runs of the
same type of experiment and to average the data. This means that after an
experiment a large number of data points are available for which an average
value has to be calculated. The choice of the unit for the averaging procedure
among the stimulus magnitude measures (i.e. among level, sound pressure, or
sound intensity) plays a crucial role. An example is given in Fig. 1.10 which
shows values of absolute thresholds for 8 subjects. In the upper part, the individual threshold values are given as dots along a scale of relative sound
intensity I/I_0. For these 8 data points the arithmetic mean, the geometric
mean, and the median are indicated by arrows.

For n data points, the arithmetic mean is calculated by

$$\frac{x_1 + x_2 + x_3 + \ldots + x_n}{n} \tag{1.7}$$

whereas the geometric mean reads

$$\sqrt[n]{x_1 \cdot x_2 \cdot x_3 \cdot \ldots \cdot x_n} \, . \tag{1.8}$$

The median just separates the data points into two equal sections, i.e. $n/2$
data points are left of the median and $n/2$ data points right of the median.
The interquartile encompasses 50% of all data points which means that 25%
of the data are outside the interquartile range on the left and 25% on the
right.

When the ratio of intensities I/I_0 is transformed into the level L, the
lower part of Fig. 1.10 is produced. For example a relative intensity of 100
corresponds to a level of 20 dB, a relative intensity of 2 to a level of 3 dB and
so forth. Because of the transformation of the scales, the arrangement of the
dots in the upper versus the lower panel of Fig. 1.10 is completely different.

In the upper part, the arithmetic mean lies between the sixth and the
seventh data point, in the lower part however between the fifth and the sixth
data point. As concerns the geometric mean, in the upper part it is situated
between the fifth and the sixth data point whereas in the lower part between
the third and the fourth data point.

The appropriate stimulus scale suitable to display psychoacoustical data
is not *a-priori* clear. Therefore it is advisable to use the median to find an
average because in contrast to the arithmetic mean or the geometric mean,
the median is resistant to transformations of the stimulus scale.

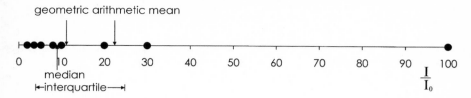

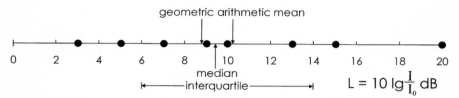

Fig. 1.10. Example of averaging 8 threshold data plotted along a scale of the linear magnitude relative intensity I/I_0 (upper panel) or along a scale of the logarithmic magnitude level L (lower panel)

Only the median and the interquartile range preserve their location relative to the data points when arranged on different scales; arithmetic and even geometric mean does not.

2. Hearing Area

This chapter addresses the hearing area and the threshold in quiet.

The hearing area is a plane in which audible sounds can be displayed. In its normal form, the hearing area is plotted with frequency on a logarithmic scale as the abscissa, and sound pressure level in dB on a linear scale as the ordinate. This means that two logarithmic scales are used because the level is related to the logarithm of sound pressure. The critical-band rate may also be used as the abscissa. This scale is more equivalent to features of our hearing system than frequency.

The usual display of the human hearing area is shown in Fig. 2.1. On the right, the ordinate scales are sound intensity in Watt per square meter (W/m^2) and sound pressure in Pascal (Pa). Sound pressure level is given

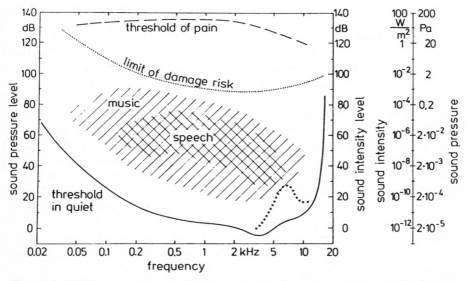

Fig. 2.1. Hearing area, i.e. area between threshold in quiet and threshold of pain. Also indicated are the areas encompassed by music and speech, and the limit of damage risk. The ordinate scale is not only expressed in sound pressure level but also in sound intensity and sound pressure. The dotted part of threshold in quiet stems from subjects who frequently listen to very loud music

for a free-field condition relative to 2×10^{-5} Pa. Sound intensity level is plotted relative to 10^{-12} W/m^2. A range of about 15 decades in intensity or 7.5 decades in sound pressure, corresponding to a range of 150 dB in sound pressure level, is encompassed by the ordinate scale. Concerning the abscissa, we must realize that our hearing organ produces sensations for pure tones within three decades in frequency ranging from 20 Hz to 20 kHz. The actual hearing area represents that range, which lies between the threshold in quiet (the limit towards low levels) and the threshold of pain (the limit towards high levels). These thresholds are given in Fig. 2.1 as solid and broken lines, respectively. These limits hold for pure tones in steady state condition, i.e. for tones lasting longer than about 100 ms.

If speech is resolved into spectral components, the region it normally occupies can also be illustrated in the hearing area. In Fig. 2.1, the range encompassed by speech sounds is indicated by the area hatched from top left to bottom right starting near 100 Hz and ending near 7 kHz. The levels indicated hold for "normal speech" as delivered for example in a small lecture hall. The components of music encompass a larger distribution in the hearing area as indicated in Fig. 2.1 by a different hatching. It starts at low frequencies near 40 Hz, and reaches about 10 kHz. Including pianissimo and fortissimo, the dynamic range of music starts at sound pressure levels below 20 dB and reaches levels in excess of 95 dB. Extreme and rare cases are ignored for the spectral distributions of music and speech displayed. It can be seen, however, that both areas are well above threshold in quiet, which is explained in more detail in Sect. 2.1.

Another high level border, very important in everyday life, is given in Fig. 2.1 as a thin dotted line – the limit of damage risk. This limit reaches quite high sound pressure levels at very low frequencies, but decreases towards levels near 90 dB in the range between 1 and 5 kHz. This limit holds for the "average person", i.e. some subjects may be more sensitive. Consequently, sound attenuation, like ear plugs, has to be offered in factories if levels about 5 dB below the level indicated by the thin dotted line are reached. This limit is valid for sounds lasting eight hours per working day and five working days per week. For shorter exposure, the sound intensity can be increased in the same way that the duration is decreased. This means that our ear may be exposed in its most sensitive frequency range to sounds with 100 dB only for about 50 minutes and to 110 dB for only about 5 minutes per day! Such exposure can easily be produced, for example by loud music played through earphones, therefore, using earphones to listen to music which produces high levels must be undertaken with care. Overexposure of our hearing system to sound initially produces temporary threshold shifts. After too many exposures this temporary threshold shift leads to a permanent shift, i.e. to a hearing loss. In this case, threshold in quiet is no longer normal but is shifted towards higher sound pressure levels and will never recover.

2.1 Threshold in Quiet

The threshold in quiet indicates as a function of frequency the sound pressure level of a pure tone that is just audible. This threshold can be measured quite easily by experienced or inexperienced subjects. The reproducibility of the threshold in quiet for a single subject is high and lies normally within ±3 dB.

The frequency dependence of the threshold in quiet can be measured precisely and quickly by Békésy-tracking. In this method, the subject uses a switch which changes the direction of the increment or decrement in sound pressure level (see Sect. 1.3). At the same time, but relatively slowly, the frequency is changed from low to high or vice versa, while the subject is changing the sound pressure level of the tone upwards and downwards via the switch according to the following rule: once the tone is definitely audible, a change of the switch reduces the level of the tone towards inaudibility. When the tone becomes definitely inaudible the switch is reversed and the level is increased towards audibility. This process continues throughout the presentation. The decrease and increase of sound pressure level is recorded as a function of time, which, because the frequency is slowly changing, also means as a function of frequency.

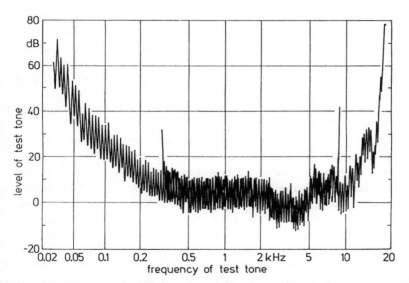

Fig. 2.2. Threshold in quiet, i.e. just-noticeable level of a test tone as a function of its frequency, registered using the method of Békésy-tracking. Note that between 0.3 and 8 kHz, the threshold is measured twice

A recording produced in this manner is shown in Fig. 2.2. A whole recording from low to high frequencies lasts about 15 minutes. The change in level must have a fine step size of less than 2 dB, otherwise clicks become audible for medium and high levels as the level increases from step to step. In order to show the reproducibility of such a tracking of threshold in quiet, two trackings, one with an upward sweep and another with a downward sweep in frequency, are indicated in Fig. 2.2 for the frequency range between 0.3 and 8 kHz. The excursions of the zigzag reach as much as 12 dB, i.e. about ±6 dB. The middle of this zigzag curve is defined as threshold in quiet. As can be seen from the registrations in Fig. 2.2, the threshold in quiet for an individual subject can be determined exactly when using this method. The frequency dependence of the threshold in quiet displayed in Fig. 2.2 is related to a certain subject, however, this frequency dependence is typical and has been recorded in a similar manner by many subjects with normal hearing.

At low frequencies, threshold in quiet requires a relatively high sound pressure level reaching about 40 dB at 50 Hz. The level at 200 Hz has already dropped to about 15 dB. For frequencies between 0.5 and 2 kHz, the threshold in quiet for the subject indicated in Fig. 2.2 remains almost independent of frequency, an effect which is relatively rare. In many cases threshold in quiet shows some excursions or small humps. In the frequency range between 2 and 5 kHz almost every subject with normal hearing exhibits a very sensitive range in which very small sound pressure levels below 0 dB are reached. For frequencies above 5 kHz, threshold in quiet shows peaks and valleys that not only vary individually but also characteristically for each subject. In many cases, threshold remains between 0 and 15 dB as long as frequency is not above 12 kHz. For even higher frequencies, threshold in quiet increases rapidly and reaches, at 16 to 18 kHz, a limit above which no sensation is produced even at high levels. This limit is dependent on the age of the subject: it is somewhere between 16 and 18 kHz at an age of 20–25 years provided that the subject has not already been exposed to sounds with levels that produce a hearing loss.

As mentioned above, each subject shows an individual frequency dependence of the threshold in quiet. From individual thresholds in quiet measured in many subjects an average threshold in quiet can be calculated. For 100 subjects, each with normal hearing, the solid curve in Fig. 2.3 indicates the median threshold of hearing. In addition to that 50% curve, data are given that encompass 10% and 90% of the subjects' individual thresholds in quiet. Thresholds below the 90% curve are usually accepted as normal. The difference between the 90% curve and the 10% curve is small for medium frequencies. Towards low and higher frequencies this difference increases. In discussing this difference, it is remarkable that the 90% curve does not follow the 50% curve in the frequency range between 3 and 8 kHz, rather it increases in the range where the 50% curve decreases. It may be that a small percentage of the young subjects already exhibit a small hearing loss in the frequency range around 4 kHz. This result points to the fact that our hearing system is most easily damaged in the frequency range between 3 and 8 kHz if it is exposed to loud sounds above the limit of damage risk. An example

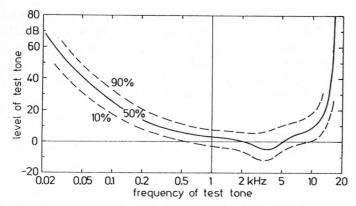

Fig. 2.3. Statistics for threshold in quiet: 50%, 90% and 10% values for threshold in quiet as a function of frequency for subjects 20 to 25 years old

of such effects is shown by the dotted line in Fig. 2.1 near threshold in quiet. It belongs to a group of students who listen frequently through earphones to loud music and indicates their median threshold in quiet in the frequency range between 3 and 12 kHz.

The sound pressure level of 0 dB and the frequency of 1 kHz are marked in Fig. 2.3 by thin lines. The difference between the crossing point of these two straight lines and the 50% curve of threshold in quiet at 1 kHz indicates that a value of 3 dB is reached at this frequency, and not 0 dB as is sometimes assumed. This latter value is reached for frequencies of about 2 and 5 kHz; between these two frequencies negative values of sound pressure level are audible for 50% of the subjects.

With increasing age hearing sensitivity is reduced, especially at high frequencies. As indicated in Fig. 2.4, threshold in quiet is shifted to a value near

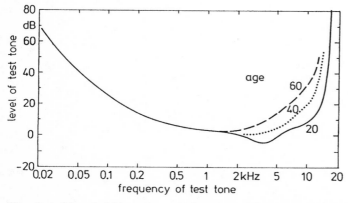

Fig. 2.4. Threshold in quiet as a function of frequency with age as a parameter

30 dB at a frequency of 10 kHz at the age of 60 years. At this age, threshold may be increased by 15 dB at 5 kHz, whereas for frequencies below 2 kHz, threshold in quiet remains almost as sensitive as for persons aged 20 years. It has to be realized, however, that this holds only for subjects who are not exposed to high noise levels during their everyday life. At the age of 40 years, the threshold shift is about half as much as that for 60 years.

3. Information Processing
in the Auditory System

In this chapter, the preprocessing of sound in the peripheral system and information processing in the neural system are addressed.

3.1 Preprocessing of Sound in the Peripheral System

Two fundamentally different regions of stimulus processing in the human auditory system can be distinguished. In the peripheral region, where the oscillations retain their original character, preprocessing occurs. In these peripheral preprocessing structures however, there are nonlinearities. The peripheral structures deliver the preprocessed oscillations to the sensory cells, which have nerve terminals that encode the mechanical/electrical stimuli into electrical action potentials. There, the second region of the hearing system begins using neural processing which finally leads to auditory sensations. This division of the hearing system into two parts can be seen in the hearing organs of all vertebrates. It should be noted that – in contrast with some other authors – we assume the first synapses to be the end of the peripheral part of the hearing system. Sometimes, especially in medicine, it is assumed that the peripheral part includes the eighth nerve.

3.1.1 Head and Outer Ear

The sound field normally assumed is that of a free, progressive, plain sound field. Any large body, such as the head of a subject, distorts this sound field. The influence of the head and the whole body of a subject in a free sound field can be measured; it is represented by the difference between the sound pressure level indicated by a small microphone in the free field (without the subject), and the sound pressure level measured in the ear canal of the subject, whose centre of the head is positioned where the microphone was located in the free field. When considering this difference, it becomes clear that the body of the subject, especially the shoulder as well as the head, outer ear and ear canal, influence the sound pressure level in front of the ear drum. Shoulders and head influence this sound pressure level most effectively at frequencies below 1500 Hz through shadowing and reflection.

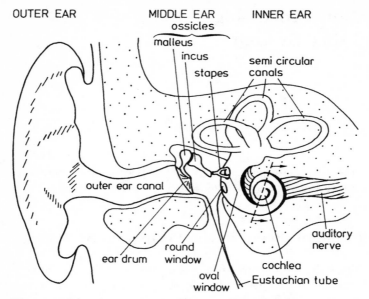

Fig. 3.1. Schematic drawing of the outer, middle and inner ear

What is generally referred to as the ear is, in fact, the outer ear shown schematically in Fig. 3.1 together with the middle and inner ear. The outer ear's function is to collect sound energy and to transmit this energy through the outer ear canal to the ear drum. The outer ear canal produces two advantages: firstly, it protects the ear drum and the middle ear from damage and secondly, it enables the inner ear to be positioned very close to the brain, thus reducing the length of the nerves and resulting in a short travel time for the action potentials in the nerve.

The outer ear canal exerts a strong influence on the frequency response of the hearing organ. It acts like an open pipe with a length of about 2 cm corresponding to a quarter of the wavelength of frequencies near 4 kHz. It is the outer ear canal that is responsible for the high sensitivity of our hearing organ in this frequency range, indicated by the dip of threshold in quiet around 4 kHz. This high sensitivity however, is also the reason for high susceptibility to damage in the region around 4 kHz.

3.1.2 Middle Ear

The sound affecting the outer ear consists of oscillations of air particles. The inner ear contains fluids that surround the sensory cells. In order to excite these cells, it is necessary to produce oscillations in the fluids. The oscillations of air particles with small forces, but large displacement, have to be transferred into motions of the salt water-like fluids with large force, but small displacements. To avoid large losses of energy through reflections, a transformation must occur in the middle ear to match the impedances of the two fluids, air outside and water inside. Impedance matching can be achieved in electrical systems with transformers. In mechanical systems, "levers" can be used, and this is precisely the task of the middle ear (see Fig. 3.1). The light but sturdy funnel-shaped tympanic membrane (eardrum) operates over a wide frequency range as a pressure receiver. It is firmly attached to the long arm of the hammer (malleus). The motions of the eardrum are transmitted to the footplate of the stirrup (stapes) by the middle ear ossicles named malleus, incus, and stapes (hammer, anvil and stirrup) which are made of very hard bone (Fig. 3.1). The stapes footplate, together with a ring-shaped membrane called the oval window, forms the entrance to the inner ear. In addition to the lever ratio of about 2 produced by the different lengths of the arms of the malleus and incus, the middle ear also produces a transformation depending on the ratio of the area of the large eardrum to that of the small footplate. This ratio is about 15. Through the lever and the area ratios, an almost perfect match between the impedances is reached in man in the middle frequency range around 1 kHz.

Normally, the middle ear space with its transforming elements is closed off from its surroundings by the eardrum on one side and the Eustachian tube on the other. However, the Eustachian tube, which is connected to the upper throat region, is opened briefly when swallowing. External influences like mountain climbing, the use of an elevator, flying, or diving can produce an extreme increase or decrease in pressure which changes the resting position of the eardrum. Consequently, the working point in the transfer characteristic of the middle ear ossicles also changes, producing a reduction of hearing sensitivity – an effect often experienced in airplanes. Normal hearing is resumed by swallowing because during a brief opening of the Eustachian tube, the air pressure in the middle ear can be equalized with that of the environment.

3.1.3 Inner Ear

The inner ear (cochlea) is shaped like a snail and is embedded in the extremely hard temporal bone (Fig. 3.2). The cochlea is filled with two different fluids and consists of three channels or scalae, which run together from the base to the apex. The footplate of the stapes is in direct contact with fluid in the scala vestibuli. Because the scala media is separated from the scala vestibuli only by the very thin and light Reissner's membrane, the two channels can be regarded, from a hydromechanical point of view, as one unit. The oscillations are transmitted to the basilar membrane through the fluids. This membrane separates the scala media from the scala tympani and supports the organ of Corti with its sensory cells. As the fluids and the surrounding bone are essentially incompressible, the fluid displayed at the oval window by the movement of the stapes must be equalized. The equalization occurs through the basilar membrane at the round window, which closes off the scala

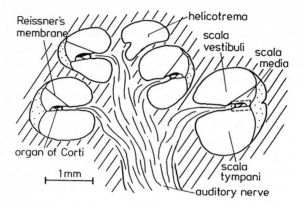

Fig. 3.2. Schematic drawing of the cross section of the inner ear

tympani at the base of the cochlea. For very low frequencies, the equalization occurs through a connection between the scalae tympani and vestibuli at the apex of the cochlea called the helicotrema.

A fluid, the perilymph, is found in the scalae vestibuli and tympani. Perilymph has a high sodium content and resembles other body fluids. It is in direct contact with the cerebrospinal fluid of the brain cavity. The fluid of the scala media, the endolymph, is in contact with the spaces of the vestibular system and has a high potassium content. Loss of the potassium ions from the scala media by diffusion is reduced by the tight membrane junctions of the cells surrounding the scala media. Any losses are rapidly replaced by an ion-exchange pump with high energy requirements found in the cell membranes of the cells of the stria vascularis, a specialized group of cells on the outer wall of the cochlea. The ion exchange in the stria vascularis generates a positive potential of 80 mV (relative to perilymph) in the scala media.

The basilar membrane separating the scala media and the scala tympani is narrow at the base but about three times wider at the apex. The cochlea forms $2\frac{1}{2}$ turns allowing a basilar membrane length of about 32 mm. The structure of the inner ear is basically the same in all mammals.

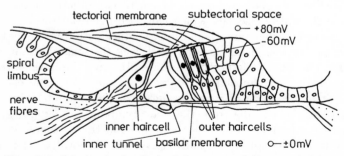

Fig. 3.3. Schematic drawing of the organ of Corti and the surrounding tissues

The function of the organ of Corti, which is located on the basilar membrane, is the transformation of the mechanical oscillations in the inner ear into a signal that can be processed by the nervous system. The organ of Corti contains various supporting cells and the very important sensory cells or hair-cells (see Fig. 3.3). The haircells are arranged in one row of inner haircells on the inner side of the organ of Corti, and three rows of outer haircells near the middle of the organ of Corti. Between the two kinds of haircells the most prominent supporting cells, the pillar cells, form the inner tunnel. The tectorial membrane covers part of the organ of Corti and is attached to the spiral limbus at the inner side of the scala media. Interestingly, the tectorial membrane contains no cells and is made up exclusively of two kinds of highly hydrated protofibrils. With a close attachment to the cells of the organ of Corti beyond the outer haircells, the tectorial membrane separates a subtectorial space from the scala media. Anatomical results demonstrate that the hairs of the inner haircells are either not attached or only weakly attached to the tectorial membrane.

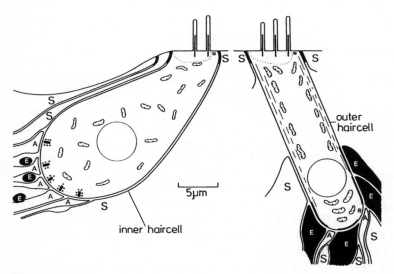

Fig. 3.4. Schematic drawing of an inner and outer haircell. Note the difference in the synaptic arrangements as outlined in the text

As can be seen in Fig. 3.4, the construction of inner and outer haircells is different. The outer haircells are thinner, pillar-shaped and (unlike the inner haircells) not tightly surrounded by supporting cells. There are obvious and regularly occurring differences in the ultrastructure of the two types of haircells. In addition, the afferent synapses of the inner haircells (going towards the brain) appear to possess the normal characteristic of chemical synapses whereas those of the outer haircells are atypical. The structural differences indicate different functions for the inner and outer haircells. In fact, more than 90% of the afferent ("A") fibres make synaptic contact with the inner haircells, with each fibre normally in contact with only one inner haircell. Each inner haircell is contacted by up to 20 afferent fibres. The rest of the afferent fibres (5% to 8%) produce a sparse innervation of the outer haircells. However, outer haircells are innervated very strongly by efferent fibres coming from the brain. Although the number of efferent ("E") fibres is only about 500, their terminals dominate the synaptic area of the outer haircells, both in terms of their size and their number. Efferent fibres travelling towards inner haircells make synaptic contact not with the cells but only with the afferent fibres departing from them. The inner haircells themselves rarely receive efferent terminals. Although the functions of the two kinds of haircells and their neural innervation are not clear in detail, it seems to be reasonable, in connection with other facts, to assume that the function of the outer haircells is restricted to the organ of Corti. The outer haircells appear to exert a large influence on the inner haircells, although there is no direct neural connection between the two systems. These influences are described in Sect. 3.1.3, part (b).

(a) Linear Passive System

Georg von Békésy, the Nobel laureate, investigated the vibratory motion of the inner ear experimentally. He regarded the scala media as a system that moves as a whole unit. According to this assumption, the displacements of the basilar membrane and Reissner's membrane are the same, i.e. for a certain location, they show the same volume displacement. Within these limitations, the displacement of the basilar membrane may be described (for high input levels or post mortem preparations) as a linear system.

The idea of von Helmholtz that low frequencies produce oscillations of the basilar membrane near the helicotrema and high frequencies near the oval window, was confirmed by the experimental results of von Békésy. The existence of travelling waves, in contrast to the previously conceived standing waves, was a new and important discovery of von Békésy. The travelling wave of the basilar membrane's vertical displacement begins with small amplitude near the oval window, grows slowly, reaches its maximum at a certain location and then rapidly dies out in the direction of the helicotrema. In Fig. 3.5b, velocities of the basilar membrane without active feedback are shown for three frequencies. The 2.5 windings of the cochlea are unwound and stretched out for this schematic drawing to a total length of 32 mm. Two curves are shown for the frequency of 400 Hz, the solid curve represents that instant where the maximum is reached, and the broken curve represents the instant a quarter period earlier. In this way, the character of the travelling wave becomes obvious; no nodes or antinodes occur as would be seen for standing waves. For all three frequencies, the envelope of the oscillations is indicated by the dotted

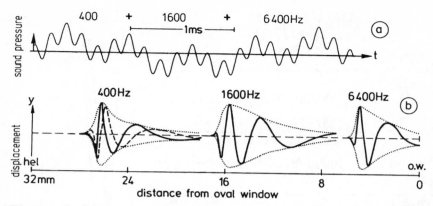

Fig. 3.5a,b. Schematic drawing of the transformation of frequency into place along the basilar membrane. In (**a**) three simultaneously presented tones of different frequencies expressed as compound time function produce travelling waves (**b**), that reach their maximum at three different places corresponding to the characteristic frequencies

lines. The amplitude gradually increases from the oval window in the direction of the helicotrema, reaches a maximum, and diminishes quite rapidly beyond this maximum. Especially striking is the clear separation of different stimulus frequencies according to the different regions of their maximum. Assuming that the three tones of 400, 1600 and 6400 Hz are presented simultaneously – as indicated in Fig. 3.5a – and are transmitted together to the oval window, a separation occurs in the inner ear according to Fig. 3.5b. Each tone causes a different region of the basilar membrane to vibrate. Thus the inner ear performs the very important task of frequency separation: energy from different frequencies is transferred to and concentrated at different places along the basilar membrane. The separation by location on the basilar membrane is known as the place principle.

An illustration using a 1-kHz tone burst presented at the oval window may show the frequency selectivity as well as the frequency-place transformation of the inner ear. In addition, the scheme indicated on the left side of Fig. 3.6 may help clarify the frequency resolution in the inner ear. A series of band-pass filters with an asymmetrical shape of their frequency response subdivides the frequency range into many sections. The centre frequencies of these filters can be correlated with places along the cochlea, indicated at the right of the figure, again in unwound form. The response near the oval window, correlated with a centre frequency at 4 kHz, is shown in the upper part of Fig. 3.6. The 1-kHz tone burst produces in this band-pass filter a short click, followed by the 1-kHz oscillation with very small amplitude; in order to be visible, this amplitude has been multiplied by a factor of four in the figure. The short click at the beginning and at the end of the time function stems from the broad-band transient which pro-

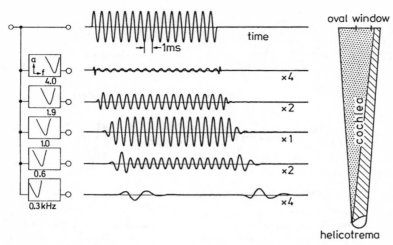

Fig. 3.6. A tone burst with a time function outlined in the upper panel produces responses at five different places on the basilar membrane, plotted in the lower five functions. Note the different amplitude scale on each row and amplitudes at different places, and the increasing delays towards the helicotrema. The drawing on the left illustrates the frequency selectivity characteristic of the five different places

duces energy in the 4-kHz range at the moment of switching on or off the 1-kHz tone burst. The second band pass (1.9 kHz centre frequency) corresponds to a place that is located further along the cochlea in the direction of the helicotrema. The size of the 1-kHz oscillation becomes much larger, but the click responses corresponding to the centre frequency of 1.9 kHz are still visible at the beginning and at the end of the burst. In this row, the amplitude is magnified by a factor of two. The amplitude of the oscillation reaches its maximum at a centre frequency of 1.0 kHz. Because the centre frequency of the band-pass filter and the frequency of the tone burst coincide, the amplitude of the 1-kHz vibration rises and decays very smoothly. The amplitude of vibration produced by the 1-kHz tone burst gets smaller and smaller for places in the cochlea located further towards the helicotrema. These places correspond to band-pass filters with centre frequencies of 0.6 and 0.3-kHz, respectively. Because the travelling wave dies out very quickly after its maximum, the 1-kHz oscillation is invisible at the 0.3-kHz centre frequency but the two clicks at the beginning and the end of the tone burst become more prominent in the two lower band-pass filters. Their frequency content corresponds to the centre frequency of the respective filter.

Another effect is also obvious from these responses: the delay time between the signal at the oval window and the response of the basilar membrane increases with increasing distance along the basilar membrane or, in other words, increases with decreasing centre frequency of the band-pass filter. This means that tones of high pitch or sound components at high frequencies produce oscillations at the entrance of the cochlea near the oval window with small delay times. Low tones, or components of sounds with low-frequency content, travel far towards the helicotrema and show long delay times. The delay increases towards the helicotrema; it reaches values of 1.5 ms for energy near 1.5 kHz and rises to 5 ms near the end of the cochlea. Figure 3.6 illustrates very clearly the two effects that occur in the inner ear treated as a linear system: frequency resolution on one hand, and the temporal effects, illustrated in delay times, on the other.

The filters indicated at the left side of Fig. 3.6 reflect a situation that is often relevant in physiological studies, where it is much easier to keep the location of observation constant and to observe the influence of frequency changes. With some approximation, such conditions can be fulfilled in special psychoacoustical measurements like psychoacoustical tuning curves. The frequency dependence found in this way is known as the frequency resolving power of the hearing system.

The displacement of the basilar membrane is the first stage in the sequence of the different auditory levels. For constant stimulus amplitude (pure tone of constant level of 80 dB SPL) the displacement at four locations, 4, 11, 20 and 29 mm from the helicotrema, is shown schematically in the upper part of Fig. 3.7 as a function of frequency on a logarithmic scale. The first temporal derivative, the velocity, often assumed to be the effective stimulus driving the inner haircells, is shown in part (b) of Fig. 3.7. Assuming a linear system, tuning curves can be constructed from part (b). Tuning curves show the amplitude of the stimulus necessary to produce a constant response magnitude at a certain place as a function of the stimulus frequency. The sound pressure levels necessary to produce a constant peak velocity of 10^{-6} m/s at the four locations are shown in part (c) of Fig. 3.7 as a function of frequency. Instead of the distance from the helicotrema, the characteristic frequency (CF) for the location is given and indicates that frequency for which an excitation is most easily produced. The curves obtained in this way are called tuning curves. The special dip indicated by plus signs in part (d) occurs at low levels only and is a consequence of the active processes discussed in the next section and which lead to much more pronounced frequency selectivity.

(b) Nonlinear Active System with Feedback

The displacements of the basilar membrane are very small. Normal conversational speech produces sound pressures in the air of about 20 mPa or sound pressure levels around 60 dB. The associated displacement of the basilar membrane is in the amplitude range of tenths of nanometers, a size that corresponds to the diameter of atoms. However, we can still hear tones which have a sound pressure 1000 times smaller. Our hearing system must use very special arrangements to produce such an extraordinary sensitivity.

The differences between the two kinds of sensory cells hint at a special construction and a special use of the two kinds of cells. Another hint concern-

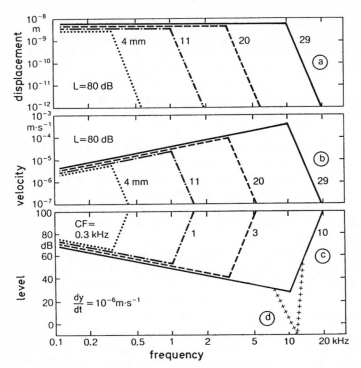

Fig. 3.7a–d. Schematic diagram illustrating the formation of tuning curves. Four different tones with frequencies of 0.3, 1, 3, and 10 kHz show characteristic places of 4, 11, 20, and 29 mm from the helicotrema, respectively. The tones are assumed to have a sound presure level of 80 dB. The displacement, y, is shown in (**a**), the velocity, dy/dt, is shown in (**b**). The sound pressure levels necessary to produce a constant velocity dy/dt at four locations as a function of frequency are shown in (**c**). Such curves are called tuning curves, however the criterion (in our case a velocity of 10^{-6} m/s) can be different from study to study. Nonlinear effects discussed in Sect. 3.1.5 increase the sensitivity to weak stimuli in the region of the characteristic frequency, so that a tuning curve marked by the crosses for CF = 12 kHz is ultimately produced

ing special arrangements at low levels is illustrated in Fig. 3.8 which shows in simplified form the tuning curve of a haircell for two levels. In Fig. 3.8, the sound pressure level of a tone necessary to produce a certain receptor potential in a haircell is plotted as a function of its frequency. The dotted curve, related to a higher voltage (10 mV), shows a shape comparable with the data outlined in Fig. 3.7c (solid): a gradual decrease in the required level with increasing frequency up to the characteristic frequency above which the required level increases strongly. The tuning curve for the lower voltage (2 mV) has a different shape. (The tuning curve shown has been shifted up-ward by 17 dB so that the two curves match at low frequencies to facilitate comparison.) For the 2-mV case (solid), the sensitivity around the character-

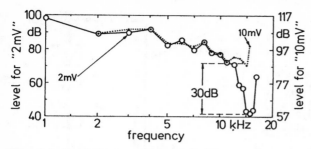

Fig. 3.8. Tuning curve of a haircell. The sound pressure level of a tone necessary to produce a certain DC receptor potential (2 mV, *circles*; 10 mV *dots*) as a function of frequency. The 10 mV curve is shifted downwards by 17 dB so that the two curves are superimposed at low frequencies. Data are replotted from Russel, I.J. and Sellick, P.M.: H. Physiol., **284**, 261 (1978)

istic frequency is much larger than expected from linear extrapolation. The level dependence of the tuning curves indicates a strong nonlinearity leading, at low levels, to an additional gain of as much as 30 dB. Very careful measurements of the displacement of the basilar membrane have shown that similar effects can already be found in such displacement patterns. These data stem from animals. Taking into account some additional peripherally located effects like cubic difference tones (see Chap. 14) or otoacoustic emissions (see Sect. 3.1.4), both observed in man, it must be assumed that our peripheral hearing system operates with some active nonlinear feedback.

How such a system acts is not yet clear in detail. However, the basic structure may be deduced from the following functional behaviour. At higher levels, the inner haircells are directly stimulated by the shearing force produced between the hairs of the sensory cells and the tectorial membrane, in response to the local velocity of the basilar membrane. For these high levels, the outer haircells are of no importance because the large displacements drive them to saturation. At low levels, however, the inner haircells are only very slightly stimulated in a direct way. An interaction between the active outer and the inner haircells is therefore assumed to be responsible for the large dynamic range of the auditory periphery, and for the sharper frequency selectivity that arises at low levels in addition to the selectivity attributable to the linear hydromechanic system of the basilar membrane. The interaction is effectively considered to be instantaneous and therefore acts by AC-components. It also includes a strong nonlinearity with a transfer characteristic showing almost symmetrical saturations.

The block diagram given in Fig. 3.9 illustrates these concepts. On the left, the inner and outer haircells are shown schematically, together with their influence on each other along the length of the basilar membrane. On the right, the functional interdependencies are outlined in simplified form for a small section of the organ of Corti. The arrows in this figure indicate the direction of the influence; they illustrate the assumption that the outer haircells ex-

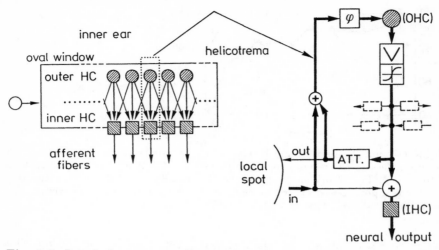

Fig. 3.9. Principal structures effecting the influence of outer haircells on inner haircells (*left part*). This structure remains the same along the organ of Corti. The right drawing shows the functional relationship between the outer (OHC) and inner (IHC) haircells for the section, as indicated by the area surrounded by dots in the left part. Note that only the inner haircells have a neural output

ert a strong influence on the inner haircells. The information contained in a stimulus is then transferred to afferent nerve fibres which terminate only on the inner haircells.

At a given location, the available stimulus affects both the inner haircells and also – through the summation point and a phase shifter – the sensitive outer haircells. The latter operate as amplifiers with a saturating characteristic, approximated by the transfer function with idealized nearly symmetrical breakpoints, corresponding to a sound pressure level of about 30 dB. Although the process is modelled in electrical terms, the influence of the outer haircells on the inner haircells could be mechanical, electromechanical, electrical, biochemical, or all or any of the above in nature. A single inner haircell may be influenced by many outer haircells in its surroundings (an effect indicated by the resistors pointing in and out laterally). However, at high levels, the amplifiers (outer haircells) are saturated and the adequate stimulus operates almost exclusively on the inner haircells.

An important part of this functional scheme is the fact that the output of the outer haircells can influence their input. Such feedback systems are very sensitive, and they have the tendency to be self-oscillating. Close to self-oscillation, the feedback loop, with appropriate phase characteristics, enhances the adequate stimulus for the outer haircells considerably. If there were no lateral spread of the feedback, this would lead to an extremely selective frequency response. Lateral spread of the feedback produces a band-pass-like frequency selectivity but still preserves the very high sensitivity as discussed in Sect. 3.1.5. Thus, in comparison with a totally passive basilar membrane, frequency selectivity is sharpened at low stimulus levels as long as saturation is not reached. The nonlinear active system feeds back to the motion of the basilar membrane so that travelling waves of distortion products are also produced (see Sect. 3.1.5 and Chap. 14). Further, the tendency of this active feedback system to oscillate sets the stage on which the four different kinds of otoacoustic emissions can be discussed (see next section).

3.1.4 Otoacoustic Emissions

Otoacoustic emissions are sounds that are produced inside the hearing system but are measured as acoustical oscillations in air. Emissions are discussed in this section in some detail because they are more and more frequently used as an efficient tool for measuring effects produced in the peripheral part of the human hearing system. Otoacoustic emissions are measured almost exclusively in the closed ear canal. The level of the emissions is very small and remains mostly far below the threshold of hearing, so that very sensitive microphones have to be used. Emissions which can be measured without any stimulation of the hearing system are called spontaneous otoacoustic emissions. Evoked emissions are produced as a reaction of our hearing system to a stimulus. For the measurement of this kind of emission, it is necessary to install not only a sensitive microphone but also a small sound transmitter in the probe. Both the microphone and the transmitter should have a broad frequency response between about 500 Hz and 4 kHz, although the emissions observed are concentrated in a frequency range between 800 Hz and 2 kHz. A schematic drawing indicating the periphery of our hearing system and an individually fitted probe is shown in Fig. 3.10. For special measurements using low-frequency signals (outlined in Fig. 3.24) sounds can be produced in an earphone (DT 48), fed through a flexible tube to the closed ear canal and monitored by an attached low-frequency microphone.

There are four kinds of otoacoustic emissions. The spontaneous otoacoustic emissions (SOAEs) are produced, as mentioned above, without any sound stimulus. The simultaneously evoked otoacoustic emissions (SEOAEs) can be measured in the closed ear canal during continuous tonal stimulation of the ear. Delayed evoked otoacoustic emissions (DEOAEs) are responses to short periodic sound impulses: these sounds may be either broad-band clicks or narrow-band tone bursts with Gaussian-shaped envelopes. After a delay which depends on the frequency of the evoking tone burst, otoacoustic emissions are measured as a response to the stimuli. Because the DEOAEs are triggered by the evoking sequence of sound bursts, the technique of time-synchronous averaging can be used to enhance the very small signal-to-noise ratio of these emissions.

The characteristics of these three kinds of emissions have been extensively studied. The distortion product otoacoustic emissions (DPOAEs), produced in the closed ear canal by stimulating with two primaries and searching for

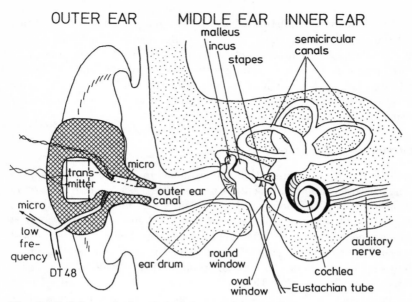

OUTER EAR   MIDDLE EAR INNER EAR

Fig. 3.10. Schematic drawing of the peripheral hearing system and the probe used to measure otoacoustic emissions, suppression-, and masking-period patterns

"ear-produced" distortion products, must be measured in more subjects before their dependencies can be outlined in general form. Only some preliminary results are discussed.

(a) Spontaneous Otoacoustic Emissions

The frequency analysis of the sound measured in the closed ear canal for a typical subject in a quiet surrounding, is shown in Fig. 3.11. Sound pressure level is plotted as a function of frequency with a resolution of a few Hertz. More than 50% of the ears of normal hearing subjects exhibit one, often several,

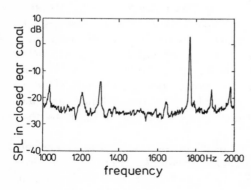

Fig. 3.11. Example of a frequency analysis of the sound pressure picked up by a sensitive microphone in the closed outer ear canal. Threshold in quiet corresponds to about 0 dB

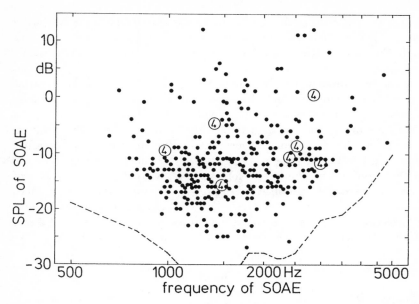

Fig. 3.12. Level of spontaneous otoacoustic emissions (SOAE) as a function of their frequency. Data stem from about 100 ears of 50 normal-hearing subjects. The (4)-symbols belong to a single ear. The dashed line indicates the noise floor measured with a bandwidth of 5 Hz

spontaneous otoacoustic emissions (SOAE). The emissions show narrow-band characteristics and have a tonal quality if amplified and reproduced acoustically. The sound pressure level of these emissions is usually between −20 and −5 dB SPL and rarely exeeds 0 dB. The levels of the spontaneous emissions measured in 50 normal hearing subjects are shown in Fig. 3.12 as a function of their frequency. For a certain subject showing several emissions, the emissions are marked by the number 4 in open circles. Searches for the number of emissions produced in single subjects using very sensitive microphones indicate that the number of emissions does not increase with increasing resolution: rather it seems that the frequency separation between neighbouring emissions does not decrease below a certain value. A careful analysis of the frequency separation between the neighbouring emissions of many subjects indicates that this distance increases with frequency. Transforming the most probable frequency separation of neighbouring emissions into critical-band rate (see Sect. 6.2) leads to a relatively strict rule: the distance between two emissions is most probably 0.4 Bark, independent of frequency. The unit "Bark" corresponds to the width of one critical band as outlined in Sect. 6.2.

Some SOAEs are not very stable in level. An emission appearing before a weekend may not be found afterwards, and vice versa. Even during the same day, spontaneous emission levels can vary distinctly. The frequencies of the spontaneous emissions, however, are very stable. Although emissions disappear, they show up again at almost the same frequency. The variation in frequency is less than 1% and in most cases it is only a few parts per thousand. Emissions with larger levels seem to be somewhat more stable than emissions with low levels.

Spontaneous emissions can be influenced by additional sounds or variations in steady-state air pressure. Tones added to the ear in which the emission occurs can reduce the amplitude of the SOAE. A reduction of 50% in the sound pressure of the emission is possible, and the frequency variation of the SOAE in this case remains very small. Plotting the sound pressure level of an additional tone (the suppressor), which is necessary to reduce the level of the spontaneous emission – for example by 6 dB – as a function of the suppressor frequency, produces a suppression-tuning curve. Such a curve is illustrated in Fig. 3.13 for a relatively large SOAE. The curve shows great similarity to the neurophysiological tuning curves discussed in Sect. 3.1.3 and characterizes the frequency selectivity of the hearing system at lower levels.

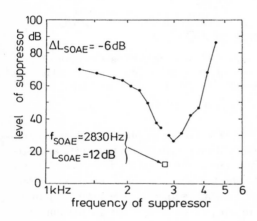

Fig. 3.13. Suppression-tuning curve, i.e. level of a suppressor tone necessary to produce a certain reduction of the level of a spontaneous otoacoustic emission, in our case 6 dB, as a function of its frequency. The square indicates the level and frequency of the spontaneous emission

In order to measure the temporal course of the effects exerted by a suppressor tone on a spontaneous emission, the bandwidth of the analysing system must be enlarged. Therefore, only emissions with a sound pressure level larger than 0 dB can be used, otherwise the signal-to-noise ratio is not sufficient. The temporal course of an emission as a reaction to the onset and offset of a suppressor is shown in Fig. 3.14. The upper part indicates the temporal course of the suppressor, the lower part the sound pressure of the spontaneous emission, both as a function of time. The reaction of the spontaneous emission does not start immediately, but with a certain delay time T_d and a certain time constant τ of about 15 ms (dotted line).

It is also possible to change the level of the spontaneous emissions periodically with a low-frequency tone. The period of these tones must be long in relation to the time constant in order to produce large effects. In discussing spontaneous emissions, it should be pointed out that normal hearing (not

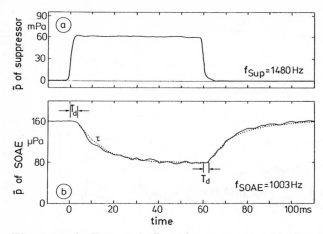

Fig. 3.14a,b. Temporal effect of a suppressor with the sound pressure versus time function, as given in (**a**), on the amplitude of the sound pressure of a large spontaneous otoacoustic emission (**b**). Exponential decay and rise times of 15 ms which match the SOAE characteristics are indicated by dotted curves

more than 20 dB hearing loss in the frequency range in question) seems to be a condition that has to be fulfilled. This means that the production of spontaneous emissions is a clear hint of good hearing capability in the frequency range in which the emission occurs. However, the non-existence of emissions is not an indication of abnormal hearing, since only 50% of normal hearing subjects show spontaneous emissions. Additionally, our experience has shown that tinnitus is not related to spontaneous emissions as long as the level of the emission is less than about 20 dB, a value which is only very rarely exceeded.

(b) Simultaneous Evoked Otoacoustic Emissions (SEOAE)

Simultaneously evoked otoacoustic emissions (SEOAE) can be identified most easily when the frequency response of the probe microphone is measured for different levels which are fed electrically into the transmitter of the probe. The uppermost curve in Fig. 3.15 shows such a response for an electrical level of 52 dB. It is almost a straight line, and indicates the frequency response of the whole probe including transmitter, microphone and ear cavity, without an active inner ear. Reducing the input level by 12 dB, the curve marked 40 dB is produced; it is – as expected – shifted 12 dB downwards. Decreasing the input level by another 10 dB produces a frequency response that is no longer a straight line, but indicates very small but consistent variations which become larger at input levels of 20 and 10 dB. Peaks and valleys are clearly visible and remain at the same frequencies even for smaller input levels of 0 and −10 dB, where the internal noise of the apparatus and of the subject increasingly influence the response. A linear system would not show such a level

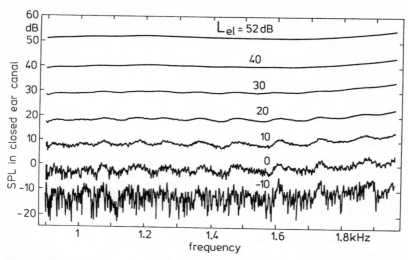

Fig. 3.15. Responses of the probe microphone for different electrical levels fed into the transmitter of the probe and picked up in the closed ear canal of a subject as a function of frequency

dependence. The ear, however, reacts in a level-dependent manner, and the small variations of sound pressure or sound pressure level are indications of SEOAEs. As shown in Fig. 3.16, such SEOAEs not only influence sound pressure, but also phase as a function of frequency. Whereas the phase response and the amplitude response are almost straight for high levels (broken lines), the phase response at low levels (solid) shows variations around the straight line. The phase response corresponds specifically to variations of sound pressure: an upward crossing of the two sound pressure curves corresponds to a peak in the phase response.

SEOAEs have been found in about 90% of ears. Spontaneous emissions (SOAE) are transformed to SEOAEs if driven by an appropriate signal about 10 to 20 dB above the SOAE level. At these levels, spontaneous emissions are synchronized to the stimulating tone. At low levels of stimulation, however, a spontaneous emission remains at its own frequency and is not synchronized to the evoking tone.

Using special apparatus and data processing, suppression tuning curves as well as temporal effects of SEOAEs can be measured. The results indicate similar responses to those already discussed for spontaneous emissions. This similarity is a hint that spontaneous emissions and simultaneously evoked emissions may be created by the same source.

(c) Delayed Evoked Otoacoustic Emissions (DEOAEs)

Delayed evoked otoacoustic emissions (DEOAEs) are responses of our hearing system to short sound impulses. Repetition rates of 20 to 50 Hz are suitable

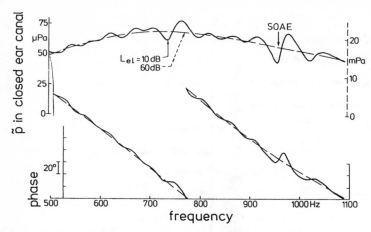

Fig. 3.16. Amplitude (*upper panel*) and phase (*lower panel*) of the sound pressure produced by a small probe transmitter and picked up in the closed ear canal of a subject for the two conditions of transmitting an electrical level of 10 dB (*solid*) and 60 dB (*broken*) to the transmitter

and correspond to an analysis time of 50 to 20 ms. Within this time, most of the delayed emissions reach their maximum. Both the time function and the level dependence of the DEOAE are important. Figure 3.17 illustrates a characteristic example. Sound pressure-time functions measured in the closed ear canal are shown on the right, where the gain is kept constant. The sensation level, i.e. the level above threshold in quiet of the burst sequence used as an evoker to produce DEOAEs, is the parameter. Because the emission is much smaller in level than the evoking bursts, the latter is far too large to be shown on a normal plot. This is especially the case (shown on the right of Fig. 3.17) if the gain is kept constant. This means that the same ordinate scale is used, although the evoking level changes. In this case, it becomes clear that the amplitude of the emission increases when the sensation level is increased from −6 to +24 dB. Above that, the emission's amplitude reaches a kind of saturation. In order to check that no unwanted distortions at high levels are produced, it is convenient to use a relative constant amplitude scale (Fig. 3.17a), instead of an absolute constant scale (Fig. 3.17b). In such a display of time functions produced by different sensation levels, it becomes clear that the relative amplitude remains equal at low levels. This means that the delayed evoked emission behaves linearly for sensation levels of evoking impulses smaller than about 18 dB. For larger sensation levels, the relative amplitude decreases, which means – as seen in Fig. 3.17b – that the absolute amplitude remains approximately constant. In the left panel, the time function of the emission changes its amplitude drastically for sensation levels

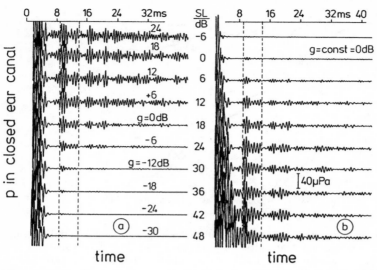

Fig. 3.17a,b. Sound pressure versus time functions for delayed evoked otoacoustic emissions produced by Gaussian-shaped tone bursts of different levels, expressed in sensation level, SL. The difference from trace to trace is 6 dB. In part (**a**), the gain of the amplifier is reduced in the same way as the sensation level is increased. This leads to a level-independent time function for the stimulus. In part (**b**), the gain is kept constant leading to increasing amplitude of the time function of the stimulus. The data are from the same subject as those of Fig. 3.15

above 18 dB while for the same conditions, the time functions of the decaying evoking bursts remain level independent. Hence, the time function of the evoking impulse is transferred correctly through the transmitting system. The equivalent RMS value of the sound pressure emitted within a certain time window is used to calculate the level of the delayed evoked emission. Such a time window is shown in Fig. 3.17 by two vertical dashed lines. Many emissions last much longer than the time window, while others are composed of a compact time function and can be described as having a certain delay.

The level calculated within the individual time window that includes the main sound pressure of the emission is plotted in Fig. 3.18, as a function of the sensation level of the evoking short burst (1 oscillation of a 2-kHz tone, in this case). At low levels, the relation between the level of the DEOAE and the evoking level is approximated by the broken 45°-line, indicating linear growth. At levels 10 to 20 dB above threshold in quiet, the emission level saturates more and more, and becomes independent of the evoking level. Different subjects and different emissions from the same subject lead to quite different levels of DEOAEs. The largest ones lie only about 5 to 10 dB below the level of the evoking sound.

The linear behaviour of delayed emissions can be demonstrated by superimposing emissions produced by single cycles. Four cycles of the evoking

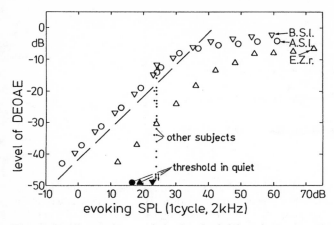

Fig. 3.18. Dependence of the level of delayed otoacoustic emissions on the level of the evoking 2-kHz, 1-cycle burst for three different subjects. Threshold in quiet of the three subjects is given by filled symbols. The dots represent data of an additional 21 subjects for an evoker level of 25 dB SPL

sound represent a tone burst, which can be thought of as being produced by four sequentially presented stimuli, each with the duration of a single cycle. The evoked emissions can be thought of as being produced in the same way. The time function of an emission produced by an evoking burst with four cycles can be compared with the time function constructed by superimposing four single emissions produced by four single evoking cycles. The result of such a construction is shown in Fig. 3.19, where the emission of the first and of the fourth single oscillation are shown in the two upper traces. The superimposed time function calculated from the responses to four single emissions is compared with the time function produced by a tone burst of four cycles. The almost exact identity of the two lower time functions in Fig. 3.19 indicates clearly that delayed evoked emissions follow linear superposition at low amplitudes.

The delay time of an emission depends on its frequency. An average of available results shows a decreasing delay time for increasing emission frequency. Figure 3.20 displays this relationship, which indicates that for low frequencies near 500 Hz a delay time of about 20 ms is measured, whereas emissions with high frequency content are produced with a much shorter delay, about 4 ms at 4 kHz. It should be mentioned, however, that many emissions show a time function that is not concentrated within a narrow time window. Long lasting emissions with partly modulated time functions often occur, and some are not completely decayed even after a delay of 80 ms. For such long-lasting emissions, the delay time shown in Fig. 3.20 holds for the first, and usually the largest, maximum in the time function of the emission. If the evoker is composed of a tone burst with a Gaussian envelope, its ef-

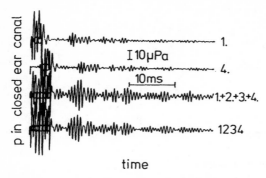

Fig. 3.19. Time functions for delayed otoacoustic emissions evoked by single oscillations of a 1.5-kHz tone with a sensation level of 8 dB ("1", "4"). Track "4" is delayed by three periods in comparison to "1". The digital summation of track "1" and track "4" together with the two tracks between lead to a time function that is denoted track "1 + 2 + 3 + 4". The emission "1234" evoked by a 4-cycle stimulus is indicated in the lowest track. Note the good agreement between the latter two time functions

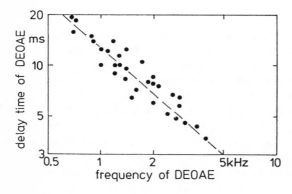

Fig. 3.20. Relationship between delay time of the delayed evoked otoacoustic emissions and their frequency region. The dashed line approximates the reciprocal value of 0.4 Bark, when expressed in frequency distance, Δf

fective frequency content is restricted to a small range, and so is that of the delayed emission.

The frequency spectrum of a stored delayed emission can be measured by analysing the periodically repeated time function of the emission without the evoking sound burst. Figure 3.21 shows such spectra for four sensation levels of the evoking impulse. The four spectra indicate very clearly that the spectral energy is not distributed continuously, but shows very distinct maxima in which most of the energy of the emission is concentrated. This effect is almost independent of the sensation level of the evoking impulse.

There are various ways of influencing the amplitude of the delayed evoked emissions with suppressing sounds. The most impressive effect is the suppression of the delayed evoked emissions by low-frequency sounds, with a period corresponding to the repetition rate of the evoking tone burst. The suppression of the delayed emissions corresponds very closely to the audibility of the

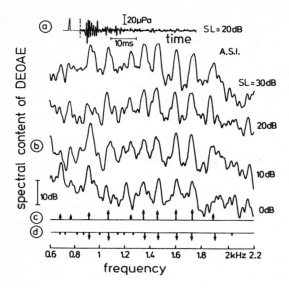

Fig. 3.21a–d. Time function (**a**) and spectra (**b**) of delayed otoacoustic emissions evoked by bursts of different sensation level. The arrows in (**c**) mark maxima in the spectra, those in (**d**) the minima in the frequency dependence of threshold in quiet of the same subject

evoking tone burst. This result points to the fact that threshold in quiet and the masked threshold are closely related to the different kinds of emissions; these relations are described in the next section.

(d) Relation Between Emissions and Threshold in Quiet

The three discussed kinds of emissions seem to be created by the same source, as can be deduced from several effects. One of these effects is that the minimal frequency distance between neighbouring spontaneous emissions, or between neighbouring maxima in the spectral composition of delayed emissions, are the same and amount to about 0.4 Bark, i.e. a little less than half a critical band. In this context, it may be interesting to realize that the reciprocal value of the delay time of the delayed evoked emissions is almost identical to this value of 0.4 Bark, as would be expected in linear circuits. Moreover, a spontaneous emission that has disappeared, has been measured as a relatively strong simultaneously evoked emission during the same day. Further, spontaneous emissions can also be measured as simultaneously evoked emissions, an effect that also indicates that the source of the three kinds of emissions may be the same.

An impressive effect, leading to the same conclusion, is the strong relation between the spectral composition of the emissions and the fine structure of the threshold measured in quiet as a function of frequency. When threshold in quiet is measured very carefully, i.e. not with continuously changing frequency but measuring point by point with a frequency distance between the points of only 2 or 3 Hz, many subjects show distinct maxima and minima.

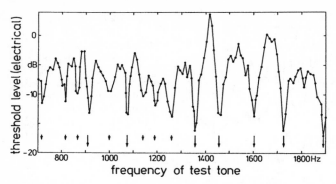

Fig. 3.22. Level of transducer voltage at threshold in quiet of a subject who produces evoked, but not spontaneous emissions. The minima are marked by arrows with two lengths corresponding to the depth of the minima. The same arrows are marked in Fig. 3.21, part (d)

The same subjects usually show also many large emissions. Figure 3.22 illustrates threshold in quiet for the same subject for whom the spectra of the delayed evoked emissions were outlined in Fig. 3.21. The minima of threshold in quiet are marked in Fig. 3.22 by arrows that are reproduced at the corresponding frequencies in panel (d) of Fig. 3.21. The pronounced minima of threshold in quiet (panel (d)) show a perfect correlation with the maxima in the spectral composition of delayed evoked emissions (panel (c)). This means that threshold in quiet shows lower values (the hearing system is more sensitive) if the spectrum of the evoked emission has a maximum.

This holds not only for delayed evoked emissions and their spectral composition but also for spontaneous otoacoustic emissions. An example is given in Fig. 3.23. The frequencies of spontaneous emissions are indicated by arrows at the lower edge together with their respective levels. The frequencies

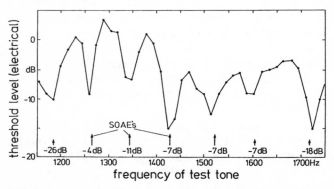

Fig. 3.23. Level of the transducer voltage at threshold in quiet for a subject who produces several spontaneous emissions (frequencies are indicated by arrows with the levels given underneath)

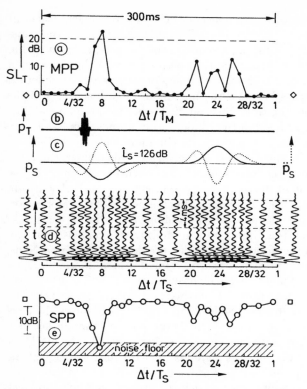

Fig. 3.24a–e. Threshold (**a**) of sequences of tone bursts (**b**) masked by a sequence of alternating Gaussian-shaped DC pressure impulses (**c**). The time functions of delayed otoacoustic emissions evoked by the tone bursts are presented in (**d**) for different times of presentation of the evoking tone burst within the period of the suppressor (**c**). The averaged RMS value of the emissions is shown in (**e**) as a function of the time at which the evoking tone burst was presented within the period of the masker. Note the mirror-like correspondence of the curves outlined in (**a**) and (**e**). Diamonds in (**a**) and squares in (**e**) correspond to data produced without masker and suppressor, respectively

of emissions correspond very closely to the minima in the frequency dependence of threshold in quiet for the same subject. This result suggests a strong correlation between hearing ability at low levels near threshold in quiet and the three kinds of otoacoustic emissions.

Another impressive example of a relationship between hearing near threshold in quiet and emissions is found in the suppression of evoked emissions by low frequency Gaussian-shaped condensation and rarefaction impulses. The time function of such a suppressor is shown by the solid curve in Fig. 3.24c. The evoking tone burst (b) is presented at a sensation level of 20 dB. The time functions of the evoked emissions are outlined in Fig. 3.24d, using a time scale pointing upwards. The horizontal position of each time function corresponds to that instant within the period of the low-frequency masker at which the evoking tone burst is presented. The time functions of the emissions evidently depend on the time of presentation of the evoking burst within the period of the suppressor. At a time that corresponds to the moment of the suppressor's rarefaction maximum, the emission disappears completely. During the period which lies between the maximum of rarefaction and the maximum of condensation, the time function of the emission remains almost unchanged. In the neighbourhood of the condensation impulse, emissions are again reduced but not as much as near the maximum of the rarefaction.

The overall behaviour is summarized in Fig. 3.24e, which shows the RMS value of the delayed evoked emission's sound pressure within the time window marked in (d) as a function of the temporal presentation of the evoker within the period of the suppressor. At the condensation maximum of the suppressor, the RMS value of the emission shows three minima. In addition, the masking effect produced on the tone burst by the alternating condensation and rarefaction impulses was measured. In order to provide direct comparisons, the same tone burst used as the evoker was used for threshold measurements. The threshold of the tone burst was also measured as a function of its position within the period of the suppressor, in this case called the masker. The resulting masking-period pattern is displayed in Fig. 3.24a. It shows the level of the test-tone burst that is just audible in the presence of the masker as a function of the time of presentation within the masker's period. The masking-period pattern outlined in Fig. 3.24a is an almost perfect mirror image of the suppression-period pattern outlined in Fig. 3.24e. This result indicates that hearing near threshold and suppression of delayed evoked emissions are very closely related: the evoked emissions disappear under those conditions for which the evoking tone burst is no longer audible.

(e) Distortion Product Emissions

When two primary tones with the frequencies f_1 and f_2 are presented to an ear, additional tones with pitches correlated to the frequencies $f_2 - f_1$ and $2f_1 - f_2$ become audible (see Chap. 14). Such audible distortion products can be cancelled by adding a tone of corresponding frequency and correct level and phase. Such cancellation, based on psychoacoustical perception, is called hearing cancellation. Based on purely objective methods and apparatus, distortion product emissions can be measured in the closed outer ear canal by a sensitive microphone. A linear set up and a very low noise floor are necessary to produce relevant data. The method of cancellation can also be used by adding a tone of the corresponding frequency adjusted in level and phase, so that the objectively measured difference tone which appears is compensated to indetectable low values.

Figure 3.25 shows an example for the distortion product at the frequency $(2f_1 - f_2) = 2 \times 1620\,\mathrm{Hz} - 1851\,\mathrm{Hz} = 1389\,\mathrm{Hz}$ and its cancellation. Two primaries are shown in the lower trace with the probe placed in a passive

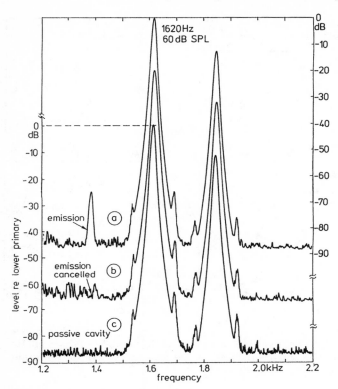

Fig. 3.25a–c. Frequency sweeps of the spectra obtained for sensation levels of the primaries $SL_1 = 60\,dB$ and $SL_2 = 50\,dB$ with the frequencies $f_1 = 1620\,Hz$, $f_2 = 1851\,Hz$ and $(2f_1 - f_2) = 1389\,Hz$ in the outer ear canal of a normally hearing subject (**a**), with the emission cancelled (**b**), and with the probe placed in a small passive cavity (**c**)

cavity instead of in the outer ear canal. It is thus shown that no measureable distortion product appeared at 1389 Hz, where the noise floor is more than 85 dB below the 1620-Hz primary. Placing the probe in the outer ear canal of a normally hearing subject, a distortion product emission at 1389 Hz is clearly indicated in the upper trace. The middle trace shows a frequency sweep with the emission cancelled by adjusting an added third tone at 1389 Hz in level and phase, so that the emission's level is reduced below the noise floor. The levels and phases needed for such cancellations are collected and plotted as a function of different parameters.

A typical example is given in Fig. 3.26 for which the sensation level, SL_1, of the lower primary at 1620 Hz, is kept constant as a parameter at 50, 60, and 70 dB, while the sensation level for cancellation (which is equal to that measured directly) is plotted as a function of the sensation level, SL_2, of the upper primary at 1800 Hz in (a) and at 1944 Hz in (b). The distortion prod-

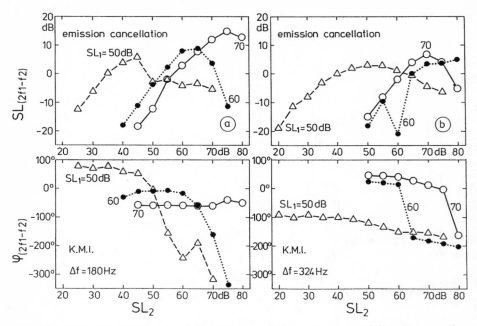

Fig. 3.26a,b. Cancellation level $SL_{(2f_1-f_2)}$ and phase $\varphi_{(2f_1-f_2)}$ of distortion product emission with $f_1 = 1620\,\text{Hz}$, $f_2 = 1800\,\text{Hz}$ and $1944\,\text{Hz}$, i.e. $(2f_1-f_2) = 1440\,\text{Hz}$ and $1296\,\text{Hz}$ or $\Delta f = 180\,\text{Hz}$ and $324\,\text{Hz}$, respectively, as a function of the sensation level SL_2 with SL_1 as the parameter (Subject K.M.l.)

uct emissions show very small levels between $-20\,\text{dB}$ (close to the noise floor) and $+15\,\text{dB}$. The dependence on level SL_2 is unusual in view of the characteristics of regular nonlinearities. Individual differencies are large. The few data available so far do not allow general statements on the characteristics of the distortion product emissions. However, it was demonstrated that abrupt changes in level and phase of the $(2f_1 - f_2)$ distortion product as a function of one of the primary levels at constant primary frequencies can occur, and that the levels needed for emission cancellation are 30 to $50\,\text{dB}$ smaller than the levels needed for hearing cancellation as discussed in Chap. 14.

3.1.5 Model of the Nonlinear Preprocessing System

A model of peripheral preprocessing has to account for all three characteristics mentioned in Sect. 3.1.3, part (b): activity, feedback with lateral coupling and nonlinearity. Such a model simulates the mechanical and electrical events in the inner ear, and explains not only the level dependent frequency selectivity but also the effects of suppression and simultaneous masking, cubic difference tone generation, the three kinds of otoacoustic emissions, and suppression- and masking-period patterns. The model is simple in its basic structure but, because of the nonlinearities, its functional behaviour is not easily understood.

The model contains linear and nonlinear networks. In order to understand the interactions between these two kinds of networks and to search for relevant approximations useful for a computer model, the model was first realized in analogue hardware. Although such a realization limits the possibility of using a large number of sections, and of creating precisely the type of nonlinearity wanted, the advantage of learning a great deal about information processing in such a system outweighs the disadvantage of some inaccuracies. The outer ear and the middle ear are ignored in the model and it is assumed that they do not produce nonlinear effects. They can therefore be treated as linear circuits, the frequency and phase response of which can easily be added at the input of the model. The hydromechanics of the inner ear and the behaviour of the outer haircells, however, play a crucial role and are responsible for the frequency-place transformation. The approximation of the complicated hydromechanic system by a one-dimensional model is rough but effective. It shows the important facts, although the subdivision of a continuous fluid medium into the sections of a hardware model necessarily creates discontinuities. In order that these artifacts be distinguishable from real facts, the number of sections is limited to about 8 per mm length of the organ of Corti, corresponding to 10 sections per critical band.

The schematic diagram shown in Fig. 3.27a indicates the structure of the electrical circuit usually described as being equivalent to the hydromechanical network in the inner ear, regarding displacement of the basilar membrane. This network, however, was transferred into the dual network (Fig. 3.27b), to enable study with voltages instead of currents. The possibility of a resonating tectorial membrane is ignored by assuming that its damping is large enough to integrate its influence into the elements of the approximation. The outer haircells are assumed in this model to act only as nonlinear amplifiers, the outputs of which strongly influence the input to the inner haircells and feed back to the vibration of the basilar membrane, not only at the same location but also at adjacent places in both directions. To account for this lateral spread, additional feedback to the next two neighbouring sections on both sides is installed, with a reduced effectiveness of half of the direct feedback for the first and one quarter for the second neighbour. This way, the additional features of lateral mechanical or electrical coupling and of two-dimensional approximations are included, but the basic structure of the model remains simple.

The information contained in a stimulus preprocessed this way is transferred to higher centres only via the inner haircells. The amplification characteristic of the outer haircells is nonlinear with an almost symmetrical saturation. At input levels that correspond to more than 10 dB SPL, the amplification characteristic becomes more and more saturated. Such behaviour is approximated in the model by diode networks. The inner haircells play a

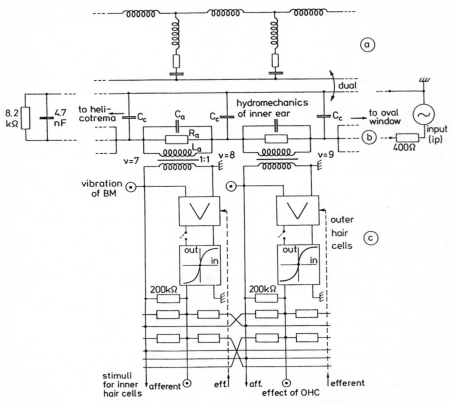

Fig. 3.27a–c. Block diagram of the hardware model realized in sections of 130 μm (corresponding to 0.1 Bark). (**a**) shows the equivalent electrical circuit of the hydromechanics of the inner ear normally used to represent basilar membrane velocity. (**b**) shows the dual circuit of (a) with driving impedance at the oval window and load impedance at the helicotrema. Part (**c**) of the model illustrates the additional nonlinear feedback circuits at each section representing the function of the outer haircells

secondary role in the model. Indeed, the inputs to the inner haircells are the outputs of the model: preprocessing in this context ends at the oscillation-to-spike-rate transformation. Finally, all sections of the model along the basilar membrane are assumed to have the same basic structure.

Details of the networks in the hardware model may be referred to in the literature. However, the nonlinear feedback loops simulating the effects of the outer haircells and outlined in Fig. 3.9, are clearly seen in Fig. 3.27c. They comprise a linear amplifier and a nonlinear symmetrical saturating device, the output of which feeds back through a large resistor to the point "vibration of BM", which is also the input to the amplifier. The lateral symmetrical coupling towards neighbouring sections through even larger resistors is also indicated.

There are two ways to illustrate the behaviour of such a model: the frequency response at a certain place, or the place response for a certain frequency. Because it is more convenient to measure, the frequency response at a certain place is more commonly used. Results of such measurements are directly comparable to neurophysiological and psychoacoustical tuning curves.

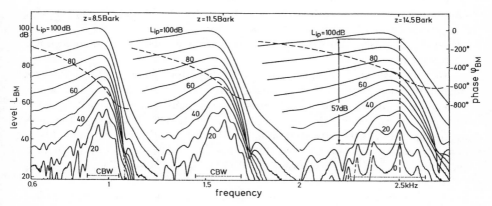

Fig. 3.28. Level response (*left-hand ordinate*) and phase response (*right-hand ordinate*) for input level of 60 dB of the output "vibration of basilar membrane" (BM) at the places 8.5, 11.5, and 14.5 Bark, corresponding to CFs of 1, 1.6, and 2.5 kHz, as a function of frequency. The parameter is the level at the input, i.e. of the driving voltage source at the 400-Ω resistor in front of the "oval window". The dotted horizontal bars correspond to the critical bandwidth

An important characteristic of the model is the level dependence of its frequency response. An example is given in Fig. 3.28 for three sections of 8.5, 11.5 and 14.5 Bark, i.e. a distance on the basilar membrane of 11.4, 15.4, and 19.4 mm from the helicotrema, corresponding to characteristic frequencies of about 1, 1.6, and 2.5 kHz, respectively. The characteristic frequency is defined as that frequency for which the frequency response reaches its highest local maximum at low input levels. The peak shifts towards lower frequencies at high input levels. The general characteristic of the frequency response is of a low-pass shape, as expected. The low-frequency slope is relatively shallow, whereas the high-frequency slope is rather steep. At lower levels, the frequency response is much more strongly peaked than at high levels, and shows several maxima due to the irregularities of the model elements. At levels below 10 to 20 dB, the shape of the response becomes almost level independent; the same holds for levels above 90 dB. This means that the model acts quasi-linearly at very low and very high levels. Another effect is the nonlinearity in the function which relates the input level L_{ip} (parameter) to the peak values of the level L_{BM} (left-hand ordinate). This compression of dynamic range may be quantified for the data at 14.5 Bark, i.e. a characteristic frequency (CF) near 2.5 kHz. For an input level difference of 100 dB (parameter), the level L_{BM} rises only 57 dB. This is best considered as an enhancement of as much as 43 dB at low levels at or in the neighbourhood of the CF.

The phase response obtained for a medium input level (60 dB) is indicated by the dashed lines for the three data sets in Fig. 3.28. The phase lag increases with increasing frequency and reaches about 700°, 600°, and 500° at the characteristic frequencies for places of 8.5, 11.5, and 14.5 Bark respectively. The general characteristic of the phase response is level independent, but the phase lag increases step-wise for low input levels at places which correspond with level maxima.

In this context, it is interesting to realize that the most probable frequency distance between neighbouring minima for threshold in quiet or between neighbouring spontaneous emissions, is closely related to the frequency distance needed to shift the phase by 180° near the CF. The data of Fig. 3.28 at 1, 1.6, and 2.5 kHz indicate about 70, 110 and 160 Hz, respectively. These spacings agree with the SEOAE data measured in the model and with the reciprocal value of the corresponding DEOAE-delay times of about 19, 9, and 6 ms.

Level responses of the model measured as a function of place in response to certain frequencies are plotted in Fig. 3.29, together with the corresponding phase-place responses for input levels of 60 dB. For a frequency of 1580 Hz, level responses are given for five different levels. The response for large input levels shows a shallow upper slope, whereas for low input levels this response is steeper, although still relatively broadly peaked, indicating that the feedback produces a band-pass-like tuning at low input levels. Phase responses at 60 dB are shown by the dashed lines in Fig. 3.29. The more the wave travels towards the helicotrema (i.e. towards lower critical-band rate), the more phase lag is established. More than 1000° can be accumulated before the phase lag levels off.

The phase lag along the length of the model varies little as a function of level for input levels larger than 70 dB. For lower levels, however, the phase response shows a ripple, related to that of the level response, indicating superimposed standing waves. As a consequence, the phase response at low critical-band rates may end up with 360° more phase lag at medium and low levels in comparison with the response at high levels.

All three kinds of otoacoustic emissions can be created in the model. Increasing the gain in the feedback loops so that oscillations occur produces the equivalent of spontaneous emissions. Such spontaneous emissions can be suppressed by adding tones leading to suppression-tuning curves in the way described for human subjects in Sect. 3.1.4. The same holds for simultaneously evoked emissions; their existence is indicated by the ripple in the frequency

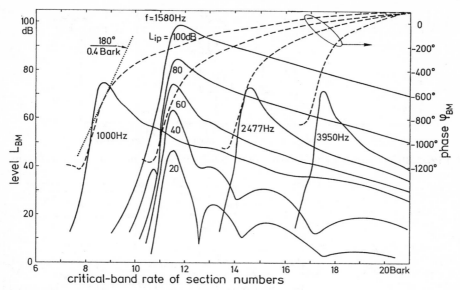

Fig. 3.29. Level response (*left-hand ordinate*) and phase response (*right-hand ordinate*) as a function of section number (expressed in critical-band rate) with input level, L_{ip}, as the parameter for 1580 Hz (otherwise $L_{ip} = 60$ dB). Input frequencies are 1000, 1580, 2477, and 3950 Hz, which produce maximal BM levels at places corresponding to 8.5, 11.5, 14.5, and 17.5 Bark for low input levels

responses plotted in Fig. 3.28 for low levels. Suppression-period patterns can also be simulated in the model as well as post-stimulus suppression effects with delayed evoked emissions.

The disadvantage of the hardware model which allows a resolution of only 10 sections per critical band, can be reduced in a computer model where the number of sections can be greatly increased, with only the available computing time limiting the number. The basic structure of a computer model acting in the frequency domain but ignoring lateral feedback coupling, is outlined in Fig. 3.30. Each section consists of a complex element due to hydromechanics ($\underline{Y}_{c\nu}$); an additional element ($\underline{Z}_{nl\nu}$) reproduces nonlinearity and feedback. Solutions in the time domain can be found using a computer model that is based on the wave-parameter-filter strategy. Such models are flexible enough to simulate lateral feedback coupling as well. The agreement between the results produced in computer models and in the analogue model is very good, especially if irregularities in the distribution of the elements along the critical-band rate are introduced. These are necessary to produce emissions that are not seen in computer models with homogeneously distributed elements.

Examples of simultaneously evoked emissions produced in the wave-parameter computer model are shown in Fig. 3.31a. Two irregularities are introduced so that the emissions occur in a frequency range similar to that of

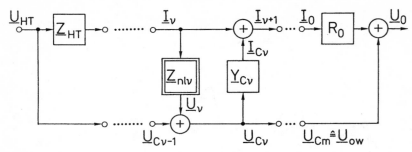

Fig. 3.30. Signal flow chart of the computer model acting in the frequency domain

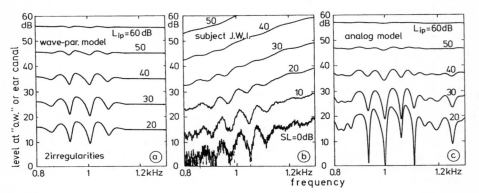

Fig. 3.31a–c. Simultaneously evoked emissions, i.e. level-frequency responses at low input levels calculated in the wave-parameter model (**a**) with two irregularities introduced near 8.5 Bark corresponding to 1 kHz. Data outlined in (**b**) are measured in the closed ear canal of a human subject, those outlined in (**c**) stem from the analogue model

the strong emissions picked up in the closed ear canal of a subject as shown in Fig. 3.31b. Similar data produced in the analogue model are outlined in Fig. 3.31c. The corresponding delayed evoked emissions are illustrated in the three panels of Fig. 3.32. A comparison of Fig. 3.31 and Fig. 3.32 indicates the previously mentioned relationship between the frequency differences of neighbouring extreme values, Δf, and the delay-time, t_d, given by $\Delta f = 1/t_d$.

The creation of delayed emissions from the contribution of many sections, can also be illustrated using the analogue model. It is the advantage of such a model that voltages equivalent to basilar membrane velocity can be picked up as easily as voltages equivalent to oval window velocity. Figure 3.33a shows the time functions of voltages picked up at several sections for a low level (30 dB). Figure 3.33b indicates the corresponding function for input levels of 70 dB. From this it becomes clear that the delayed emissions are not produced by early responses of the basilar membrane in response to the evoking tone burst. Rather, the decaying oscillations of some parts of the basilar membrane

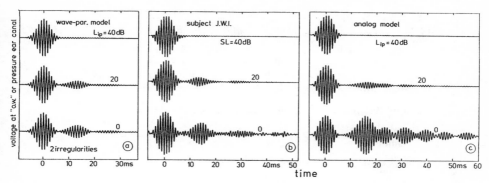

Fig. 3.32a–c. Delayed evoked emissions, i.e. pressure-time responses at low input levels calculated in the wave-parameter model with two irregularities (**a**), measured in the closed ear canal of a human subject (**b**) and picked up from the analogue model (**c**). The data correspond to those outlined in Fig. 3.31

are responsible for the production of the long delay of the delayed emissions. The effect can be seen more clearly by inspecting the contributions of each section separately. The contributions cancel each other out in the early part of the delay. For the delay corresponding to that of the DEOAEs, however, the time functions of the contributions are nearly in-phase and are added together to produce the delayed emission. Figure 3.33b indicates that the model does not show delayed emissions at high input levels although large oscillations of the basilar membrane are clearly present.

Suppression of delayed emissions by low-frequency tones can also be demonstrated in the hardware model. Careful studies of the behaviour of the hardware model under such conditions show that the very large low-frequency components drive the nonlinearity in the feedback loops into saturation. Thus the feedback gain is strongly reduced, the decay is faster and therefore the emission is drastically reduced. Moreover, the model gives a simple but effective explanation of the behaviour measured in masking-period patterns and in suppression-period patterns, in that the second derivative of the suppressor's time function (dotted in Fig. 3.24c) can be assumed to be the source of masking and suppression.

The data measured in man and that recorded from the models suggest that:
a) the cochlea acts in a way similar to that of the models; b) the three kinds of emissions stem from the same source; c) the phase response of the cochlea's hydromechanics is somehow responsible for the frequency distance between neighbouring spontaneous emissions and between extreme values of either evoked emissions or thresholds in quiet; d) the long delay of delayed evoked emissions is due to many decaying contributions from various places along the basilar membrane, which cancel each other out just after the evoking stimulus but add up to the delayed emission later; and e) the

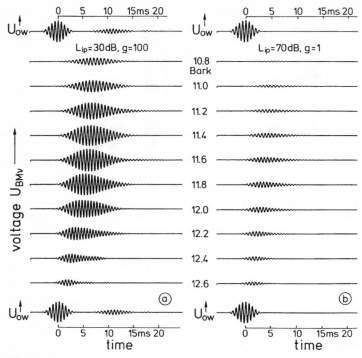

Fig. 3.33a,b. Oval window velocity equivalent voltages (*top* and *bottom*) and basilar membrane velocity equivalent voltages (places 10.8 to 12.6 Bark) from the analogue model as responses to Gaussian-shaped 1.5-kHz tone bursts with input levels of 30 dB (*left*) and 70 dB (*right*). The gain, *g*, is reduced as much as the level is increased

double-peaked shape of the suppression-period patterns produced by high-level, low-frequency sounds reflects the nearly symmetrically shaped saturating nonlinearity of the feedback loops, which approximate the function of the outer haircells in the model.

3.2 Information Processing in the Nervous System

The entire flow of information from the inner ear to the brain in man runs through approximately 30000 afferent auditory nerve fibres. These fibres differ in spontaneous activity and in the range of their frequency response. The distribution of frequencies that occurs peripherally in the cochlea is preserved in these fibres, i.e., the characteristic frequency of a fibre is determined by that part of the basilar membrane where it innervates an inner haircell. Together with the fact that the nerve fibres tend to maintain their spatial relations to one another, this results in a systematic arrangement of frequency responses according to location in all centres of the brain, and is called tonotopic organization. Effects already seen in the peripheral preprocessing are also found in nerve fibres. At low sound levels for example, a stimulus containing several frequencies stimulates several separated small groups of nerve fibres. At higher levels, however, the fibres are less selective, which means that a single tone stimulates many fibres at different locations. The dynamic range of the haircells, and therefore of the fibres, is assumed to be about 40 to 60 dB. Despite these restrictions we can perceive intensities over a range of more than 100 dB because of two factors: firstly, the peripheral preprocessing produces some dynamic compression, and secondly, at higher intensities different nerve fibres become stimulated according to their sensitivity. At low frequencies, the fibres respond according to the instantaneous phase of the motion of the basilar membrane. At high frequencies, above about 3.5 kHz, this phase synchronization disappears.

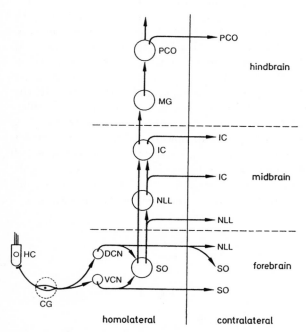

Fig. 3.34. Highly simplified schematic diagram of the afferent (*ascending*) neural connections in the brain from an inner haircell in the organ of Corti. Afferent fibres from the other side of the brain (contralateral) and efferent (*descending*) fibres are not included. HC haircell; CG cochlear (*spiral*) ganglion; DCN dorsal cochlear nucleus; VCN ventral cochlear nucleus; SO superior olivary nucleus; NLL nucleus of the lateral lemniscus; IC inferior colliculus; MG medial geniculate; PCO primary cortex

In order to obtain information about the location of the sound source in space, it is necessary, even in lower centres of the brain, that a significant exchange of information from the two ears occurs. Through a comparison of patterns related to intensity, phase and latency from both ears, the information concerning source location can be extracted. Figure 3.34 shows a highly simplified schematic diagram of the afferent neural pathways to and within the brain from an inner haircell of the organ of Corti. Although afferent fibres from the other side of the brain (contralateral) and efferent fibres are not included, the diagram is complicated despite high simplification. The higher the centres of the auditory system, the more complex are the cell constructions and the cell responses. The sensory cells seem to become more specialized at higher neural levels. Some of them react to interaural delays, some to very small level differences between the ears, some to frequency modulated tones and others to amplitude changes. Our knowledge of the whole system may be simplified and summarized as follows: firstly, there is a large interchange of information between the two sides of the auditory nervous system in lower centres, so that information for sound localization can be analysed in early parts where temporal information is more accurately determined, and secondly, many components of the stimulus are analysed separately, although only some of the brain cells are specialized while others show generalized responses. The increasing complexity of the cell responses in higher centres can be traced to the combination of the various separated components. In this way, the brain can analyse relatively simple signals, for example time differences, in lower centres but analyses complex stimuli in areas with complex, multimodal responses. It should be realized, however, that our understanding of information processing, especially in higher centres of the brain, is still incomplete. From this point of view, psychoacoustics, which describes the relationship between the lowest level (the stimulus) and the highest possible level (the sensation), is an attractive way of assessing our hearing system.

4. Masking

The masking of a pure tone by noise or by other tones is described in this chapter. Both psychoacoustical tuning curves and temporal effects in masking are addressed, effects related to the pulsation threshold are described, and finally, models of masking are developed.

Masking plays a very important role in everyday life. For a conversation on the pavements of a quiet street, for example, little speech power is necessary for the speakers to understand each other. However, if a loud truck passes by, our conversation is severely disturbed: by keeping the speech power constant, our partner can no longer hear us. There are two ways of overcoming this phenomenon of masking. We can either wait until the truck passed and then continue our conversation, or we can raise our voice to produce more speech power and greater loudness. Our partner then can hear the speech sound again. Similar effects take place in most pieces of music. One instrument may be masked by another if one of them produces high levels while the other remains faint. If the loud instrument pauses, the faint one becomes audible again. These are typical examples of simultaneous masking. To measure the effect of masking quantitatively, the masked threshold is usually determined. The masked threshold is the sound pressure level of a test sound (usually a sinusoidal test tone), necessary to be just audible in the presence of a masker. Masked threshold, in all but a very few special cases, always lies above threshold in quiet; it is identical with threshold in quiet when the frequencies of the masker and the test sound are very different.

If the masker is increased steadily, there is a continuous transition between an audible (unmasked) test tone and one that is totally masked. This means that besides total masking, partial masking also occurs. Partial masking reduces the loudness of a test tone but does not mask the test tone completely. This effect often takes place in conversations. Because partial masking is related to a reduction in loudness, it will be discussed in Chap. 8.

Masking effects can be measured not only when masker and test sound are presented simultaneously, but also when they are not simultaneous. In the latter case, the test sound has to be a short burst or sound impulse which can be presented before the masker stimulus is switched on. The masking effect produced under these conditions is called pre-stimulus masking, shorted to "premasking" (the expression "backward masking" is also used). This effect is not very strong, but if the test sound is presented after the masker is switched off, then quite pronounced effects occur. Because the test sound is presented after the termination of the masker, the effect is called post-stimulus masking, shorted to "postmasking" (the expression "forward masking" is also used).

4.1 Masking of Pure Tones by Noise

Different kinds of noises are common in psychoacoustics. As described in Sect. 1.1, white noise represents a broad-band noise most easily defined in physical terms. The spectral density of white noise is independent of frequency; it produces no pitch and no rhythm. The frequency range of white noise in auditory research is limited to the 20 Hz to 20 kHz band. Besides white noise, there exist noises such as pink noise, in which high frequencies are attenuated. Another important broad-band noise discussed in this section is called uniform masking noise. Strong frequency dependence in the spectral density of noise leads to narrow-band noise and to low-pass or high-pass noise. If masking effects on the slopes of such noises are sought, then care has to be taken to produce slopes of the attenuation of the noise as a function of frequency, that are at least as steep as the frequency selectivity of our hearing system.

4.1.1 Pure Tones Masked by Broad-Band Noise

White noise is defined as having a frequency-independent spectral density. Figure 4.1 shows threshold level as a function of the frequency of the test tone, in the presence of a white noise with several different density levels.

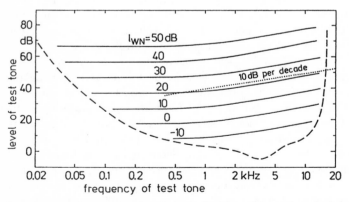

Fig. 4.1. Level of test tone just masked by white noise of given density level l_{WN}, as a function of the test-tone frequency. The dashed curve indicates the threshold in quiet

Threshold in quiet, described in Chap. 2, is indicated by the broken line. Although white noise has a frequency-independent spectral density, the masked thresholds, indicated by solid lines, are horizontal only at low frequencies. Above about 500 Hz, the masked thresholds rise with increasing frequency. The slope of this increase corresponds to about 10 dB per decade, illustrated by the dotted line. At low frequencies, the masked thresholds lie about 17 dB above the given density level. Thus numbers representing the values of spectral density, l_{WN}, indicate that even negative values of the density level produce masking. Increasing the density level by 10 dB shifts the masked threshold upwards by the same 10 dB. This interesting result indicates the linear behaviour of masking produced by broad-band noises. At very low and very high frequencies, masked thresholds are the same as the threshold in quiet. It is interesting to note that the strong individual differences in the dependence of threshold in quiet on frequency almost completely disappear when thresholds masked by broad-band noises are measured – an effect that is based on the ear's frequency selectivity representing masker and test tone within the same band.

For some measurements, a masked threshold independent of frequency over the entire audible frequency range is required. Such a masking curve can be produced by a special noise with a density level that depends on frequency. Such a noise represents a mirror image of the frequency dependence of the masked threshold for white noise. The attenuation of a network, which has to be put in series with a white-noise generator to produce such a uniform masking noise, is shown in the upper panel of Fig. 4.2. The resulting noise is called uniform masking noise, because it produces – as shown in the lower

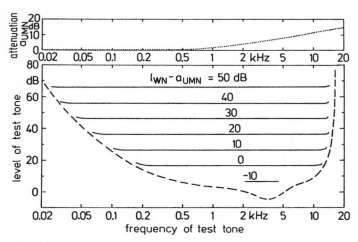

Fig. 4.2. Level of test tone just masked by uniform masking noise of given density level as a function of the frequency of the test tone. The upper curve (*dotted line*) shows the attenuation needed to produce uniform masking noise from white noise

panel of Fig. 4.2 – a masked threshold that is independent of frequency. In this case, the parameter is given as the density level of white noise from which the attenuation of the network is subtracted. Because this attenuation is zero at frequencies below about 500 Hz, the masked thresholds indicated in Fig. 4.2 are the same as those shown in Fig. 4.1 for this low frequency range.

4.1.2 Pure Tones Masked by Narrow-Band Noise

In this context, narrow-band noise means a noise with a bandwidth equal to or smaller than the critical bandwidth (about 100 Hz below and $0.2\,f$ above 500 Hz, as outlined in Chap. 6). It is more meaningful when narrow-band noise is used to give data in terms of the total level of the noise instead of its density level. Using the equations given in Sect. 1.1, it is easy to transform the density level into the total level once the bandwidth is known. Figure 4.3 shows the thresholds of pure tones masked by critical-band wide noise at centre frequencies of 0.25, 1, and 4 kHz. The level of each masking noise is 60 dB and the corresponding bandwidths of the noises are 100, 160, and 700 Hz, respectively. The slopes of the noises above and below the centre frequency of each filter are very steep (more than 200 dB/octave), in order to exceed the frequency selectivity of our hearing system. The frequency dependence of the threshold masked by the 1-kHz narrow-band noise is very similar on the axes of Fig. 4.3 to that produced by the 4-kHz narrow-band noise. The frequency dependence of the threshold masked by the 250-Hz narrow-band noise, however, seems to be broader. A second effect is also noticeable: the maximum of the masked threshold shows the tendency to be lower for higher centre frequencies of the masker, although the level of the narrow-band masker is 60 dB at all centre frequencies. The difference between the maximum of the masked thresholds and the horizontal dashed

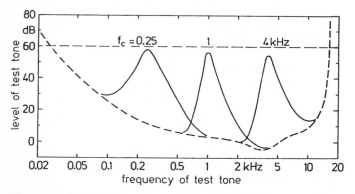

Fig. 4.3. Level of test tone just masked by critical-band wide noise with level of 60 dB, and centre frequencies of 0.25, 1, and 4 kHz. The broken curve is again threshold in quiet

line in Fig. 4.3, indicating the 60-dB test-tone level, amounts to 2 dB for 250-Hz, 3 dB for 1-kHz, and 5 dB for 4-kHz centre frequency. Ascending from low frequencies, masked thresholds show a very steep increase, and after reaching the maximum, a somewhat flatter decrease. The increase amounts to about 100 dB per octave. This steep rise indicates the need for very steep filters, otherwise the frequency response of the filter and not that of our hearing system is measured.

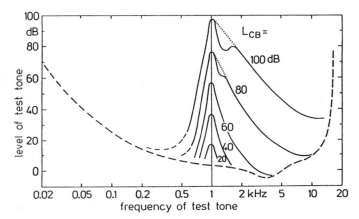

Fig. 4.4. Level of test tone just masked by critical-band wide noise with centre frequency of 1 kHz and different levels as a function of the frequency of the test tone

Figure 4.4 shows the dependence of masked threshold on the level of a noise centred at 1 kHz. All masked thresholds show a very steep rise from low to higher frequencies before the maximum masking is reached. The slope of this rise seems to be independent of the level of the noise masker, and the maximum always is reached 3 dB below the level of the masking noise. Beyond the maximum, the masked thresholds decay towards lower levels quite quickly for low and medium masker levels. At higher masker levels, however, the slope towards high frequencies becomes increasingly shallow. Therefore, the frequency dependence of the masked threshold is level-dependent or nonlinear. The nonlinear rise of the upper slope of the masked threshold with masker level is an interesting effect which plays an important part both in masking and in other auditory phenomena. The dips indicated in Fig. 4.4 for masker levels of 80 and 100 dB stem from nonlinear effects in our hearing system, which lead to audible difference noises created by interaction between the test tone and the narrow-band noise. With increasing test-tone level, the subject reaches threshold by listening for anything additional; in this case, it is the difference noise and not the test tone that is heard. The latter only becomes audible when the test-tone level is increased to the values indicated by the dotted lines.

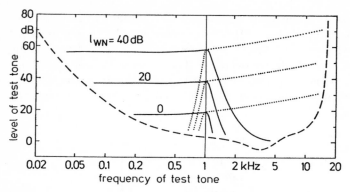

Fig. 4.5. Level of test tone just masked by low-pass noise (*solid curves*) and high-pass noise (*dotted curves*) for different density levels of the noises as a function of the frequency of the test tone. The cut-off frequencies of the high-pass and low-pass noise are 0.9 and 1.1 kHz, respectively

4.1.3 Pure Tones Masked by Low-Pass or High-Pass Noise

The masking of pure tones by white noise limited by a steep low-pass filter (solid lines) or a very steep high-pass filter (dotted lines) with cut-off frequencies of 1.1 and 0.9 kHz, respectively, is shown in Fig. 4.5. The parameter, as for white noise, is the density level. The masked thresholds decrease at the cut-off frequency not with the steepness of the attenuation of the noise, but in the form shown in Fig. 4.4 for the masking of narrow-band noise. Below the cut-off frequency of the low-pass noise, the masked thresholds are the same as found using white noise as masker. The same holds true for frequencies of the test tone above the cut-off frequency of the high-pass noise. There, the masked threshold increases with the test-tone frequency by about 10 dB per decade. This means that masking on the slopes produced by band-limited noises can be approximated by the sound pressure levels of the masker falling within the critical band at the cut-off. The slopes found with narrow-band maskers show up again in the masked thresholds produced by low-pass (solid) and high-pass noises (dotted). This result indicates that the masked thresholds produced by narrow-band noises and shown in Figs. 4.3 and 4.4, play an important part in describing masking effects of noise maskers with different spectral shapes.

4.2 Masking of Pure Tones by Tones

In this section the masking of pure tones by pure tones and tonal complexes is discussed.

4.2.1 Pure Tones Masked by Pure Tones

Although the stimuli needed to study the masking of pure tones by pure tones are simple, such masking experiments have many difficulties especially at medium and higher levels of the masker. Figure 4.6 shows the threshold of a test tone as a function of its frequency when masked by a 1-kHz masker at a level of 80 dB. As in all the measurements described in the preceding sections, the subject responds as soon as the presence of the test tone produces some sensation in addition to the sensation of the steady-state-masker (detection of anything). An effect that appears to be quite dominant in this case is that beats are audible when the frequency of the test tone is in the neighbourhood of the 1-kHz masker. For example, a test tone presented at a frequency of 990 Hz and a level of 60 dB produces a beating quality at 10 Hz. The subject listening to such a beating tone hears something different from the steady-state masker and therefore responds, although the criterion is very different from hearing an additional tone. Considering the whole frequency range from 500 Hz to 10 kHz it is clear that beating becomes audible in two regions around 2 and 3 kHz in addition to the region around 1 kHz.

In addition to the problem of beats, another difficulty arises for inexperienced subjects. At test-tone frequencies near 1.4 kHz, the subject indicates audibility of an additional tone at the relatively low test-tone level of 40 dB. A careful examination of these results and discussions with experienced subjects show that inexperienced subjects do not hear the test tone at that frequency and level, but a difference tone near 600 Hz. This difference tone is produced through nonlinear distortions that originate in our own hearing system. The threshold of this difference tone is not the threshold of the test tone we are seeking. The test tone with its appropriate pitch is only detected at levels

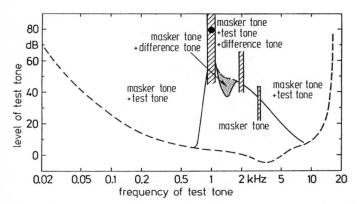

Fig. 4.6. Level of test tone just masked by a masking tone (1 kHz, 80 dB) as a function of the test-tone frequency. The different areas are characterized by the different sensations. The cross-hatched areas, for example, characterize regions of beating

above about 50 dB. Only experienced subjects can differentiate between the threshold of the difference tone and the threshold of the test tone.

To explain this complicated situation, different regions in the plane outlined in Fig. 4.6 are marked with indications of which sounds are heard by the subject. Below threshold in quiet of the test tone (broken line), nothing but the masker is audible. Below about 700 Hz, increasing the level of the test tone above threshold in quiet produces a region in which the masker tone and the test tone are audible. At frequencies between about 700 Hz and 9 kHz, the 80-dB 1-kHz masker produces a region in which only the masker tone is audible, even though threshold of the test tone in quiet is much lower. Areas of audible beats are marked by hatching. The region in which only the masker tone and the difference tone (but not the test tone) are audible is marked by stippling. Above masked threshold of the test tone at frequencies between 1 and 2 kHz, difference tones are also audible. All these results indicate that thresholds of tones masked by tonal maskers are far more difficult to measure than thresholds of tones masked by noise.

None the less, with well-trained subjects and some special equipment to reduce the audibility of the difference tones, thresholds of the test tones masked by tonal maskers can be measured or at least estimated. The region of beats cannot be avoided but one data point, where the frequency of the test tone is identical to that of the masker, can be measured. For the point shown, the test tone was 90° out of phase with the masker. Figure 4.7 shows average results from many subjects using this method. Individual differences are larger for such measurements relative to those obtained with noise maskers. In contrast with the results shown in Fig. 4.4, the data in Fig. 4.7 indicate a clear tendency for the slope towards lower frequencies to become less steep with decreasing masker level. On the other hand, slopes towards higher frequencies become shallower with *increasing* level of the masker. The pronounced

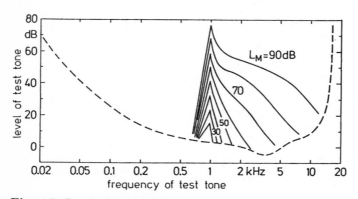

Fig. 4.7. Level of test tone masked by 1-kHz tones of different level as a function of frequency of the test tone. The shape of curves in the neighbourhood of 1 kHz can only be estimated

maximum in the neighbourhood of the masker tone occurs at similar frequencies to those found for masking by narrow-band noise. However, the peak of masked threshold is reduced with tonal maskers.

The different behaviour of the high and low frequency slopes at low levels produces an effect that is somewhat unexpected. At low levels, a greater spread of masking towards the lower frequencies than towards the higher frequencies occurs. At high levels, this behaviour is reversed, so that a greater spread of masking is found towards higher frequencies than towards lower frequencies. While the effect at higher levels is well known from masking with narrow-band maskers, the effect at low levels is rather unexpected. Between these two level ranges, i.e. near a masker level of about 40 dB, the masking patterns are approximately symmetrical. This effect is found at all frequencies for which it is sensible to distinguish between different low- and high-frequency slopes. Figure 4.8 illustrates the findings in more detail. The sensation level of the test tone, i.e. the level above threshold in quiet, is used as the ordinate and is indicated by solid lines. The dotted lines show exactly the same data with an inverse frequency scale (upper abscissa) mirrored at 1 kHz. This superposition illustrates the inversion of the masking characteristic with increasing level. At a 20-dB masker level, more spread of masking towards lower frequencies occurs; at 40 dB, masking is nearly symmetrical and more spread of masking towards higher frequencies shows up at 60 dB.

The spread of masking towards higher frequencies shows a strong dependence on masker level as already indicated in Fig. 4.4. This effect can be

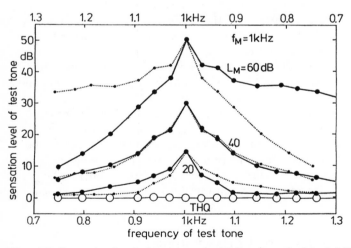

Fig. 4.8. Sensation level of test tone (*solid lines*) masked by 1-kHz tones of different level as a function of the frequency of the test tone. Threshold in quiet in this case is a horizontal line at 0 dB. The dotted lines are the same data as the solid lines with an inverted frequency scale (*upper abscissa*) mirrored at 1 kHz. This superposition allows the inversion of the masking characteristic with increasing masker level to be readily seen

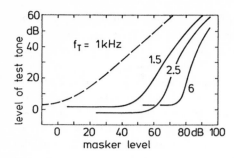

Fig. 4.9. Level of test tone masked by a masker at 1 kHz as a function of the masker level. The parameter is the frequency of the test tone

illustrated more clearly if the abscissa and the parameter of Figs. 4.7 and 4.8 are exchanged. In this way, Fig. 4.9 is created with the level of the test tone again as ordinate, but with the frequency of the test tone as the parameter and the level of the masker as the abscissa. In such a display, an identical increment of masker level and test-tone level would produce a 45° line, which is only approximated by the data for the 1-kHz test tone 90° out of phase with the masker (broken line). However, the increment in this case is a little less over the whole range of masker level. The failure to produce a 45° line exactly is called the near-miss of Weber's law, which describes the audibility of an increment in level for tones. The higher the test-tone frequency, the more the slopes of the rising curves deviate from the 45° slope. The solid lines in Fig. 4.9 represent test-tone frequencies above the masker frequency. The lines remain flat at low masker levels at the threshold in quiet but rise more and more steeply with increasing test-tone frequency. Instead of a slope of 1, the curve for the test-tone frequency of 6 kHz shows a slope as high as 3. Hence the increase in threshold level of the test tone is three times larger than the increase in masker level. The data given in Fig. 4.9 represent average values. Individual data for single subjects sometimes yield slopes as steep as 6. This means that an increment in masker level of 1 dB can produce an increment in the masked threshold of the test tone of up to 6 dB.

The results displayed in Figs. 4.7 to 4.9 also hold for other masker frequencies if appropriate scales are chosen. Here, the effects shown with narrow-band noise maskers appear again: except at frequencies of the masker below 500 Hz where the masked thresholds as a function of the test-tone frequency appear to be broader, the shape of the curves may be predicted by shifting the whole curves in Figs. 4.4, 4.7 and 4.8 horizontally until the maximum appears at the masker frequency.

4.2.2 Pure Tones Masked by Complex Tones

Pure tones appear relatively rarely in nature. Only some bird songs and the sounds produced by a flute can be considered to be pure tones. Most of the instrumental sounds in music are composed of a fundamental tone and many harmonics. The difference in timbre produced by different musical instru-

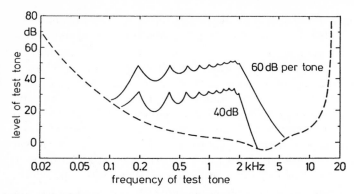

Fig. 4.10. Level of test tone masked by ten harmonics of 200 Hz as a function of the frequency of the test tone. The levels of the individual harmonics of equal size are given as the parameter

ments depends on the frequency spectra of their harmonics. Whereas a flute produces primarily one single component, the fundamental, a trumpet produces many harmonic partials and therefore elicits a much broader masking effect than a flute. Figure 4.10 shows thresholds of pure tones, masked by a complex tone composed of a 200-Hz fundamental frequency and nine higher harmonics, all with the same amplitude but random in phase. The masked thresholds are given for sound pressure levels of 40 and 60 dB of each partial. On the logarithmic frequency scale, the distance between the partials is relatively large at low frequencies, but becomes very small between the ninth and tenth harmonic. Accordingly, the dips between the harmonics become smaller and smaller with increasing frequency of the test tone. In the frequency range between 1.5 and 2 kHz, the maxima and the minima can hardly be distinguished. At frequencies above the last harmonic, in our case 2 kHz, the masked thresholds are flatter towards higher frequencies at higher levels of the masking complex. At frequencies one to two octaves above the highest spectral component, masked thresholds approach threshold in quiet. In music, many complex tones, each composed of many harmonics, are used at the same time. This means that the corresponding masking effect can be assumed to produce shapes similar to those outlined in Fig. 4.10. However, the minima between the lines become even smaller because the density of the lines is higher.

It should be noted here, that non-random phase conditions of the components can lead to temporal envelopes of the sound that can be described as impulsive. Consequently, temporal effects in masking may become a crucial factor in determining masked thresholds. Effects of this kind are discussed in Sect. 4.4.

The masking patterns produced by narrow-band noise maskers or by pure-tone maskers show differences despite the same level and the same (centre) frequency: prominent differences occur with respect to the level dependence of the slope towards lower frequencies. It is possible to approximate noise by a relatively small number of equal-sized pure tones, the frequencies of which are spread randomly within the bandwidth of the "noise". Thus, it may be reasonable to measure masking with an increasing number of tones and to compare the effects with the masking effects produced by narrow-band noise. Figure 4.11 gives an example for a centre frequency of 2 kHz (left) and an overall level of 70 dB. The critical bandwidth at that frequency is about 330 Hz. The approximation of the critical-band wide noise starts with just one tone at 2 kHz, continues with two tones at 1910 and 2100 Hz, or 1840 and 2170 Hz, and ends with five tones at frequencies of 1840, 1915, 2000, 2080, and 2170 Hz (right). The corresponding masking produced at the low frequency side is illustrated in Fig. 4.11. Again, the test-tone level is shown as a function of frequency (upper scale) or of critical-band rate (lower scale). The masking effect produced by the tone or the combination of tones is shown by open symbols connected with solid lines.

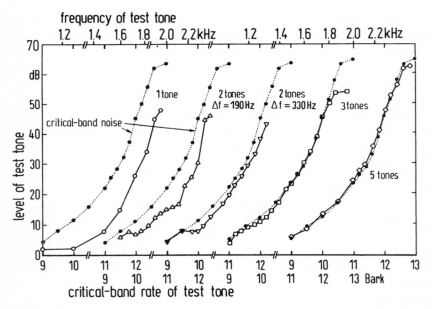

Fig. 4.11. Level of test tone just masked by a number of tones within the critical band around 2 kHz as a function of the frequency (*upper scale*) and critical-band rate (*lower scale*) of the test tone. The total level of the maskers is kept constant at 70 dB. The dotted curve indicates data measured with critical-band wide noise as the masker. These data should be compared with the data produced by the different number of tones (*solid lines* and *symbols*). The more tones that are used to approximate the narrow-band noise, the better the coincidence of the corresponding curves. Note the shift of the abscissa scale corresponding to each set of curves

The data displayed in Fig. 4.11 show very clearly that a single masker tone is an inappropriate approximation of a narrow-band noise masker. Two tones produce masking effects relatively close to those produced by narrow-band noise, provided the distance between the two tones is chosen in such a way that the tone frequencies correspond closely to the lower and higher cut-off frequency of the narrow-band noise. However, there still remain differences of up to 7 dB between the two masking curves. An approximation of the narrow-band noise by five tones produces an almost identical masking curve. The remaining differences are within the accuracy of measurement.

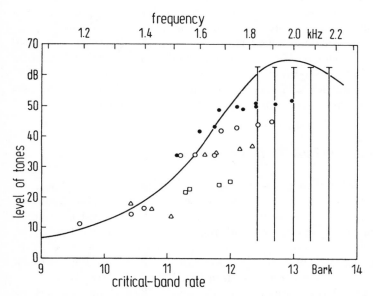

Fig. 4.12. Masking pattern and difference tones. The solid curve indicates the level of a test tone just masked by five tones, each with an SPL of 63 dB near 2 kHz (*vertical lines*), as a function of frequency (*upper scale*) or critical-band rate (*lower scale*). The symbols indicate levels and frequencies of estimated difference tones of odd order, n ($n = 3$: *dots*; $n = 5$: *open circles*; $n = 7$: *triangles*; $n = 9$: *squares*)

Because five tones produce the same masking as a narrow-band noise, but one single tone produces a much steeper masking slope, it may be possible to find the reason for this difference using the five-tone complex. Using a special procedure, which is explained in detail in Chap. 14, the level of the difference tones produced by the five-tone complex can be estimated. The difference tones of odd order play the most important role. The level of all these difference tones, as estimated through subjective measurements, is displayed in Fig. 4.12. The difference tones of third, fifth, seventh, and ninth order are indicated by different symbols. The frequencies of the five-tone complex are indicated by vertical lines. The threshold of a pure tone masked by the five-tone complex is given as a solid curve. A comparison between frequency and level of the difference tones on the one hand, and the masked threshold on the other, suggests that the masked threshold in the frequency range of 1300 to 1700 Hz is due to the difference tones which also produce masking. Therefore, it can be assumed that the frequency selectivity of the ear remains the same, irrespective of whether a narrow-band noise or a tone is used as the masker. However, the internally produced nonlinear components, either difference tones or difference noises (in the case of narrow-band noise as the masker) change the physical stimulus into an internal stimulus which is broader and therefore produces more masking at the low-frequency side of the masker. At the high-frequency side, this effect does not appear to play a role because masking is already spread much more towards higher frequencies. Therefore, it can be assumed that the frequency selectivity measured with a pure tone as the masker, although level dependent, is the largest that is possible. Masking produced by a narrow-band noise is somewhat less selective at the low-frequency side. Due to the appearance of distortion products (in this case continuous spectra) it is almost level independent.

4.3 Psychoacoustical Tuning Curves

Masking effects of tones by tones can be plotted in different ways. There are four variables, the frequency and level of the test tone and the frequency and level of the masker. Normally the threshold level of the test tone in the presence of a masker of given level and frequency is plotted as a function of test-tone frequency. An example of such a masking pattern obtained with the method of tracking is outlined in Fig. 4.13a together with threshold in quiet. It can be compared with the data given in Fig. 4.7, which have already been discussed in detail. Psychoacoustical tuning curves follow a different pattern. There, the masker level needed in order to mask a test tone of a given low level and frequency is plotted as a function of the masker frequency. Such a curve can also be measured using the tracking method. In this case, the subject listens to the test tone but varies the level of the masker to make the test tone audible and then inaudible. The result, displayed in Fig. 4.13b, indicates an inverted curve (called tuning curve) relative to the classical masking pattern of Fig. 4.13a. Two aspects are typical for tuning curves as shown in Fig. 4.13b: the slope towards low frequencies is shallower than the slope towards higher frequencies, and the minimum is reached at a masker frequency a little bit above the frequency of the test tone (indicated by an asterisk).

The interchange of parameters and ordinates changes the characteristics of masking curves, the best known being classic masking curves. They show the just-audible level of the test tone masked by a pure-tone masker, as a function of the frequency of the test tone with the level of the masker as a parameter. These curves depend on the frequency of the masker as shown in

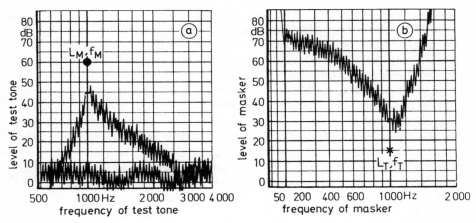

Fig. 4.13a,b. Examples of the use of the tracking method to measure continuously threshold in quiet and masked threshold, as a function of the frequency of test tone (**a**) and psychoacoustical tuning curve (**b**). The level of the test tone as a function of its frequency when masked by the indicated tonal masker (*dot*) is shown in (**a**). The level of the masker is the ordinate in (**b**), whereas frequency of the masker is the abscissa; the level of the test tone and its frequency are kept constant at the values indicated by the asterisk

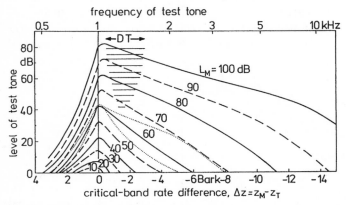

Fig. 4.14. Normalized classic masking patterns: level of test tone just masked by a masker tone, the level of which is indicated. The abscissa is either the frequency of the test tone (*upper scale*) or the critical-band rate difference between masker and test tone (*lower scale*). The two thin dotted curves indicate extreme individual deviation of two subjects from the average for the masker level of 60 dB. The hatched area indicates the range within which difference tones (DT) are audible and disturb the measurements

Fig. 4.3. The dependence on masker frequency can be avoided if the critical-band rate is used as the abscissa instead of a logarithmic frequency scale. The critical-band rate will be discussed in detail in Sect. 6.2 but its advantage in normalizing the classic masking patterns will be used here; the abscissa will be transformed from the frequency of the test tone into the critical-band rate difference, Δz, between the critical-band rate z_M of the masker frequency and the critical-band rate z_T of the test-tone frequency. Averaged data of such a pattern are plotted in Fig. 4.14, on the assumption that the threshold level for the test tone in quiet is always adjusted to 0 dB. A transformation of this set of classic masking patterns into a set of data similar to those displayed in Fig. 4.9 is shown in Fig. 4.15. These data can be read either as the test-tone level just masked by the masker level, or as the masker level necessary to just mask a test tone of level L_T. The critical-band rate difference, Δz, is the parameter. The typical characteristics of masking mentioned above are again recognized in Fig. 4.15. The curves with the positive and those with the corresponding negative values of the parameter Δz cross each other (marked by open circles) for masker levels near 35 to 40 dB. This means that the masking curve is shaped symmetrically for these masker levels. For masker levels above 40 dB, negative values of Δz lead to values of the test-tone level which are larger in relation to those reached for positive values of Δz. For example, a chosen masker level of 60 dB leads to 27 dB test-tone level for $\Delta z = -2$ Bark, but to 8 dB for +2 Bark. At masker levels below 40 dB, the conditions are reversed indicating that masking spreads more towards lower frequencies than towards higher frequencies. Figure 4.15 represents a set of

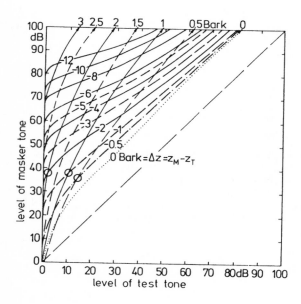

Fig. 4.15. The same data as in Fig. 4.14 with masker level as the ordinate, test-tone level as the abscissa and the difference, Δz, between critical-band rate of masker z_M and that of test tone z_T as the parameter. Such a set of curves can be used to construct psychoacoustical tuning curves outlined in Fig. 4.16. Open circles indicate points where equal positive and negative parameter values Δz cross each other

data out of which classic masking patterns as well as psychoacoustical tuning curves can be constructed.

Psychoacoustical tuning curves can be read from Fig. 4.15 most easily by keeping the test-tone level constant at a relatively small value. Although the test-tone level should not exceed 20 dB for psychoacoustical tuning curves, corresponding to neurophysiological tuning curves, it is of interest to have the whole set of tuning curves available. Figure 4.16 shows such a set of psychoacoustical tuning curves in a normalized fashion, i.e. with the critical-band rate difference, Δz, as the abscissa. The masker level necessary to just mask a test tone of a level L_T (parameter) and a frequency, f_T, expressed in z_T, is plotted as a function of masker frequency, f_M, expressed in critical-band rate difference, $\Delta z = z_M - z_T$, i.e. the difference between the critical-band rate of the masker and that of the test tone. As shown in Fig. 4.13b, such tuning curves can be measured directly by experienced subjects using a Békésy-tracking method. All the curves plotted in Fig. 4.16 are asymmetric, with a tendency towards larger asymmetry at higher test-tone levels. At low frequencies, i.e. for large negative values of Δz, the tuning curves seem to run parallel. The same holds for positive values of Δz. However, for small levels of the test tone, the tuning curves show a more pronounced dip towards lower masker levels that deepens as the test-tone level becomes lower. Comparing Figs. 4.14 to 4.16, it is clear that the classic masking curves and the psychoacoustical tuning curves represent identical data that can be transformed from one to the other by simply interchanging ordinates and parameters.

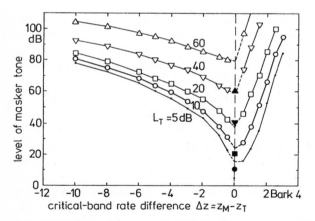

Fig. 4.16. Normalized psychoacoustical tuning curves, i.e. level of the masker tone necessary to just mask a test tone of the level indicated (parameter), as a function of the frequency of the masker expressed as the difference between critical-band rate of the masker and the test tone. The filled symbols indicate the level of the test tone used to produce the corresponding tuning curves

4.4 Temporal Effects

Masking in steady-state condition with long-lasting test and masking sounds, was described in previous sections. However, the transmission of information in music or speech implies a strong temporal structure of the sound. Loud sounds are followed by faint sounds and vice versa. In speech, the vowels generally represent the loudest parts whereas consonants are relatively faint. A plosive consonant is a typical example of a sound that is often masked by a preceding loud vowel. The effect occurs not only because of the reverberation of the room in which speech is received, but also in free-field conditions, because of the temporal effects of masking which characterize our hearing system.

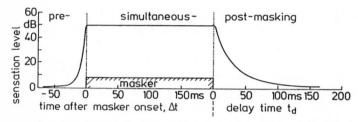

Fig. 4.17. Schematic drawing to illustrate and characterize the regions within which premasking, simultaneous masking and postmasking occur. Note that postmasking uses a different time origin than premasking and simultaneous masking

To measure these effects quantitatively, maskers of limited duration are presented and masking effects tested with short test-tone bursts or short pulses. Further, the short signal is shifted in time relative to the masker, as illustrated in Fig. 4.17, where a 200-ms masker masks a short tone burst with a duration as small as possible and negligible in relation to the duration of the masker. In such a case, it is advantageous to use two different time scales: in the first, the value Δt corresponds to the time relative to the onset of the masker – both positive and negative values exist. The second time scale starts at the end of the masker. This time is often called delay time and indicated by t_d. It is convenient to use as the ordinate not the sound pressure level of the test-tone burst, but the level above the threshold of this sound. This level is referred to as the sensation level. Three different temporal regions of masking relative to the presentation of the masker stimulus can be differentiated. Premasking takes place during that period of time before the masker is switched on. In this period, negative values of Δt apply. The period of pre-stimulus masking is followed by simultaneous masking when the masker and test sound are presented simultaneously. In this condition, Δt is positive. After the end of the masker, post-stimulus masking, normally called postmasking, occurs. During the time scale given by positive delay time, t_d, the masker is not physically existent; nevertheless, it still produces masking.

The effect of postmasking corresponds to a decay in the effect of the masker and is more or less expected. Premasking, however, represents an effect that is unforeseen because it appears during a time before the masker is switched on. This does not mean, of course, that our hearing system is able to listen into the future. Rather, the effect is understandable if one realizes that each sensation – including premasking – does not exist instantaneously, but requires a build-up time to be perceived. If we assume a quick build-up time for loud maskers and a slower build-up time for faint test sounds, then we can understand why premasking exists. The time during which premasking can be measured is relatively short and lasts only about 20 ms. Postmasking, on the other hand, can last longer than 100 ms and ends after about a 200-ms delay. Therefore, postmasking is the dominant non-simultaneous temporal masking effect.

4.4.1 Simultaneous Masking

Both threshold in quiet and masked thresholds depend on the duration of the test sound. These dependencies have to be known in order to discuss temporal effects in non-simultaneous masking, because very short test sounds are needed for such measurements. Two dependencies can be differentiated: one is the dependence of thresholds on the duration of a single test sound, and the other is the dependence on repetition rate of repeated short test sounds. Figure 4.18 illustrates the first of these dependencies for bursts of a sinusoidal test tone. Threshold in quiet is shown by dotted lines and thresholds for tone bursts masked by uniform masking noise of 40 and 60 dB sound pressure level are indicated by solid lines. Both dependencies, that as a function of duration and that as a function of repetition rate, are identical for threshold in quiet and for masked threshold. The dependence on duration shows a constant

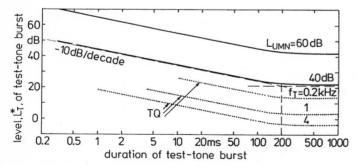

Fig. 4.18. Level of just-audible test-tone bursts, L_T^*, as a function of duration of the burst in quiet condition (TQ, *dotted curves*, for three frequencies of test tones) and masked by uniform masking noise of given level (*solid curves*). Note that level, L_T^*, is the level of a continuous tone out of which the test-tone burst is extracted. Broken thin lines mark asymptotes

test-tone threshold for durations longer than 200 ms corresponding to that
of long-lasting sounds. For durations shorter than 200 ms, threshold in quiet
and masked threshold increase with decreasing duration at a rate of 10 dB
per decade. This behaviour can be described by assuming that the hearing
system integrates the sound intensity over a period of 200 ms. The frequency
dependence of this effect is indicated by different curves for threshold in
quiet. With uniform masking noise, masked threshold becomes independent
of frequency (Fig. 4.2) and the two solid lines hold for all frequencies within
the audible range. The dependence on repetition rate shows similar functions
predicted by the assumption that the hearing system integrates the sound
intensity over 200 ms. The 200-ms interval corresponds to a repetition rate
of 5 Hz. Therefore, the thresholds are independent of repetition rate for repetition rates smaller than 5 Hz, but decrease for larger repetition rates until
finally the steady-state condition is reached. This occurs for 5-ms tone bursts
at a repetition rate of 200 Hz.

Reducing the duration of a tone burst produces a widening of its spectrum.
This effect limits the shortest useable duration to a value at which the spectral
width corresponds to that of the critical band. A quick rise time that occurs
with a narrow spectrum is produced by Gaussian rise and fall. Therefore
Gaussian-shaped tone bursts are usually used to measure temporal effects.

If sound bursts extracted from uniform exciting noise are used as the
test sound, the threshold in quiet as a function of the duration of the test
sound depends on duration in a way similar to that shown for tone bursts
(see dotted line in Fig. 4.19). Thresholds for noise bursts masked by uniform
masking noise, however, show a somewhat different dependence indicated in
Fig. 4.19 by solid lines. Approaching from steady-state condition, i.e. from
large durations, the masked threshold increases somewhat more gradually
as duration decreases in comparison with the increase measured with tone
bursts.

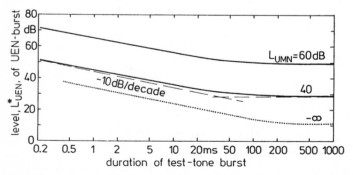

Fig. 4.19. Level of uniform-exciting noise bursts masked by continuous uniform
masking noise of a given level as a function of noise-burst duration. The dotted
curve corresponds to threshold in quiet. Broken thin lines mark asymptotes

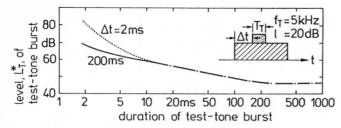

Fig. 4.20. Level of test-tone bursts presented 2 ms (*dotted*) or 200 ms (*solid*) after the onset of a white-noise masker with 20-dB density level, as a function of the duration of the 5 kHz test-tone burst. Note the difference between the two delay times at short durations (overshoot effect)

The difference between the dependencies for masked threshold and for threshold in quiet becomes understandable from the following fact: threshold in quiet is effectively measured for a small frequency range only, namely that for which the hearing system is most sensitive. This means that threshold for broad-band noise in quiet is, in effect, measured with a narrow-band test sound with a frequency content around 3 kHz. Hence, threshold in quiet for the broad-band noise is similar to that for a narrow-band centred near the most sensitive frequency range of the hearing system. Components of the noise that lie outside the most sensitive frequency range seem to be ignored.

The dependence on repetition rate shows similar effects as already discussed for tones. Masked threshold remains constant for very small repetition rates up to about 10 Hz. For larger repetition rates, it decreases again and finally reaches steady-state condition.

Within the range of simultaneous masking, additional effects exist shortly after the masker is switched on. The inset of Fig. 4.20 shows the temporal structure used for such an experiment. A high-frequency test tone (5 kHz) is used to produce short test-tone bursts. The masker burst is obtained from white noise and is long, relative to the duration of the test-tone burst. For a masker density level of 20 dB, the masked threshold is indicated in Fig. 4.20 as a function of the duration of the test-tone burst. If the test-tone burst is presented 200 ms after the onset of the masker, then no difference occurs in the dependence of the threshold on duration relative to what is known for the steady-state condition of the masker. For a very small value of Δt, i.e. 2 ms, the masked threshold for short durations shows a strong deviation from the expected behaviour: it rises by up to 10 dB at 2 ms. Such increments can be measured only for durations shorter than about 10 ms.

This effect (sometimes called the overshoot effect) also depends on the spectral content of masker and test sound. The effect disappears if the two sounds have similar spectra. Figure 4.21 gives examples for two different bandwidths of the masker. First, the masker is white noise (as used in Fig. 4.20), and the results are indicated by the dotted line. Second, the solid curve shows the results for a masker centred at 5 kHz and having the width

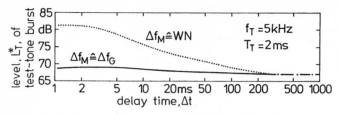

Fig. 4.21. Overshoot effect, i.e. level of 5-kHz test-tone burst of 2-ms duration as a function of the delay time. The masker has a density level of 25 dB at 5 kHz and is either only one critical band wide (*solid*) or broad-band white noise (*dotted*)

of the critical band. The delay time, Δt, between the onset of the masker and the onset of the 2-ms test-tone burst, is the abscissa. The overshoot effect for broad-band maskers is largest at very small Δt. It diminishes for larger values of Δt and disappears completely after about 200 ms. The overshoot effect does not appear for narrow-band maskers. The effects measured with broad-band noise may reach up to 10 dB and sometimes even more for very short values of T_T. It should be mentioned, however, that the individual differences are quite large – sometimes in the order of 10 dB.

4.4.2 Premasking

The temporal region in which premasking appears was introduced in Fig. 4.17. Premasking is mostly measured with single masker bursts, and reproducible results can only be obtained from trained subjects. Even for these subjects, the reproducibility of the results is worse than in simultaneous masking or post-masking.

It has not been possible so far to decide conclusively whether premasking depends on the duration of the masker. The dependence of premasking on the level of the masker can be characterized in such a way that premasking lasts about 20 ms in any condition. This means that the threshold remains unchanged until Δt reaches a negative value of 20 ms. After that, threshold rises and reaches the level found in simultaneous masking near the time at which the masker is switched on. Premasking alone plays a relatively secondary role, because the effect lasts only 20 ms and therefore is usually ignored. However, premasking and postmasking are discussed together in Sect. 4.4.4 in connection with temporal masking patterns.

4.4.3 Postmasking

A Gaussian-shaped condensation impulse with a duration of only 20 μs produces a spectral shape that corresponds to that of white noise. It is similar to that of a Dirac impulse. Postmasking produced by a white-noise masker can therefore be measured without spectral influences using such a brief Gaussian

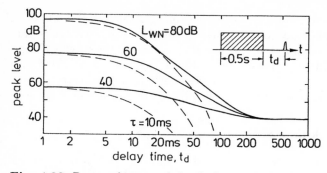

Fig. 4.22. Postmasking: peak level of a just-audible 20-μs Gaussian pressure impulse as a function of the delay time (*abscissa*) after the offset of white noise maskers of given levels. The broken curves indicate the form of exponential decay on the logarithmic abscissa

impulse as a test sound. The peak value of this Gaussian impulse expressed in level is plotted as the ordinate in Fig. 4.22, which shows the level necessary to just reach threshold in postmasking as a function of delay time t_d from the end of the masker. The parameter in the figure is the overall level of the white-noise masker. The solid lines indicate results that show almost no decay for the first 5 ms after the masker is switched off. There, the values correspond to those measured in simultaneous masking. After about 5 ms delay time, the threshold in postmasking decreases and, at about 200 ms delay time, reaches the threshold in quiet. The broken lines in Fig. 4.22 illustrate an exponential decay with a time constant of 10 ms. Comparison of the broken lines and the solid lines indicates that the effect of postmasking cannot be described as an exponential decay.

Postmasking depends on the duration of the masker. Figure 4.23 shows a typical result measured using, as a test sound in this case, a 2-kHz tone burst of 5-ms duration. Again, the time at which the test-tone burst is presented after the end of the masker is plotted as the abscissa. The level of the test-

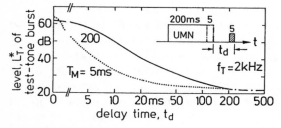

Fig. 4.23. Postmasking depends on masker duration: levels of just-audible test-tone bursts are indicated as a function of their delay time after the offset of maskers of 200-ms and 5-ms duration. The level of the uniform masking noise is 60 dB; duration of the 2-kHz test tone is 5 ms. Note that delay time in this case is the time between the end of masker and the end of test tone

tone burst is the ordinate. For a masker duration of 200 ms, the solid curve indicates postmasking comparable to that displayed in Fig. 4.22. Quite different from that is the postmasking produced by a masker burst, which lasts only 5 ms, as indicated by the dotted line in Fig. 4.23. In this case, the decay is initially much steeper. This means that postmasking depends strongly on the duration of the masker and therefore is a highly nonlinear effect.

4.4.4 Temporal Masking Patterns

Temporal masking patterns can be produced by maskers with a pronounced structure of the temporal envelope. Amplitude modulated tones or noises, frequency modulated tones, or the envelope fluctuation of narrow-band noise represent typical examples for sounds producing time-variant temporal masking patterns.

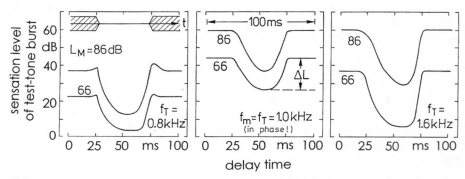

Fig. 4.24. Temporal masking patterns produced by a 10-Hz square wave amplitude-modulated 1-kHz tone. The sensation level of a 3-ms test-tone burst of given frequency is plotted as a function of the delay time, in fractions of the 100-ms masker period corresponding to 10 Hz. The level of the 1-kHz masker is the parameter; each panel corresponds to a different test-tone frequency, f_T

Simultaneous masking, postmasking, and premasking constitute the three parts of a temporal masking pattern. All three masking effects depend on the spectrum of the masker. For a broad-band noise masker, the frequency selectivity of the masking effect plays a secondary role; for narrow-band maskers, frequency selectivity becomes dominant. At the upper and lower frequency slopes of the masking pattern, masked threshold is much lower than in the frequency range of the narrow-band noise masker. The form of the temporal pattern remains similar for all frequencies. Figure 4.24 shows the temporal masking pattern produced by a 1-kHz masker which is square wave modulated with a Gaussian rise-fall time of 2 ms. The modulation frequency amounts to 10 Hz, so that the period of 100 ms is subdivided into 50 ms with the masker tone on and a 50-ms pause. In the figure, the temporal scale is chosen so that zero time corresponds to the middle of the masker's on-time. Consequently, the pause is in the middle of each diagram. The small inset at the top of the left panel illustrates the condition. The sensation level of the 3-ms test-tone burst of given frequency, f_T, is used as the ordinate. Test-tone frequency changes across the three panels of the figure, and the level of the masker is the parameter. For test-tone frequencies of 0.8 and 1 kHz, the shape of the solid lines indicating the temporal masking pattern is quite similar and shifted by about the difference in level between the two maskers. However, the distance between the two temporal masking patterns, especially within the pause, is larger for the test-tone frequency of 1.6 kHz indicated in the right panel. This corresponds to the nonlinear level dependence of the upper slope of masking which was discussed in Sect. 4.2.1. All temporal masking patterns indicate the relatively steep rise of premasking and the relatively slow decay of post-masking. However, the decay of postmasking is much shorter than might be expected from Fig. 4.22. This difference can be understood from the fact explained in Fig. 4.23; postmasking depends on the duration of the masker. For the temporal masking pattern displayed in Fig. 4.24, the duration of the masker is only 50 ms, i.e. much shorter than the 200 ms of Fig. 4.23, and the decay of postmasking is therefore steeper.

Figure 4.25 shows temporal masking patterns of a uniform masking noise, rectangularly modulated with 5 Hz (solid, lower abscissa) or 100 Hz (dotted, upper abscissa).

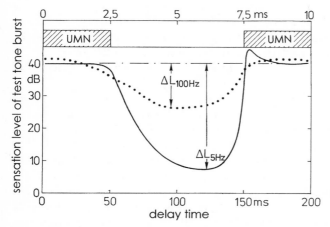

Fig. 4.25. Temporal masking patterns produced by uniform masking noise, square-wave modulated with modulation frequencies of 5 or 100 Hz. Test tone impulses of 3 ms duration at 3 kHz test tone frequency

For low modulation frequency (5 Hz) the temporal gap in the uniform masking noise is well resolved by the hearing system and the difference ΔL between the maximum and the minimum in the temporal masking pattern amounts to more than 30 dB. At higher modulation frequency (100 Hz), on the other hand, the level difference ΔL is significantly reduced in particular because of the effects of post masking. With some simplification it can be stated that short temporal gaps in a masker are "bridged" by post masking effects.

Sounds modulated periodically in frequency produce temporal masking patterns in a way similar to amplitude-modulated sounds. A typical example is that of a sinusoidally frequency-modulated pure-tone masker. In this case, the threshold has to be measured with short test-tone bursts presented at specific times during the period of the frequency modulation, and the frequencies of the test-tone bursts have to be chosen to cover at least the total range of frequency deviation. The parameters to be varied are the temporal position of the test-tone burst within the period of the modulation, the frequency of the test-tone burst, the frequency of modulation and the level of the masker. The value to be measured is always the level of the just-audible test-tone burst.

Sinusoidal frequency modulation (±700 Hz) of a 1500-Hz masker is used in the outlined example. The relationship between the instantaneous frequency of the frequency-modulated tone as a function of the temporal position within the period of the modulation, divided into eight segments, and the frequencies of the test-tone burst (horizontal dashed lines) are shown in Fig. 4.26. The corresponding critical-band rates of the test-tone frequencies are also indicated. The period of the modulation begins at the highest instantaneous frequency (2200 Hz), reaches its minimal frequency in the middle of the period and crosses the centre frequency of 1500 Hz at the temporal positions of 2/8 and 6/8 of the period. At the bottom, the time function of the test-tone burst with a frequency of 1200 Hz at a temporal position of 2/8 of the period is shown for a modulation frequency of 8 Hz (corresponding to a period of 125 ms).

Figure 4.27a shows temporal masking patterns produced by a frequency-modulated 1500-Hz tone with an SPL of 50 dB for eight different test-tone frequencies. The parameter in each of the eight panels is the modulation frequency. The key is in the bottom right panel. The data indicate clearly that the ear is able to follow the temporal course of the masking effect produced by the frequency-modulated tone up to frequencies of modulation of about 8 Hz. For higher modulation frequencies, the pattern loses its structure. At the highest modulation frequency of 128 Hz, the pattern is almost totally flat. This is also true for the higher SPL (70 dB) of the frequency-modulated masker indicated in Fig. 4.27b. The patterns at a test-tone frequency of 700 Hz with either 50 or 70 dB maskers show only one peak within the period. The peak is produced by the slope of masking below the lowest instantaneous frequency of the masker. At a frequency of 900 Hz, the peak of the temporal masking pattern becomes somewhat broader, because the frequency of the frequency-modulated masker reaches this test-tone frequency twice within

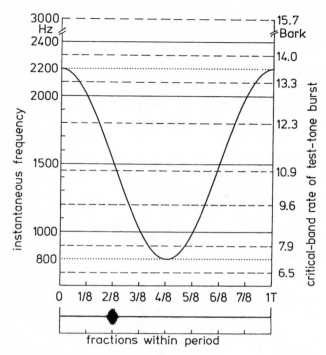

Fig. 4.26. Instantaneous frequency of a sinusoidally ($\pm$700 Hz) frequency-modulated 1.5 kHz tone as a function of time expressed in units of one eighth the period of the masker. The dashed horizontal lines indicate frequencies (*left scale*) and critical-band rates (*right scale*) of the test-tone burst used to measure masked threshold. A 1200-Hz 5-ms long test-tone burst is indicated at a temporal position of $2T/8$ in the lower part

a short temporal distance (see Fig. 4.26). At test-tone frequencies of 1200 and 1450 Hz, the pattern shows two peaks for lower modulation frequencies, indicating that the masker's instantaneous frequency surpasses the frequency of the test-tone burst twice within the period of modulation. Corresponding effects are shown for higher test-tone frequencies in Fig. 4.27. It can also be seen that – as expected – masking at higher levels extends further towards higher frequencies than it does at lower levels.

For some applications, it may be interesting to see these data plotted as a function of the frequency of the test-tone burst or of its critical-band rate. The masking patterns produced by the 70-dB frequency-modulated masker are replotted in Fig. 4.28 for two modulation frequencies, 0.5 Hz (upper panels) and 8 Hz (lower panels). The parameter is the position of the test-tone burst within the period of the masker. The two left panels indicate the masked threshold at the time incidence of the masker's extreme positions: for position

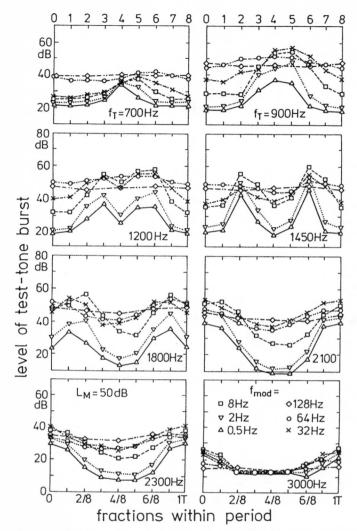

Fig. 4.27a,b. Temporal masking patterns, i.e. level of just-audible test-tone burst as a function of its temporal position within the interval, given in eighths of the masker period (FM tone 1.5 kHz ± 700 Hz). The parameter in each panel of given test-tone frequency is the frequency of modulation indicated by different symbols shown in the key in the lowest right panel of part (**a**). Masker-tone level is 50 dB in part (**a**) and 70 dB in part (**b**)

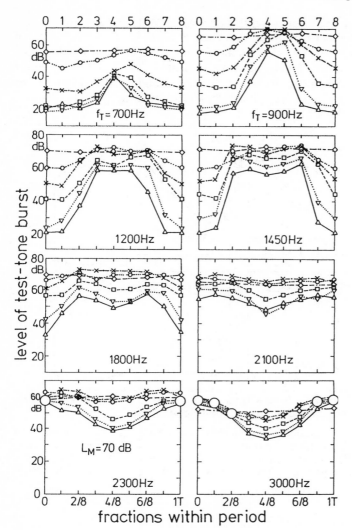

Fig. 4.27b. Caption see opposite page

"0" (2200 Hz) and for position "4" (800 Hz). The two right panels indicate the data for the zero-crossing of the frequency modulation, i.e. the 1500-Hz point, which is reached at the temporal positions "2" and "6". The patterns for the extreme values ("0" and "4") show the expected values. The same is true for the values of the zero-crossing ("2" and "6") at the 0.5-Hz modulation frequency. Closer inspection, however, shows that postmasking already influences these data, so that the "2" curve remains below the "6" curve for lower critical-band rates and vice versa for higher critical-band rates. The two curves "2" and "6" should be identical if premasking and postmasking were to behave similarly. Because postmasking lasts much longer than pre-masking, these differences are to be expected. For a modulation frequency of 8 Hz, this effect becomes more pronounced leading to differences of almost 20 dB in masked threshold for the positions indicated by "2" and "6" at lower critical-band rates. The difference is not as large at higher critical-band rates because of the relatively shallow upper slope of the masking pattern.

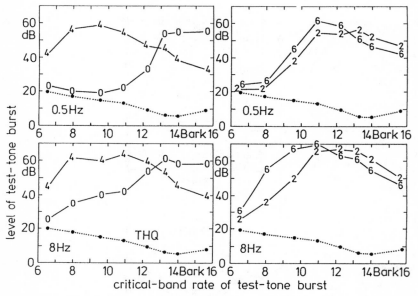

Fig. 4.28. Level of just-audible test-tone burst as a function of its critical-band rate. The parameter is the temporal position of the test-tone burst (given in eighths of the masker period). Modulation frequency of the 1.5 kHz, ±700-Hz FM masker is 0.5 Hz in the upper and 8 Hz in the lower two panels. Threshold in quiet (THQ) is indicated by dots

For many sensations which depend on temporal effects, the difference in masking within the period of the masker plays an important role. Figure 4.29 indicates masked thresholds for the positions of the extreme values of frequency deviation as well as for the zero-crossings as a function of the modulation frequency. The upper row holds for a masker level of 70 dB, the lower row for 50 dB. The three frequencies correspond to a low value of 900 Hz, a value somewhat above the centre frequency (2100 Hz), and a frequency of 3000 Hz, which is above the high extreme frequency value of 2200 Hz. The difference between the lowest and the highest curve within each graph indicates the level difference that masking follows within each period of the masker. This difference clearly diminishes for high modulation frequencies. The modulation frequency of 128 Hz corresponds to a period of about 8 ms. The 5-ms

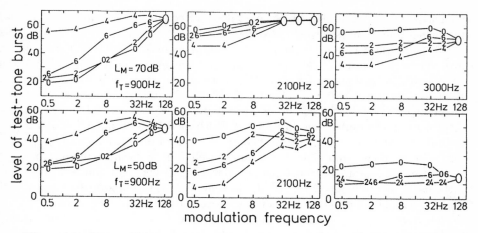

Fig. 4.29. Just-audible test-tone burst level as a function of the modulation frequency of the 1.5-kHz, ±700-Hz FM masker. The parameter is the temporal position of the test-tone burst given in eighths of the masker's period. The frequency of the test-tone burst is indicated in each panel. Masker level is 70 dB in the upper and 50 dB in the lower row

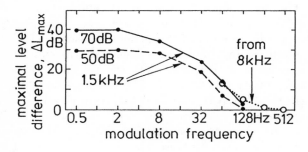

Fig. 4.30. Maximal difference of just-audible test-tone level within the period of the 1.5-kHz, ±700-Hz FM masker for 50-dB and for 70-dB masker level as indicated. The dotted part stems from data produced by an 8-kHz, ±2-kHz FM masker

duration of the test-tone burst is already larger than half of the period in this case. This means that threshold differences at this high modulation frequency may have been reduced by this inadequate relation. Additional measurements were produced at an 8-kHz centre frequency of the masker with a deviation of ±2 kHz. In this condition, test-tone bursts of only 1-ms duration still have a spectral width within the critical bandwidth. Therefore, measurements up to 512 Hz modulation frequency can be made. Such data indicate that a modulation frequency of 250 Hz produces a very small difference between maximal and minimal masking within the period, about 1 dB or less. Using this result together with the data given in Fig. 4.27, the maximal difference, $\Delta L_{\max}$, between thresholds within one period as a function of the modulation frequency is shown in Fig. 4.30 for 70 and 50 dB masker levels. The figure indicates that the hearing system can follow the modulation very well up to modulation frequencies of almost 8 Hz. For higher modulation frequencies, the value ΔL_{max} diminishes and reaches zero close to 250 Hz.

All these data can be summarized in the conclusion that the masking level versus critical-band rate versus time pattern for frequency-modulated sounds can be estimated for slow frequency changes up to a modulation frequency of about 8 Hz, directly from the steady-state transformation of masking pattern into the corresponding excitation versus critical-band rate pattern using the hints given in Sect. 6.4. With increasing frequency of modulation, the effect that has already been measured for amplitude modulation occurs for frequency-modulated maskers as well. This is because premasking (small time constant) and postmasking (larger time constant) influence the patterns. Their influence leads to the flattening of the temporal pattern of masking at high modulation frequencies above 250 Hz. In this region of modulation frequency, the sound acts as a steady-state sound, i.e. the temporal masking patterns disappear.

Many sounds that produce masking do not show any period. This is especially true for noise maskers. In order to be able to measure masking elicited by such noises, a special noise having a long period of several seconds has to be produced. Such noises are generated by repeated random sequences of pulses. The period of such noises is not audible to the subjects if it is longer than several seconds. Using a test-tone burst presented at a certain moment within a period of this artificial noise, a masking effect similar to that produced in temporal masking patterns can be measured. For a 200-ms long part of such a masker, the threshold of a 3-kHz test-tone burst of 2-ms duration is shown in Fig. 4.31 as a function of time. The masker in this case is a repetition noise at 4 kHz with a bandwidth of only 32 Hz and a period of 2 s. The solid line in Fig. 4.31 shows the logarithm of the envelope of sound pressure as a function of time. The value is indicated as peak level of the narrow-band noise masker on the left-hand ordinate. Because of the 32-Hz bandwidth of the masking noise, maxima and

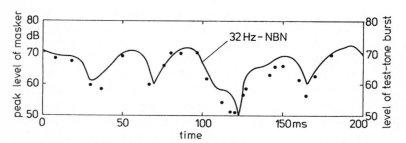

Fig. 4.31. Temporal masking pattern produced by a 32-Hz wide narrow-band noise centered at 4 kHz. The time course of its peak sound pressure level is indicated by the solid curve (*left ordinate scale*). The level of a 3-kHz test-tone burst of 2 ms duration just masked by this repeated 2 s long narrow-band noise measured at certain delay times is indicated by dots (*right ordinate scale*)

minima in the envelope of the masker are clearly visible. The average distance in time between these maxima or minima corresponds according to the formula in Chap. 1 to about 50 ms.

Masked threshold was measured on the low-frequency slope using 3-kHz tone bursts. Under this condition, peak levels of the envelope of the masker almost coincide with the sound pressure level of the test-tone burst needed to be just audible (indicated by dots and the right-hand ordinate scale). A comparison between the solid line and the dots, which represent the masked thresholds, shows agreement between the two values not only by average, but also with the temporal structure of the envelope of the masker. The narrower the bandwidth of the noise, the better the masking follows the temporal envelope of the masker. For broader noise bands, the envelope changes quickly and therefore postmasking fills up most of the valleys. Figure 4.31 indicates clearly that masking follows not only periodical changes, but also statistical changes.

4.4.5 Masking-Period Patterns

For measurements of masking-period patterns, the masking sound is the time-function of the masker's sound pressure itself and *not* that of its temporal envelope as with temporal masking patterns. In masking-period patterns, the sound pressure level or the sensation level of just-audible triggered sequences of short, high-frequency test-tone bursts are measured as a function of their temporal spacing throughout the period of the masker. Special, individually adapted probes as well as special microphones are necessary to measure temporal masking patterns in case of maskers with very low frequency (Fig. 3.10). Masking-period patterns seem to be a psychoacoustical equivalent of neurophysiologically measured period histograms. It is possible to measure such patterns for low-frequency maskers up to about 200 to 500 Hz, depending on the subject. The frequency of the test-tone burst can be chosen between a few hundred Hz and a few kHz, but the duration of the test-tone burst must be short in relation to the period of the masker. The upper limit of test-tone frequency is reached when the low-frequency masker no longer masks the test tone.

The spectral distributions and the time functions of the masker and test signal may illustrate the boundary conditions. Figure 4.32a indicates the spectral distribution of masking produced by a 100-Hz sinusoidal masker with a sound pressure level of 90 dB. The heavy dots indicate the frequency range in which masking occurs. The broken line marks threshold in quiet. The spectral content of the series of test-tone bursts with a centre frequency of 2.5 kHz, a duration of 1 ms, and a Gaussian rise/fall time of 0.5 ms (the repetition rate is identical to masker frequency) is indicated by the thin dotted curve. From the curves it becomes clear that a masking-period pattern can be measured only in that frequency range in which the masker produces a masking effect. The second boundary condition can be seen clearly in Fig. 4.32b where the

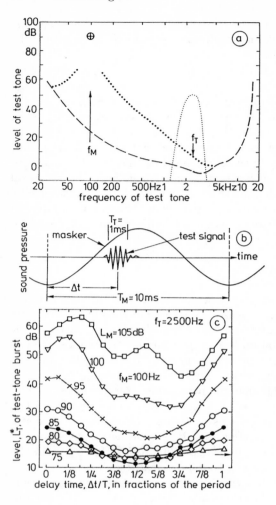

Fig. 4.32a–c. General spectral **(a)** and temporal **(b)** conditions for producing low-frequency masking-period patterns **(c)**. In (a), the spectral distribution of the masking effect produced by a 90-dB, 100-Hz masker is shown by the dotted curve above threshold in quiet (*broken*). The thin dotted curve indicates the spectral distribution of the test signal (series of 2.5-kHz tone bursts of 1-ms duration and Gaussian rise/fall time of 0.5 ms; repetition rate is also 100 Hz). The temporal condition is indicated in (b) for the period of 10 ms. The delay time, Δt, of the tone burst is given in relation to the instant of maximal rarefaction of the masker. The test-tone burst is presented periodically once within the period of the masker. The masking-period pattern in (c) shows the level of the test-tone burst needed to be just masked as a function of delay time in fractions of the masker's period with the masker level as the parameter

time functions of the masker and of the test signal are indicated. The test signal has to be short in relation to the masker period, i.e. shorter than one eighth of the masker period.

Masking-period patterns produced by these maskers and test tones are indicated in Fig. 4.32c with masker level as the parameter. Below 75 dB of the 100-Hz masker, no masking occurs. For 85 dB, masking is larger for that part of the period of the masker during which rarefaction occurs. A rarefaction maximum is used as reference ($\Delta t = 0$) for the time scale. It is interesting to note that in the region of the condensation maximum, which is in the middle of the pattern, masked threshold becomes even lower than the threshold in quiet indicated by the arrow on the right ordinate of Fig. 4.32c. This is an unexpected effect which has been confirmed for a large number of subjects.

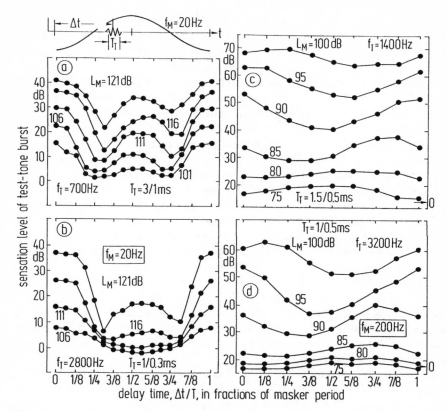

Fig. 4.33a–d. Masking-period patterns, i.e. sensation level of test-tone bursts as a function of their delay time in fractions of the masker period for 20-Hz maskers (*left*) and 200-Hz maskers (*right*). Test-tone frequency, duration and rise time of the test-tone bursts are indicated in each panel. The masker level is the parameter

For a masker level of 95 dB, the masking-period pattern is shifted towards higher levels of the test-tone burst. At 105 dB, the masking-period pattern is shifted even higher and shows two peaks within the period of the masker.

More data on masking-period patterns are given in Fig. 4.33. On the left, two sets of data are drawn for a masker frequency of 20 Hz which has a period of 50 ms. Test-tone frequencies of 700 Hz (upper panel) and 2800 Hz (lower panel) are used. Both sets of curves show very similar temporal behaviour. The highest threshold within the pattern is always reached for rarefaction maximum. A second maximum is sometimes reached near the condensation maximum, although this second maximum appears only at higher masker levels. Two clear minima are observed between the two maxima near the zero-crossings of the time function of the masker. Because similar data could be produced for this 20-Hz masker using test-tone frequencies of 350 Hz as well as 1400 Hz, it can be concluded that the cochlear duct vibrates in phase from a place corresponding to a CF of 350 Hz down to the end of the basal turn. This conclusion is not so evident for a masker frequency of 200 Hz, the masking-period patterns of which are displayed in the two right panels of Fig. 4.33. Because the frequency of 200 Hz corresponds to a period of only 5 ms, the test-tone burst duration has to be quite short. A test-tone frequency of 1400 Hz is already a little bit low in order to produce a short tone burst, the spectral width of which is smaller than the corresponding critical band. Under these conditions, phase shifts occur as a function of test-tone frequency (compare the upper and lower right-hand panels) and as a function of masker level (parameter). All the data clearly show a close relationship between the time function of masking-period patterns and that of the masker.

In order to obtain more insight into this relationship, very low-frequency maskers with non-sinusoidal temporal functions have been used to produce masking-period patterns: Gaussian impulses alternating in sign, their first, and their second integrals. The time functions of these maskers are plotted in Fig. 4.34 together with the corresponding masking-period patterns. In order to produce masking of at least 20 dB somewhere within the period of the masker, large masker levels are needed. For the twice-integrated Gaussian impulses (Fig. 4.34a), almost no masking occurs during the first half of the period although a masker level as high as 140 dB is used, but a very distinct maximum appears in the second (rarefaction) half. Indeed, the maximum coincides with the minimum of the masker's time function, i.e. with the peak value of the rarefaction.

The data measured for the integrated Gaussian impulse (Fig. 4.34b) with a masker peak level of 131 dB, show two distinct peaks even though no extreme values of the masker's time function occur at these positions. The alternating Gaussian masker itself (Fig. 4.34c) produces a masking-period pattern that reaches a maximum in the first half of the period at the same time at which the masker's time function reaches its most negative value. Three other maxima in the masking-period pattern are produced during the second half of the masker period. The middle maximum corresponds to that moment at which the masker's time function reaches its most positive value. For the two neighbouring maxima, however, there exist no corresponding maximal values in the masker's time function. These three data sets indicate that the masking-period pattern's time dependence cannot always be directly related to the masker's time function. However, it is obvious that positive values of the second derivative of the masker's time function seem to play the dominant role in producing masking-period patterns of tone bursts masked by periodical low-frequency maskers. Note that the same effect is evident in suppression-period patterns as outlined in Fig. 3.24.

4.4.6 Pulsation Threshold

If two sounds are presented alternately, it may occur that one of the sounds is perceived as being continuous despite the fact that in reality it is gated on and

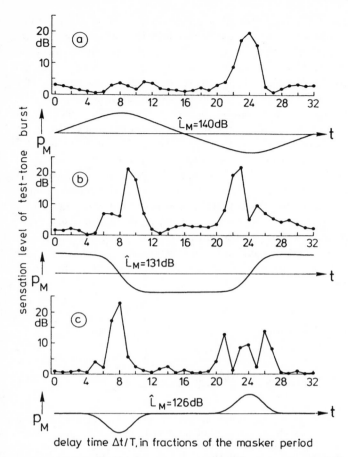

delay time Δt/T, in fractions of the masker period

Fig. 4.34a–c. Masking-period patterns produced by maskers, the sound pressure versus time functions of which are indicated below each pattern. The period of each masker is 300 ms, the time function of the masker in (**b**) is the first derivative and that of the masker in (**c**) the second derivative of the masker shown in (**a**). The level above threshold in quiet of a 1.4-kHz test-tone burst with a duration of 2 ms and a rise/fall time of 0.5 ms is indicated as a function of the delay time in 32 fractions of the masker period

off at periodic intervals. The corresponding temporal pattern of such stimuli is shown in Fig. 4.35. The solid curves represent the temporal envelope of one sound called the "masker", the dashed curve represents the temporal envelope of the test tone. The durations of "masker", T_M, the gap between "masker" and test tone, t_g, and the duration of the test tone, T_T, are measured at 70% of the respective maximum amplitude. The rise and fall time, T_{rG}, of the Gaussian-shaped gating signal is measured between 10% and 90%. When the sequence of "masker" and test tone is presented at a reasonable level for both sounds, a sequence of two pulsed sounds is perceived. However, if the level of the test tone is decreased to a specific value, then the test tone starts to sound continuous and is no longer heard as pulsed. This particular value of the test-tone level is called the pulsation threshold. It should be mentioned that if the test-tone level is decreased further, the test tone continues to be heard as a continuous tone until it reaches absolute threshold and disappears. In the pulsation-threshold technique, the "masker" does not really mask the test tone, but gives rise to the continuity phenomenon. Therefore, in this section the first sound is always denoted "masker" in quotation marks in order to point to this effect.

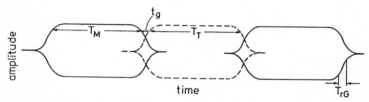

Fig. 4.35. Stimulus pattern for the measurement of pulsation thresholds. The temporal envelope of the "masker" (*solid*) and of the test tone (*dashed*) are shown; the "masker" duration, T_M, the test-tone duration, T_T, the duration of the gap between "masker" and test tone, t_g, and the rise time of the Gaussian-shaped gating signal, T_{rG}, are also indicated

In a manner comparable to the masking patterns determined by masking described in Sect. 4.1.2, pulsation patterns can be produced using the pulsation-threshold technique. An example of a pulsation pattern with a critical-band wide noise at 2 kHz as "masker" is shown in Fig. 4.36. The level L_T of the test tone at pulsation threshold is given as a function of both test-tone frequency (top scale) and critical-band rate (bottom scale). While the duration t_g of the gap between "masker" and test tone and the rise time of the Gaussian-shaped gating signal were kept constant, the duration of the "masker" and test tone were varied. In Fig. 4.36, triangles represent a pulsation pattern for durations of "masker" and test tone $T_M = T_T = 30$ ms, the circles stand for $T_M = T_T = 110$ ms and the squares represent the data for $T_M = T_T = 300$ ms (medians of 8 subjects). In all experiments, the level of the "masker" was kept constant at 70 dB.

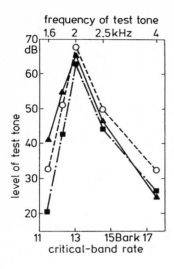

frequency of test tone

level of test tone

critical-band rate

Fig. 4.36. Pulsation patterns of critical-band wide noise "masker" centred at 2 kHz. "Masker" level L_M = 70 dB; temporal gap between "masker" and test tone t_g = 5 ms; rise/fall time of Gaussian-shaped gating signal T_{rG} = 10 ms; duration of "masker" T_M and test tone T_T = 30 ms (*triangles*), 110 ms (*circles*), and 300 ms (*squares*)

The results plotted in Fig. 4.36 show that the shape of the pulsation pattern depends slightly on the choice of the duration of "masker" and test tone. In particular, the steepness of the lower slope of the pulsation pattern decreases somewhat with decreasing duration of "masker" and test tone. Consequently, an attempt was made to develop an "optimum" stimulus pattern for the measurement of pulsation thresholds. The arguments hinge on the results displayed in Fig. 4.37, where the left panels show results for narrow-band noise centred at 2 kHz (13 Bark) and the right panels show data for a pure tone at 2 kHz. Masking patterns are indicated by open triangles and solid lines, pulsation patterns by filled triangles and dashed lines.

Whereas the measurement of masking patterns for narrow-band noise represents an easy task for the subject, masking patterns of pure tones are confounded by additional effects such as beats, roughness and combination tones (see Sect. 4.2.1). The following strategy was therefore adopted: for a critical-band wide noise at 2 kHz with 60 dB SPL, both the masking pattern and the pulsation pattern were determined. The temporal features of the stimulus pattern for the pulsation threshold measurement shown in Fig. 4.35 were then varied, until the resulting pulsation pattern matched to the corresponding masking pattern as closely as possible. Figure 4.37a shows that the masking pattern and the pulsation pattern coincide very closely (average deviation is less than 1 dB) when the stimulus pattern was as follows: "masker" duration T_M = 100 ms, test-tone duration T_T = 100 ms, temporal gap between "masker" and test tone t_g = 5 ms, and Gaussian-shaped gating signal T_{rG} = 10 ms. This "optimum" stimulus pattern for the measurement of pulsation threshold was used in all further experiments on pulsation patterns outlined in this chapter. A comparison of solid and dashed curves in Fig. 4.37c reveals that not only at 60 dB SPL but also at 20, 40, and 80 dB, masking

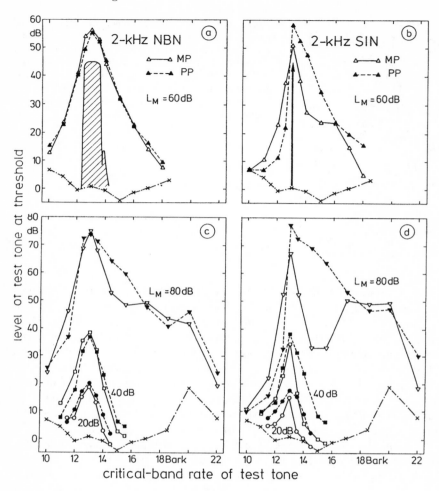

Fig. 4.37a–d. Comparison of masking patterns and pulsation patterns produced by the following maskers ("maskers"): *left panels:* critical-band wide noise at 13 Bark (2 kHz); *right panels:* pure tone at 13 Bark. *Open triangles* and *solid lines:* masking patterns, *filled triangles* and *broken lines:* pulsation patterns, *crosses* and *dash-dotted lines:* threshold in quiet. (**a**) critical-band wide noise with 60 dB SPL; the hatched area indicates the spectral distribution of the critical-band wide noise, (**b**) pure tone at 60 dB SPL, the arrow indicates the spectral position of the masker tone, (**c**) critical-band wide noise, and (**d**) pure tones at 20, 40, and 80 dB SPL

pattern and pulsation pattern of narrow-band noise are quite similar. Only at 80 dB SPL does the upper slope of the masking pattern show the influence of nonlinearities (see Sect. 4.2.1) that are not present in the pulsation threshold pattern.

The right panels of Fig. 4.37 enable a comparison of masking patterns and pulsation patterns for a pure-tone masker. In contrast with the data obtained with a noise masker (Fig. 4.37a), the data plotted in Fig. 4.37b indicate large differences between masking patterns and pulsation patterns, even for a configuration where the masking pattern and pulsation pattern of a narrow-band noise nearly coincide. Figure 4.37d also shows that at 20 and 40 dB and, in particular, at 80 dB, large differences between masking pattern and pulsation pattern occur. However, the pulsation patterns also show the "tilt" as seen in masking patterns of pure tones: at low masker level the lower slope of the pattern is shallower than the upper slope and at high masker level the lower slope is steeper.

A comparison of the data plotted in Fig. 4.37a and Fig. 4.37b reveals that the patterns for the pure-tone masker show a steeper lower slope than the patterns for the narrow-band noise masker, particularly for the pulsation patterns. An experiment was conducted to determine whether this effect also occurs at frequencies different from 2 kHz. In Fig. 4.38, pulsation patterns are given for maskers at 0.4, 1, 2, and 4 kHz. Triangles and solid lines represent pulsation patterns for critical-band wide "maskers", circles and dashed lines represent pulsation patterns for pure-tone "masker". The squares connected

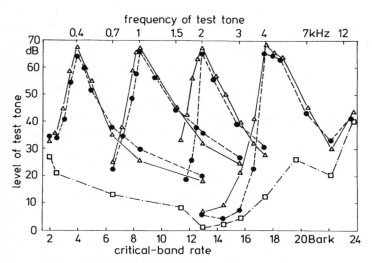

Fig. 4.38. Comparison of pulsation patterns of critical-band wide noise versus those of pure tones. "Masker" level $L_M = 70$ dB. "Masker" frequency $f_M = 0.4$, 1, 2, and 4 kHz; *triangles* and *solid lines:* critical-band wide "masker"; *circles* and *dashed lines:* pure tone "masker"; *squares* and *dashed-dotted lines:* medians of threshold in quiet of the eight observers involved

by dash-dotted lines indicate the median threshold in quiet for the eight observers involved. For both the critical-band wide noises and the pure tones, an SPL of 70 dB was chosen.

The results displayed in Fig. 4.38 indicate that lower slopes of pulsation patterns for pure tones, compared with critical-band wide noise, differ for maskers at 2 kHz and to a lesser degree for maskers at 4 kHz. At 0.4 and 1 kHz, the pulsation patterns of critical-band wide "masker" and pure tone "masker" yield similar patterns. This also holds for the upper slopes of the pulsation patterns at 2 and 4 kHz. By and large, the pulsation patterns shown in Fig. 4.38 greatly resemble the masking patterns of critical-band wide noise shown in Fig. 4.3. This similarity only holds, however, if the temporal sequence used for measuring pulsation threshold consists of $T_M = T_T = 100$ ms and $t_g = 5$ ms.

4.4.7 Mixed Spectral and Temporal Masking

In many cases of complex maskers, both spectral and temporal effects appear. It is not possible to discuss this in detail because there are many varying parameters. Three examples may indicate possible side effects and show what our hearing system may pick up in complex masking experiments.

The basis of masking is always the same: the ratio (spectrally and temporally) between the excitation level versus critical-band rate versus time patterns produced by the masker, and that produced by the test sound is ultimately responsible for whether the test sound is heard or not.

A frequency-modulated pure-tone masker produces constant SPL. However, its excitation level pattern moves up and down along the critical-band scale, with an extension corresponding to its frequency deviation and a temporal speed corresponding to its modulation frequency. If a long lasting tone is used as a test sound, the minimal excitation level of the masker produced within the modulation period at the critical-band rate of the test tone determines threshold. When a short test-tone burst, repeatedly presented with a rate corresponding to modulation frequency, is used then the temporal relation between the FM of the masker and the incidence of burst presentation within the FM period plays an additional role, as has been shown in Fig. 4.26.

This effect of FM seems to be reasonable. However, FM-like sounds can also be produced by determining the phases of a large number of harmonically related tonal components of a complex masker, following special rules. Changing the phases from random values to this special condition without changing the level of the partial tones may be thought as a strategy to uncover our ears' phase sensitivity. Actually, such phase changes are resolved by the spectral and temporal resolution ability of our hearing system.

Gaussian noise can be simulated by a large number of low-level tonal components with random phase. Such approximations have to be adopted with care when used as maskers, because of the non-Gaussian amplitude distribution produced in the case of too few lines and/or non-arbitrary phase conditions. The same holds for noises produced by "statistically" chosen sequences of "0"s and "1"s. Such noises often produce masking effects different from those produced by Gaussian noise because of the different amplitude distribution and the different temporal and spectral pattern within narrow-frequency bands. Such patterns are important for the excitation level versus critical-band rate versus time pattern, the basis of masking.

4.5 "Addition" of Masking

The effect of masking by certain well-defined maskers has been described in Sects. 4.1 and 4.2. In many cases, there is not one masker (for example a sinusoidal tone at a lower frequency) but also a second masker which may, for example, be a narrow-band noise centred at 1 kHz. At lower frequencies, masking will be defined by the low-frequency masker. For frequencies around 1 kHz, masking will be defined by the narrow-band noise. There is a region, however, for which the masking effect produced by the low-frequency tone and the masking effect produced by the narrow-band noise, each of them presented separately, produce an equal shift of threshold. This may be, in the case mentioned, in the neighbourhood of 800 Hz. The effect expressed as "addition of masking" can be formulated by a question: does threshold masked by the two noises remain the same as threshold masked by one sound alone, or does it rise by 3 dB (intensity addition) or 6 dB (sound pressure addition), or does it rise by some other value? Results of corresponding measurements using simultaneous masking can be summarized relatively clearly. Similar measurements using nonsimultaneous masking, for example post-masking, produce effects that cannot be generalized easily. It seems that bandwidth plays an important role with a number of other variables if temporal parameters are involved.

4.5.1 "Addition" of Simultaneous Masking

In order to characterize the thresholds produced by one of the maskers, only one number will be used as an index: the masked threshold will be indicated by the level of the test tone at a certain frequency (L_1 or L_2), and the increment in masked threshold produced by adding the second masker will be indicated by the incremental value, ΔL_T. One of the most important variables is the difference between the two thresholds (L_1 and L_2) produced by each of the maskers. This difference in masked threshold will be called $\Delta L_{T(1-2)}$. It must be expected that the increment in threshold, ΔL_T, will be largest when the difference between the masked thresholds produced by each masker is zero, i.e. when each masker produces the same masked threshold. If the threshold produced by one masker is much larger than that produced by the other, then the masking produced by the maskers together is likely to be dominated by the masker that produces larger masking.

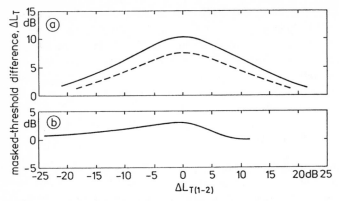

Fig. 4.39a,b. Difference in masked thresholds, ΔL_T, of test tones masked either by two maskers (a 105-Hz tone and a critical-band wide noise at 1 kHz) together, or by the masker which produces the larger masking, is shown as a function of the difference $\Delta L_{T(1-2)}$ between masked thresholds produced by the two maskers alone. Part (**a**) holds for data produced at the slopes (*upper* or *lower*) of masking patterns (*solid*: averaged curve for masked thresholds higher than 15 dB above threshold in quiet, *dashed*: 5 to 15 dB above threshold in quiet). Part (**b**) holds for data produced within the critical band of the noise masker

The increment ΔL_T of masking when two maskers are presented, in relation to that produced by only one masker, is indicated in Fig. 4.39 as a function of the level difference of the masked thresholds produced by each of the two maskers separately. Two curves are presented, the broken one corresponding to masked thresholds only 5 to 15 dB above the threshold in quiet, and the solid curve corresponding to masked thresholds more than 15 dB above threshold in quiet. The expected maximum at a level difference $\Delta L_{T(1-2)} = 0$ shows up clearly in both cases. However, the maximum ΔL_T value reached is much larger than 3 dB and very nearly a factor of 2 larger than 6 dB! For measurements well above threshold in quiet 12 dB are often obtained when $\Delta L_{T(1-2)} = 0$. The decrement of the curve towards negative or positive values of $\Delta L_{T(1-2)}$ has a slope that is shallower than expected. The masked threshold increment still reaches about 6 to 8 dB, even when the difference between the masked threshold produced by each masker is as much as 10 dB. The data in Fig. 4.39a apply to masked thresholds being determined at the slopes of excitation. This means that masked threshold is not determined by what is called the critical-band level of the maskers, i.e. main excitation, but by the excitation slopes, upper or lower. If measurements are performed in a frequency range that is determined as main excitation (in our case in the neighbourhood of 1 kHz, the centre frequency of the narrow-band masker), the increment in masked threshold reaches only 3 dB when the two maskers produce the same masking (see Fig. 4.39b). This means that in such cases the addition of masking of these maskers seems to be closely correlated to intensity addition.

Besides measurements using two maskers, others have been performed with up to four additional maskers. In this case, it is most effective if each of the four maskers produces the same masked threshold, which should be at least 20 dB above threshold in quiet. A masked threshold increment of up to 21 dB can then be achieved when the four maskers are presented simultaneously (compared with the 6 dB that would result from the addition of intensity). These effects again show clearly that "addition of masking" follows rules that are not equivalent to the addition of intensity, but may be describable by the addition of specific loudnesses.

4.5.2 "Addition" of Postmasking

As discussed previously, post-masking depends on the duration of the masker. When using narrow-band maskers, post-masking also depends on bandwidth. Therefore, many variables besides the levels of the two maskers influence the addition of the effects produced by masker "1" and masker "2". One example that produces relatively large effects will be discussed. Both the maskers are relatively long (about 500 ms); the test-tone duration is 20 ms and Gaussian-shaped, with a 2-ms rise time. The delay time between the end of the masker and the end of the test tone is kept constant at 46 ms. A narrow-band noise centred at 2.8 kHz is used as masker M1. The frequency of the test-tone burst is the same as the centre frequency of masker M1. Masker M2 is a sinusoidal tone at 3.22 kHz (Fig. 4.40a) or a broad-band noise (Fig. 4.40b). The spectral configurations are illustrated by the insets. Masked threshold is plotted in both cases as a function of the bandwidth of masker M1. The curves represent medians of 16 data points produced by eight subjects. The arrows on the left mark the post-masking threshold produced by masker M2 alone. The left-most data points in the two figures belong to a sinusoidal masker M1. Proceeding from this masker to a narrow-band masker with 30-Hz bandwidth results in an increase of masked threshold of about 12 dB for masker M1 alone. Above a bandwidth of 100 Hz, masked threshold decreases to an almost constant value for bandwidths larger than 660 Hz. If the two maskers are presented simultaneously, post-masked thresholds remain almost independent of the bandwidth of masker M1. This leads to a large decrement of post-masked threshold when the second masker M2 is added. This is in complete contrast to data for simultaneous masking. The decrement of masking with the addition of two maskers for post-masking, depends strongly on the parameters chosen.

The examples indicated in Fig. 4.40 show the most pronounced effect. The data indicate that this astonishing effect may depend on the time structure of masker M1, which produces a pronounced temporal pattern for narrow bandwidth below 600 Hz. Therefore, a model describing these results must be based on spectral and temporal features of the two maskers so that the distinctly audible fluctuations produced for a narrow-band masker M1 are taken into account. These fluctuations are strongly reduced when a second masker

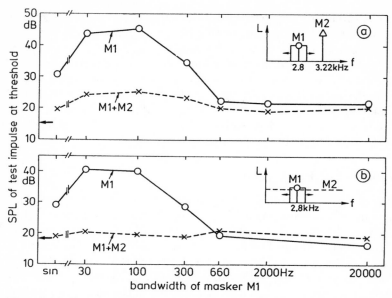

Fig. 4.40. Post-masked thresholds for test-tone bursts centered in frequency at masker M1, a band-pass noise, as a function of this masker's bandwidth. Solid lines represent masked thresholds produced by masker M1 alone, dashed lines those produced by the two maskers, M1 and M2, presented simultaneously. The arrows indicate post-masked thresholds produced by masker M2 alone. Masker M1 is centered at 2.8 kHz with a critical-band level of 40 dB. Masker M2 is a pure tone at 3.22 kHz with an SPL of 60 dB for part (**a**) and a broad-band noise of 20 kHz bandwidth and an SPL of 60 dB in part (**b**)

M2, a pure tone or a white noise is added. Consequently, in addition to the spectral effect, a pronounced temporal effect leading to a reduction of audible fluctuations is introduced, which finally produces a decrease of post-masked threshold by as much as 13 to 24 dB. The latter effect depends strongly on the masker's bandwidth. Large effects appear when M1 is a narrow-band noise, whereas almost no effect occurs when M1 is a broad-band noise.

4.6 Models of Masking

Masking effects fall not only into different temporal regions, such as simultaneous masking and nonsimultaneous masking, but can also be described either psychoacoustically or – at least for simultaneous masking – using the cochlear pre-processing model. Masking also has a relatively close relation to just-noticeable variations. An example may illustrate this: if a white noise from one generator is presented as a masker and another white noise from another generator is used as the test sound to be masked, two incoherent sounds are produced. It is also possible to interpret this sound sequence as a rectangularly amplitude-modulated white noise. This effect may come even closer to what we call modulation if the two sounds originate from the same noise generator and are therefore coherent. In this case, the equivalence of the sequence of sounds to that of rectangular amplitude modulation becomes obvious. Thus a suitable model of masking has close relation to a suitable model of just-noticeable variations, as outlined in Sect. 7.5.1. It may be helpful to read Chaps. 6–8 before reading the next two sections.

4.6.1 Psychoacoustical Model of Simultaneous Masking

The model for just-noticeable sound variations with durations longer than 200 ms (see Sect. 7.5.1) postulates that a variation is audible for conditions in which an increment of excitation level larger than 1 dB at any place along the critical-band rate scale is produced. Because masking is very closely related to excitation (see Chap. 6), this postulation has to include masking effects in the steady-state condition, e.g. masking of pure tones by long-lasting broadband noises. The fluctuation of noises that occur with the small bandwidths produced in the early stages of the hearing system through its frequency selectivity must also be considered. At low frequencies, for example, the corresponding critical bandwidth is only 100 Hz wide, whereas at high frequencies near 10 kHz, the bandwidth is more than 2 kHz wide. Consequently, noise at high frequencies produces a steady-state condition, whereas noises at low frequencies produce a strong temporal modulation which hinders the process of hearing the interrupted test tone used in masking procedures.

An example may illustrate the use of the model where white noise is the masker and a sinusoidal tone is the test sound. White noise is a broadband noise that produces only main excitations. At high frequencies, the fluctuation of the noise is not effective because the critical bandwidth is very large. Therefore, the logarithmic threshold factor is 1 dB, the just-noticeable relative increment in intensities is 0.25, and the corresponding masking index is -6 dB. If we assume a test-tone frequency of 9 kHz, the corresponding critical bandwidth is 2000 Hz. Because main excitation level is identical to critical-band level, the masked threshold can be calculated using the equation $L_T = L_G + a_v = l_{WN} + 10 \log (\Delta f_G / \text{Hz}) + a_v$. For the example given above, and for a masker density level $l_{WN} = 30$ dB, the level of the test tone at masked threshold is calculated to be $L_T = (30 + 10 \log 2000 - 6)$ dB $= (30 + 33 - 6)$ dB $= 57$ dB. This value fits well with the data outlined in Fig. 4.1.

Similar calculations can be done at the low-frequency region. There, the masking index is only -2 dB, the just-noticeable relative intensity increment is 0.65 and the logarithmic value of the 1-dB increment at high frequencies changes to little more than a 2-dB increment at these low frequencies. For a test-tone frequency of, for example, 300 Hz, the critical bandwidth is 100 Hz, leading to a test-tone level at masked threshold $L_T = (30+20-2)$ dB $= 48$ dB. This value also agrees with the measured data. Such an agreement is not surprising because the excitation level versus critical-band rate pattern was produced from masked threshold pattern, and the model for masking uses this relationship in the reverse direction.

Sometimes it is also interesting to use, instead of the increment in level, the increment in relative specific loudness (see Chap. 8). The excitation is then transformed to specific loudness using (8.7). In this case, the logarithmic incremental value of 1 dB corresponds to the relative increment of $\Delta N'/N' = 0.06$ at high frequencies (see also (8.2)) and to 0.13 at low frequencies. Such relationships can be used not only for steady-state conditions in simultaneous masking but also for postmasking effects.

Near threshold in quiet, masked threshold becomes dependent not only on masker level but also on level of threshold in quiet. In this case, the masker noise and the internal noise responsible for threshold in quiet have to be added. Using such an assumption, the threshold of tones masked by broadband noises of lower levels changes from the masked threshold into threshold in quiet very smoothly. Threshold in quiet as such can be picked up from this model assuming an internal noise as the masking sound.

Simultaneous masking depends on duration of the test sound in such a way that the product of intensity and duration is constant for durations smaller than 200 ms. For larger durations, threshold is independent of duration. That effect can also be described as an intensity weighting by a time constant with a characteristic value of 200 ms.

When, instead of broad-band noises, narrow-band noises or tones are used as maskers, slope excitation is also created. Because slope excitation is based on the masking patterns produced by narrow-band sounds, the facts described above apply even more: the model of masking is based on the excitation versus critical-band rate pattern; therefore, it is obvious that this model must describe spectral masking in steady-state condition perfectly.

4.6.2 Psychoacoustical Model of Non-simultaneous Masking

The effect of post-stimulus masking (postmasking) is dominant in non-simultaneous masking; premasking plays a relatively secondary role in comparison. The effect of postmasking is shown in Figs. 4.22 and 4.23. The astonishing effect that postmasking depends on the duration of the masker is clearly outlined in Fig. 4.23. This effect can be modelled more easily in the specific loudness versus critical-band rate versus time pattern described in detail in Sect. 8.7.1. If the duration-dependent decay of postmasking is transformed into an excitation-time pattern, which is transformed further into a specific loudness versus time pattern, then the decay can be simulated by two time constants. One is very quick and effective in the case of short maskers while the other acts more slowly and is seen with a masker duration longer than 100 ms. This behaviour can be simulated by a relatively simple network, which contains two resistors, two capacitors and a diode. Its input is specific loudness derived directly from the instantaneous direct excitation, i.e. the

level within the critical band in question. The transfer function from excitation level to specific loudness is given in (8.7). The input-time functions
of almost rectangular shape, when switching the masker on and off quickly,
are transferred into specific loudness versus time functions that correspond
closely to the postmasking plotted on an ordinate scale of specific loudness.
Postmasking threshold can be achieved using such a model by assuming that
postmasking threshold is reached for the relative increment of specific loudness $\Delta N'/N' > 0.06$ at any time. This means that a deviation from the
normal decay of specific loudness by more than 6% is audible. Postmasking
effects can be usefully approximated in this way.

4.6.3 Masking Described in Cochlear Active Feedback Models

Simultaneous masking can be described in the cochlear preprocessing model
by suppression effects. The nonlinear active feedback loop, as discussed in
Sect. 3.1.5, is not yet in its final stage. However, it can be used as an approximation which may be improved later with more reliable data.

It can be shown that simultaneous masking appears within the peripheral
cochlear processes before the signals are transformed into neural information
by the first synapses. Nonsimultaneous masking, however, was found not to
be measurable within the peripheral process of the cochlea. For this reason,
we can describe only simultaneous masking using cochlear models.

Figure 3.27 shows the model of peripheral nonlinear active preprocessing.
It reveals clearly that a large masker drives the nonlinear transfer function
installed in the feedback loop into saturation. This way, the feedback loop
is automatically switched off more and more by increasing levels, leading to
a decrease of the gain at and near the characteristic frequency of the place
in question. Furthermore, suppression occurs when a second tone, the test
tone, is added. This test tone occurs in a condition within the feedback loop
that is determined by the amplitude of the masker. Taking into account that
a certain increment in excitation level at or near the characteristic place of
the test tone is necessary to just hear the test tone, masking curves can be
measured in the analogue model or calculated in the computer model. Results
of such measurements using the analogue model are indicated in Fig. 4.41.
The criterion used is that threshold is reached when ever a 1-dB increment
is produced at any place along the critical-band rate scale by the addition of
the test tone. The parameters are adapted so that they correspond closely to
the level range within which masking of tone by tone is usually outlined.

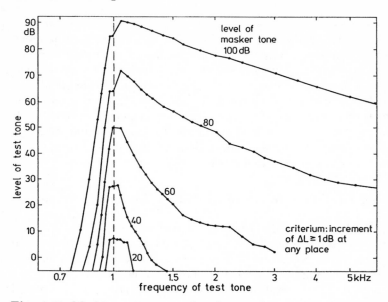

Fig. 4.41. Masking patterns produced using the nonlinear active feedback model: the level of the test tone necessary to produce, at any place along the basilar membrane, an excitation increment of 1 dB as a function of the test-tone frequency. The parameter is the level of the 1-kHz masking tone

5. Pitch and Pitch Strength

In this chapter the pitch of pure tones, complex tones and noise bands is addressed, and models for spectral pitch and virtual pitch are developed. In addition, the pitch strength of various sounds is assessed.

5.1 Pitch of Pure Tones

5.1.1 Ratio Pitch

The pitch of pure tones can be measured by different procedures. One possibility is that the subject is presented with a pure tone of frequency f_1 and has to adjust the frequency $f_{1/2}$ of a second tone in such a manner that the second tone produces half the pitch of the first tone. If, for instance, a pure tone of 440 Hz is used as sound 1 and a pure tone of variable frequency as sound 2, and the subject listening alternately to sounds 1 and 2, adjusts sound 2 to produce half the pitch elicited by sound 1, the average setting for the second tone is a frequency of 220 Hz. This means that at low frequencies, the halving of the pitch sensation corresponds to a ratio of 2 : 1 in frequency. This result at low frequencies is expected particularly from musically trained subjects. At high frequencies, however, some unexpected effects occur. If a frequency of 8 kHz is chosen for f_1, subjects produce for the sensation of "half pitch" not a frequency of 4 kHz, but a frequency of about 1300 Hz. Although there exist large individual differences, the value of 1300 Hz, on average, could be confirmed in many experiments. Measurements at other frequencies above 1 kHz confirm the tendency observed: for the perception of "half pitch", a ratio of the corresponding frequencies larger than 2 : 1 is necessary. This relation is shown in Fig. 5.1 by the solid curve. The frequency f_1 is given at the upper abscissa, the frequency $f_{1/2}$ at the left ordinate. The broken curve indicates the ratio of 2 : 1 between frequency f_1 and frequency $f_{1/2}$. Up to a frequency of about 1 kHz, the broken line and the solid line coincide; significant deviations occur at higher frequencies. The example that 1300 Hz represents "half pitch" of 8 kHz is indicated in Fig. 5.1 by arrows and the thin broken lines. In the same way that the sensation "half pitch" is determined, the sensation "double pitch" can be measured. For both types of experiments, a method of constant stimuli is frequently used. Moreover, at high frequencies, narrow noise bands are sometimes used as stimuli instead of pure tones. The data given in Fig. 5.1 as solid lines represent, with the appropriate interchange of axes, an average of the measurements for "half pitch" and "double pitch".

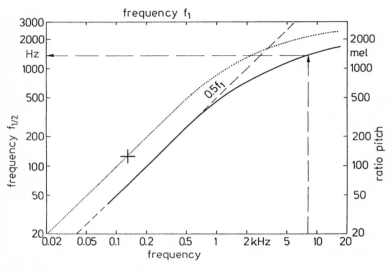

Fig. 5.1. Frequency and ratio pitch. The relationship between the frequency f_1 and the frequency $f_{1/2}$ producing "half pitch" (*solid*). Ratio pitch as a function of frequency (*dotted*). *Cross:* reference 125 Hz = 125 mel. *Dashed lines with arrows:* indication that 1300 Hz corresponds to the "half pitch" of 8 kHz

Ratios can be determined from experiments with halving and doubling of sensations, but not absolute values. To get absolute values it is necessary to define a reference point for the sensation "ratio pitch" as function of frequency. For the results plotted in Fig. 5.1, it is advisable to choose the reference point at low frequencies where the frequencies f_1 and $f_{1/2}$ are proportional, and to assume as the constant of proportionality the factor 1. In this way, the dotted line in Fig. 5.1 in the frequency region below 500 Hz, is produced by shifting the solid line by a factor of 2 towards the left. As a reference frequency, 125 Hz is chosen and marked in Fig. 5.1 by a cross. The dotted line in Fig. 5.1 then indicates that the numerical value of the frequency is identical to the numerical value of the ratio pitch at low frequencies. Because ratio pitch determined this way is related to our sensation of melodies, it was assigned the unit "mel". Therefore, a pure tone of 125 Hz has a ratio pitch of 125 mel, and the tuning standard, 440 Hz, shows a ratio pitch with almost the same numerical value. However, at high frequencies, the numerical value of frequency and that of ratio pitch deviate substantially from another. As indicated by the dotted line in Fig. 5.1, a frequency of 8 kHz corresponds to a ratio pitch of 2100 mel, and a frequency of 1300 Hz corresponds to a

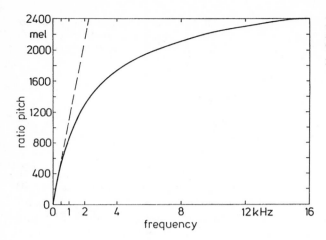

Fig. 5.2. Ratio pitch as a function of frequency. The abscissa and ordinate are linear

ratio pitch of 1050 mel. Thus the experimental finding that a tone of 1300 Hz produces "half pitch" of an 8 kHz comparison tone is reflected in the numerical values of the corresponding ratio pitch, because 1050 mel are half of 2100 mel.

Figure 5.2 shows on linear scales the relationship between ratio pitch and frequency. The proprotionality of frequency and ratio pitch at frequencies up to about 500 Hz becomes clear from the steep increase at the start of the curve near the origin. At higher frequencies, the curve bends more and more and reaches a ratio pitch of only 2400 mel at a frequency near 16 kHz. A comparison of the dependence of ratio pitch on frequency as illustrated in Fig. 5.2 with the dependencies displayed in Figs. 6.9 and 7.9, indicates that all three curves show great similarity. This important correlation will be discussed in more detail in Chap. 7.

5.1.2 Pitch Shifts

The pitch of pure tones depends not only on frequency, but also on other parameters such as sound pressure level. This effect can be assessed quantitatively by comparing the pitches of pure tones at different levels. The pitch of a pure tone at level L and frequency f_L is measured by matching its pitch to that of a tone with level of 40 dB and a frequency of $f_{40\,dB}$.

If a 200-Hz tone is alternatively presented at levels of 80 dB and 40 dB, the louder tone produces a lower pitch than the softer tone. However, if the same experiment is performed with pure tones at 6 kHz, the effect is reversed: the 6-kHz tone with 80 dB SPL produces a higher pitch than the 6-kHz tone at 40 dB SPL. Hence, frequency alone is *not* sufficient to describe the pitch produced by a pure tone, because the pitch sensation depends somewhat on level although much less than on frequency.

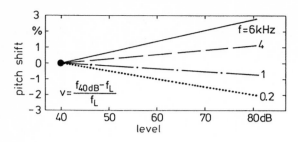

Fig. 5.3. Pitch shift of pure tones as a function of level for the four test frequencies indicated

The dependence of the pitch of pure tones on the level is displayed in Fig. 5.3, where the pitch shift, v, is given as a function of level when 40 dB is used as the standard. As indicated in the inset of Fig. 5.3, the pitch shift is calculated as the difference in frequency of a pure tone at 40 dB SPL and that of another pure tone at the level L under the condition in which the pure tones of different levels produce the *same* pitch. In Fig. 5.3, average values for the pitch shifts obtained from many subjects and for many repetitions are given; pitch shifts of single individuals can deviate substantially from the average data given in Fig. 5.3. The results displayed in Fig. 5.3 suggest that for an increase in sound pressure level of 40 dB, the pitch of pure tones is shifted on average by not more than about 3%. This relatively small effect can be neglected in many cases. If, however, the pitch of a pure tone has to be known very precisely, then both the frequency and the level have to be given.

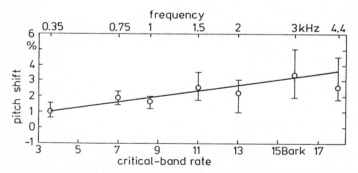

Fig. 5.4. Pitch shift of test tones partially masked by uniform masking noise as a function of frequency and critical-band rate. Overall masker level 60 dB, test-tone level 50 dB

Pitch shifts of pure tones can also occur if additional sounds that produce partial masking are presented. Pitch shifts produced by a broad-band noise masker are shown in Fig. 5.4, and are given as a function of both frequency and critical-band rate of the pure tones, the level of which is 50 dB. For uniform masking noise with 60 dB SPL as the partial-masking sound, the tones

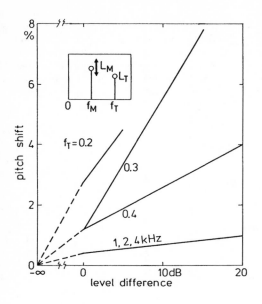

Fig. 5.5. Pitch shift of pure tones elicited by a pure tone of lower frequency, as a function of the level difference between the tones. The parameter is the test-tone frequency

show a pitch shift between 1 and about 3%. For partial masking with broadband sounds, the pitch shift of pure tones tends to increase with increasing frequency, even at low frequencies.

With narrow-band sounds as partial maskers, larger pitch shifts can be obtained. In Fig. 5.5, the pitch shift of pure tones, partially masked by pure tones of lower frequency, is given as a function of the level difference between the partial-masking tone and the test tone. In all cases, the partial-masking tone was presented 1 octave *lower* than the test tone, i.e. the frequencies of the test tone and partial-masking tone are in a ratio of 2 : 1. The loudness level of the unmasked test tone was kept constant at 50 phon.

The results displayed in Fig. 5.5 show pitch shifts up to 8% at low frequencies near 300 Hz, and a pitch shift of only 1% at high frequencies between 1 and 4 kHz, due to the octave ratio of partial-masking tone and test tone. For a 4-kHz test tone partially masked by a 3-kHz tone or a narrow-band noise, a pitch shift up to about 6% can be achieved.

In Fig. 5.6, pitch shifts of pure tones are displayed when the partial-masking tone is presented 1 octave *higher* than the test tone. In this case there are negative pitch shifts which, because of the octave ratio, are more pronounced at low frequencies (100 Hz) than at mid-frequencies (500 Hz).

With some simplification, the following rule can be proposed: partial masking produced by sounds which are lower in frequency than a test tone yields positive pitch shifts, whereas partial-masking sounds higher in frequency than the test tone produce negative pitch shifts. In terms of the corresponding excitation patterns, the pitch of pure tones is shifted away from the spectral slope of the partial-masking sound.

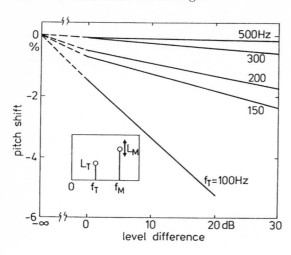

Fig. 5.6. Pitch shift of pure tones elicited by a pure tone of higher frequency, as a function of the level difference between the tones. The parameter is the test-tone frequency

5.2 Model of Spectral Pitch

For pure tones, a model of spectral pitch can be based on the corresponding masking patterns. As shown in Sect. 5.1, the pitch of pure tones depends not only on their frequency, but also somewhat on their level and the presence of partial-masking sounds. The sensation "pitch of pure tones" can be described by that frequency (a physical value) of a pure tone at 40 dB SPL, which produces the same pitch as the pure tone in question. This value is called "frequency pitch", H_F. The total pitch shift v of a pure tone can be devided into two components: one component, v_L, due to the level dependence of pitch, and a second component, v_M, due to the partial masking, where

$$v = v_L + v_M \; . \tag{5.1}$$

The frequency pitch H_F of pure tones given in pitch units (pu) can be calculated according to the formula:

$$H_F = (f_T/\text{Hz})\,(1 + v)\,\text{pu} \; , \tag{5.2}$$

where f_T is the frequency of the tone in Hz. An examination of (5.2) reveals that the frequency pitch of pure tones is easily obtained once the corresponding pitch shifts are known. Therefore, in the following, pitch shifts are calculated on the basis of masking patterns.

The basic features of the model of spectral pitch of pure tones are explained in Fig. 5.7 which shows the masking pattern of a masking sound M, and the masking pattern of the test tone T, for which the spectral pitch has to be calculated. In addition, the threshold in quiet, THQ, is given as a dashed line. The spectral pitch of the test tone can be described on the basis of the corresponding masking patterns using three magnitudes. The first magnitude "1" is denoted ΔMPTM and represents the difference between the masking

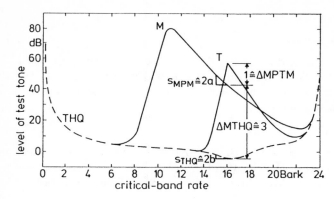

pattern of the test tone (T) and the masking pattern of the masker (M) at the critical-band rate of the test tone. The second magnitude comprises two components: the component "2a" represents the steepness s_{MPM} of the masking pattern produced by the masking sound in a region 1 Bark below the critical-band rate of the test tone. The other component "2b" represents the steepness of the threshold in quiet, also in a region one critical band below the critical-band rate of the test tone. The third magnitude "3" is denoted ΔMTHQ, and represents the difference between the masking pattern of the masker at the critical-band rate of the test tone, and the threshold in quiet at this critical-band rate.

As a first example, the pitch shift produced by variations in the level of pure tones will be described by the model of spectral pitch. In this case, the pattern shown in Fig. 5.7 is considerably simplified because only the masking pattern of the test tone (T) has to be considered since no masker (M) is present. Thus, in this case, the magnitude "3" does not exist, and the magnitude "1" encompasses the whole region between the top of the masking pattern of the test tone (T) and the threshold in quiet at the critical-band rate of the test tone. Furthermore, in this special case only the component "2b" has to be considered. Thus, the pitch shift of a pure tone caused by its level can be calculated on the basis of the height of its masking pattern above absolute threshold, and the steepness of absolute threshold in a region 1 Bark below the critical-band rate of the test tone. The corresponding formula reads as follows:

$$v_L = 3 \times 10^{-4}(\Delta MPTM/dB) \cdot (e^{0.33s_{THQ}/(dB/Bark)} - 1.75) , \qquad (5.3)$$

where v_L, according to (5.1), is the pitch shift due to level. The calculation of the pitch shifts of a pure tone caused by a partial-masking sound is somewhat more complicated, since magnitude "3" has also to be taken into account. The value of the pitch shift v_M can be calculated as the product of three factors, each correlated with one of the three magnitudes denoted "1", "2", and "3" in Fig. 5.7, and an additional constant.

The result is described by the formula:

$$v_\text{M} = 1.6 \times 10^{-3} \times g_1 g_2 g_3 \ . \tag{5.4}$$

The function g_1 describes the dependence of the pitch shift of the pure tones on the differences between the masking pattern of the test tone and the masking pattern of the masker at the critical-band rate of the test tone. The corresponding formula is given by

$$g_1 = -0.033 \Delta \text{MPTM}/\text{dB} + 1.37 \ . \tag{5.5}$$

The second function, g_2, depends on the steepness of the masker's masking pattern and the steepness of the threshold in quiet, both in the region one critical band below the critical-band rate of the test tone. The corresponding formula reads:

$$g_2 = 9\text{e}^{-0.15(\Delta s + 12)} - \text{e}^{-0.15(\Delta s + 24)} + 0.4 \ , \quad \text{with}$$
$$\Delta s = (s_\text{MPM} - s_\text{THQ})/(\text{dB/Bark}) \ . \tag{5.6}$$

The third function, g_3, represents the difference between the masking pattern of the masking sound and the threshold in quiet at the critical-band rate of the test tone. It is most easily described by distinguishing three different level ranges as follows:

$$g_3 = \begin{cases} 0 & \text{for } \Delta \text{MTHQ} < 15\,\text{dB or} > 70\,\text{dB} \ ; \\ 0.2 \Delta \text{MTHQ}/\text{dB} - 3 & \text{for } 15\,\text{dB} \le \Delta \text{MTHQ} \le 60\,\text{dB} \ ; \\ -0.9 \Delta \text{MTHQ}/\text{dB} + 63 & \text{for } 60\,\text{dB} < \Delta \text{MTHQ} \le 70\,\text{dB} \ . \end{cases} \tag{5.7}$$

While the patterns displayed in Fig. 5.7 may qualitatively elucidate the "philosophy" behind the model of spectral pitch, its quantitative realisation given in the formulae above was optimised by fitting as many data from the literature as possible.

5.3 Pitch of Complex Tones

Complex tones can be regarded as the sum of several pure tones. If the frequencies of the pure tones are integer multiples of a common basic or fundamental frequency, the resulting complex tone is called an harmonic complex tone. Despite the fact that complex tones comprise several pure tones, complex tones occur much more frequently in daily life than pure tones. The vowels of human speech, for example, or the sounds produced by many musical instruments are harmonic complex tones.

The pitch of complex tones can be assessed by pitch matches with pure tones. Although complex tones contain many pure tones, they do *not* usually produce many pitches, rather one single or perhaps one prominent pitch. Basically, the pitch of complex tones corresponds to a frequency close to the frequency difference between the components in the case of harmonic complex tones, i.e. the fundamental frequency. Closer inspection, however, reveals

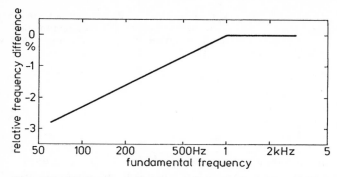

Fig. 5.8. Pitch of complex tones. Relative frequency difference between the fundamental frequency of a complex tone and a pure tone of the same pitch, as a function of fundamental frequency. Overall level of complex tone 50 dB, of pure tone 60 dB

slight but systematic deviations from this rule. An example is given in Fig. 5.8, where the relative frequency difference between a pure tone of matching pitch and the fundamental frequency of an harmonic complex tone is given as a function of the corresponding fundamental frequency. Starting with the fundamental frequency, the complex tone contains all harmonics with equal amplitude up to 500 Hz, and a spectral weighting with −3 dB/octave for higher frequencies. The overall level for the harmonic complex tone is 50 dB and the level of the matching tone usually 60 dB, but at frequencies below 100 Hz it is 70 dB. The results displayed in Fig. 5.8 show that for fundamental frequencies below 1 kHz, the relative frequency difference becomes increasingly negative with decreasing fundamental frequency. For example, the difference at 60 Hz amounts to almost −3%, i.e. a pure tone, with a frequency of 58.2 Hz produces the same pitch as the harmonic complex tone with a 60 Hz fundamental frequency. At 400 Hz fundamental frequency, the relative frequency difference amounts to about −1%, i.e. a pure tone of 396 Hz produces the same pitch as a complex tone with 400 Hz fundamental frequency. For frequencies above 1 kHz, the frequency of the pure tone and the fundamental frequency of the complex tone of the same pitch are equal.

The pitch of harmonic complex tones depends on level. Figure 5.9 shows the pitch shift of a complex tone with a 200-Hz fundamental frequency as a function of its level. An increasing negative pitch shift shows up with increasing level of the complex tone. Similar behaviour was described in Fig. 5.3 for pure tones at low frequencies. This indicates that the pitch of a complex tone is based on the spectral pitch of its lower components. This result is in line with the data shown in Fig. 5.8, which indicates that a pure tone has a frequency *lower* than the fundamental frequency of a complex tone of same pitch. One explanation, suggested by the data displayed in Fig. 5.6, is that the fundamental of a complex tone is shifted towards lower frequencies by the second harmonic.

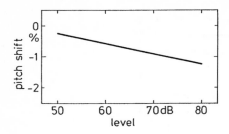

Fig. 5.9. Pitch shift of complex tones as a function of level. Fundamental frequency 200 Hz, level of matching tone 50 dB

If the lower harmonics are removed from a complex tone, the pitch hardly changes. This means that the pitch of the (incomplete) harmonic tone without fundamental frequency usually corresponds closely to the pitch of its fundamental. The effect where the "residual" higher harmonics of a complex tone produce a pitch that corresponds to the low (fundamental) frequency has been termed residue pitch, low pitch, or virtual pitch.

Not all complex tones from which the lower harmonics have been removed elicit a virtual pitch. Rather, specific combinations of fundamental frequency and of the frequency of the lowest component must occur in order to produce a virtual pitch. An existence region of virtual pitch, as displayed in Fig. 5.10, can be defined. It shows the fundamental frequency of a complex tone as a function of the lowest component, below which all spectral components were removed. In order to produce a virtual pitch, spectral components within the shaded area of Fig. 5.10 have to be presented.

The results displayed in Fig. 5.10 indicate that a complex tone with its lowest frequency component above 5 kHz does not produce a virtual pitch whatever the fundamental frequency. At low fundamental frequencies, i.e. close spacing of the spectral lines, this limit is attained at even lower frequencies. For example a complex tone with 50 Hz fundamental frequency produces a virtual pitch only if the frequency of the lowest spectral line does not exceed 1 kHz. This means that only the lower harmonics up to the 20th harmonic can produce the sensation of virtual pitch.

The pitch of incomplete harmonic tones can be assessed in the same manner as the pitch of complete harmonic complex tones. As with the data displayed in Fig. 5.8 for complete complex tones, incomplete complex tones also show negative values of the relative frequency difference, which, however, are larger by a factor of about two. The dependence of the pitch of incomplete complex tones on level also can be deduced from the level dependence of the lowest spectral component. Therefore, if the lowest component is around 3 kHz, then the pitch shift *increases* with level, in contrast with the data displayed in Fig. 5.9 for complete complex tones.

Instead of removing some of the spectral components of a harmonic complex tone by high-pass filtering, spectral components can be rendered inaudible by presenting a low-pass noise of sufficient level and steep spectral slope. In this case, the lower frequency components of the complex tone are com-

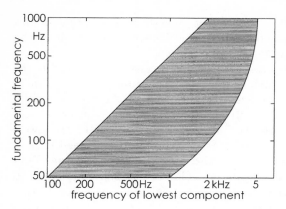

Fig. 5.10. Existence region for virtual pitch. Fundamental frequency as a function of the lowest frequency component. *Shaded area*: Spectral region in which spectral components of incomplete line spectra have to be contained in order to produce a virtual pitch

pletely masked, and the frequency of the lowest component of the incomplete complex tone starts near the cut-off frequency of the low-pass filter.

So far, harmonic complex tones for which the component frequencies are integer multiples of the fundamental frequency have been discussed, where the frequency spacing of the components is equal to the fundamental. However, all spectral components of a harmonic complex tone can be shifted by some amount, to produce an inharmonic complex tone. With inharmonic tones, from which the lower components have been removed, rather ambiguous virtual pitches can sometimes be produced. Figure 5.11 gives an example for a complex tone with a fundamental frequency of 300 Hz, which is filtered by an octave-band filter with a centre frequency of 2 kHz. Since this filter shows a lower limiting frequency of 1.4 kHz and an upper limiting frequency of 2.8 kHz, the pass band contains the 5th to 9th harmonic of the complex tone with frequencies of 1500, 1800, 2100, 2400, and 2700 Hz. In this case, the pitch produced by the residue tone corresponds to the frequency of a matching tone of 290 Hz (circle in the middle of Fig. 5.11).

If the five frequency components of the filtered harmonic complex tone are shifted by a constant amount, say 100 Hz, towards lower frequencies, then the lowest component will have a frequency of 1400 Hz, the next component

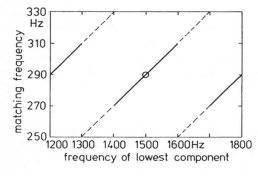

Fig. 5.11. Pitch of inharmonic complex tones. Matching frequency as a function of the frequency of the lowest component. Complex tone composed of five pure tones with 300 Hz line spacing

will be located at 1700 Hz, the following at 2000 Hz and so forth. In this case, the frequencies of the components are no longer integer multiples of the fundamental frequency of 300 Hz, and the whole complex is called an incomplete inharmonic complex tone, with a pitch corresponding to 270 Hz. Hence, a shift of all spectral components by 100 Hz towards lower frequencies yields a decrease of 20 Hz in the matching frequency. As displayed in Fig. 5.11, a shift of all spectral components by 100 Hz towards higher frequencies leads to a matching frequency of 310 Hz. For an upward shift of 150 Hz, the lowest component has a frequency of 1650 Hz. As indicated by the dashed lines in Fig. 5.11, the pitch of the inharmonic complex tone in this case becomes faint and ambiguous, and pitch matches are possible both at 260 and 320 Hz. If the spectral lines of the filtered harmonic complex tone are shifted further towards higher frequencies, the pitch again becomes more prominent and less ambiguous. Figure 5.11 indicates that the pitch of such inharmonic complex tones shows a sawtooth like shape: starting from a situation in which the frequency of the lowest component is, say, 1400 Hz, the pitch first increases with increasing shift of the spectral components, then becomes ambiguous and, near where the frequency of the lowest component is 1650 Hz, jumps down to a lower value only to increase again for the further increasing shift of the spectral components.

5.4 Model of Virtual Pitch

A sophisticated model of virtual pitch has been elaborated by Terhardt. The basic features of this model will be illustrated in this section. In general, the model is based on the fact that the first six to eight harmonics of a complex tone can be perceived as separate spectral pitches. These spectral pitches form the elements from which virtual pitch is extracted by a type of "Gestalt" recognition phenomenon. A visual analogue for the model of virtual pitch is illustrated in Fig. 5.12. The word "pitch" displayed on the left side is produced by thin border lines as an analogue of a complex tone containing all the relevant harmonics. On the right side of Fig. 5.12, the letters are only indicated by parts of their borders analogous to a complex tone from which some of the basic features, the lower harmonics for example, have been removed. The two parts of Fig. 5.12 are meant to illustrate the "philosophy" of the virtual pitch concept: from an incomplete set of basic features (incomplete border lines or incomplete spectral pitches) a complete image (the word "pitch" or the virtual pitch) is readily deduced by a mechanism of "Gestalt" recognition.

  **Fig. 5.12.** Visual analogue of the model of virtual pitch

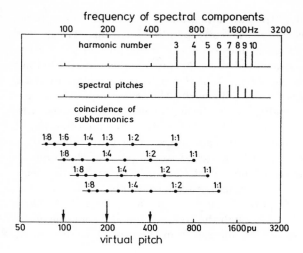

Fig. 5.13. Illustration of the model of virtual pitch based on the coincidence of subharmonics, derived from the spectral pitches corresponding to the spectral lines of the complex tone

The model of virtual pitch can be illustrated using Fig. 5.13 which, for didactical reasons, includes some simplifications. The influence of pitch shifts, for instance, is neglected at this stage. In the upper part of Fig. 5.13, a complex tone with a fundamental frequency of 200 Hz and from which the first two harmonics have been removed is displayed schematically. Both the harmonic number and the frequency of the spectral components are given. In the first stage, spectral pitches are derived (neglecting pitch shifts) from the spectral components, and a spectral weighting with a maximum around 600 Hz is applied. Next, subharmonics are calculated for each spectral pitch present. Finally, the coincidence of the subharmonics of each spectral pitch is evaluated.

For example, (again neglecting pitch shifts) the spectral component at 600 Hz is first transformed into a spectral pitch at 600 pitch units (pu). Starting from this value, the first eight subharmonics which occur at 300 pu, 200 pu, 150 pu, 120 pu, 100 pu, 85.7 pu, and 75 pu are calculated. In Fig. 5.13, each of these subharmonics is indicated by a dot, and the corresponding ratio is given in numbers. The same procedure is performed with the next spectral component at 800 Hz. In this case, we start from a spectral pitch at 800 pu, get the first subharmonic at 400 pu, the next at 266.7 pu, the next at 200 pu and so on. The same procedure is then applied for the spectral pitches at 1000 pu and at 1200 pu. In this way, an array of "yardsticks" containing dots representing the respective subharmonics is obtained. From this array, virtual pitch is deduced as follows: a scanning mechanism simply counts the number of dots that are contained in a narrow "pitch window", which is shifted like a cursor from left to right. At 200 pu in Fig. 5.13, four dots are found in the window. A large number of coincident subharmonics indicates a strong virtual pitch and therefore this spot is marked by a long arrow on the virtual pitch scale. Near 100 pu and 400 pu two dots are found in the window, therefore two small arrows are plotted at the corresponding locations. The largest number of coincidences of subharmonics occurs near 200 pu and the virtual pitch of the complex tone is calculated to be 200 pu as indicated by the long arrow. However, near 100 pu and 400 pu, candidates for the calculated virtual pitch also occur, but with less weight. This means that the complex tone produces a virtual pitch corresponding to 200 pu with some octave ambiguities (100 and 400 pu) in both directions. Such octave ambiguities are often found in experiments on virtual pitch. In our case, however, a pure tone with a frequency a little below 200 Hz will be matched to the pitch of the complex tone with the spectrum shown at the top of Fig. 5.13. To realize the complete model of virtual pitch, many further details have to be assessed and the model is implemented on a computer. Such a model was optimised for many published experimental results. Although the computer model contains many refinements, its basic features can be inferred from the procedure illustrated in Fig. 5.13.

5.5 Pitch of Noise

Noise with steep spectral slopes can elicit pitch sensations. The pitch corresponds closely to the cut-off frequency of the filter for both low-pass and high-pass noise. This result is illustrated in Fig. 5.14 where the frequency of a pure tone matched to the pitch of filtered noise is plotted as a function of cut-off frequency. Circles illustrate pitch matches for low-pass noise, triangles indicate pitch matches for high-pass noise. Very steep filters with a spectral slope of at least 120 dB/octave were used in both cases. The dashed

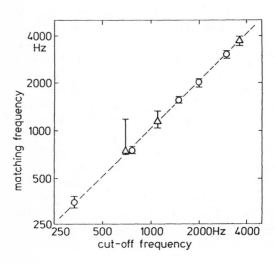

Fig. 5.14. Pitch of low-pass and high-pass noise. Frequency of matching tone as a function of cut-off frequency. *Circles:* Low-pass noise, *triangles:* high-pass noise. *Dashed line:* equality of cut-off frequency and matching frequency

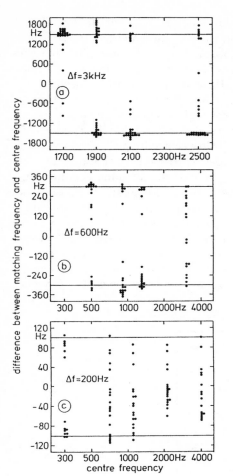

Fig. 5.15a–c. Pitch of band-pass noise. Difference between matching frequency and centre frequency of the noise as a function of centre frequency. Dots indicate individual pitch matches, solid lines represent the cut-off frequencies of the band-pass noises with bandwidths 3 kHz (*upper panel*), 600 Hz (middle panel) and 200 Hz (*lower panel*)

line in Fig. 5.14 corresponds to equality of matching frequency and cut-off frequency, indicating an almost perfect approximation. Data for low-pass noises (circles) show small interquartile ranges, but larger interquartile ranges are found for high-pass noises (triangles) especially at lower frequencies. Only at cut-off frequencies above about 800 Hz, high-pass noise elicits a pitch that is relatively faint, whereas low-pass noise produces a more distinct pitch at all cut-off frequencies. This effect on pitch strength will be discussed in more detail in Sect. 5.7.

From the data of low-pass and high-pass noise, it might be expected that band-pass noise produces two pitches corresponding to the upper and lower cut-offs. Results of corresponding experiments are shown in Fig. 5.15a where the difference between the frequency of the matching tone and the centre frequency of the band-pass noise is shown as a function of this centre frequency.

A constant bandwidth of 3 kHz was maintained throughout the experiment so that the upper and lower cut-off frequency of the band-pass noise can be indicated by the solid lines in Fig. 5.15a. As expected, most pitch matches correspond to the respective cut-off frequencies, i.e. fall near the solid lines. However, at a centre frequency of 1700 Hz, most pitch matches occur at the upper cut-off of 3200 Hz; no pitch is elicited at the lower cut-off frequency of 200 Hz, in line with the data obtained for high-pass noise which produces a pitch sensation only for cut-off frequencies above about 400 Hz. Figure 5.15b shows results for a 600-Hz bandwidth. For low centre frequencies of the band-pass noise, the pitch sensations elicited again correspond to the spectral edges (solid lines). However, at a centre frequency of 3 kHz, pitch matches no longer occur at the spectral edges but are distributed throughout the pass band. This tendency is more often seen with a bandwidth of 200 Hz, as displayed in Fig. 5.15c. Only at the 300-Hz centre frequency do two pitches corresponding to the spectral edges show up. With increasing centre frequency, however, the pitches correspond more and more to the centre of the noise band.

Figure 5.16 shows results on the pitch of very narrow noise bands. The difference between the frequency of the matching tone and the centre frequency

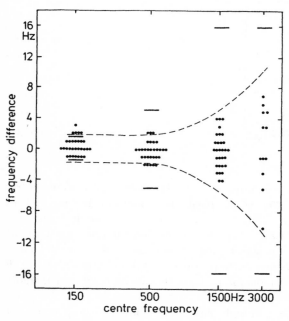

Fig. 5.16. Pitch of narrow-band noise. Frequency difference between matching tone and centre frequency as a function of the centre frequency. *Dots:* individual pitch matches; *thick bars:* cut-off frequencies; *dashed lines:* just-noticeable frequency variations for pure tones

of the corresponding narrow-band noise is plotted as a function of centre frequency. Bandwidths of 3.16, 10, 31.6, and 31.6 Hz were chosen at centre frequencies of 150, 500, 1500, and 3000 Hz, respectively. The cut-off frequencies of these narrow-band noises are indicated in Fig. 5.16 by thick bars. The dots again represent individual pitch matches of five subjects. The dashed lines represent the just-noticeable frequency deviations for pure tones as described in Sect. 7.2. In Fig. 5.16, almost all pitch matches (dots) lie within the dashed lines. Therefore, within the accuracy of pitch matches (dashed lines), narrow-band noises at different centre frequencies elicit only one pitch sensation corresponding to their centre frequency.

Summarizing the results in a simplified form, it can be stated that band-pass noise produces pitches that correspond to the frequencies of the spectral edges. If the spectral edges are close together, the two edge pitches fuse to a single pitch corresponding to the centre frequency of the narrow-band noise.

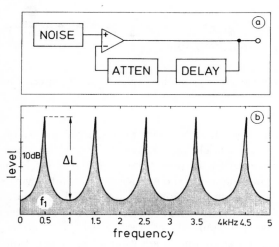

Fig. 5.17a,b. Schematic diagram of circuit (**a**) for the production of peaked ripple noise (**b**)

A noise with spectral peaks (rippled noise) can be produced by feeding back delayed versions of a noise to the input. Figure 5.17a shows a schematic diagram of the circuit used. In Fig. 5.17b, the spectral distribution of a noise with strong peaks, called peaked ripple noise, is displayed. The frequency f_1 of the first spectral peak is determined by the delay time; the level difference, ΔL, between maxima and minima in the spectrum is governed by the attenuation. For the spectrum displayed in Fig. 5.17b, a delay of 1 ms and an attenuation of 0.4 dB was chosen. Such noises produce distinct pitch sensations.

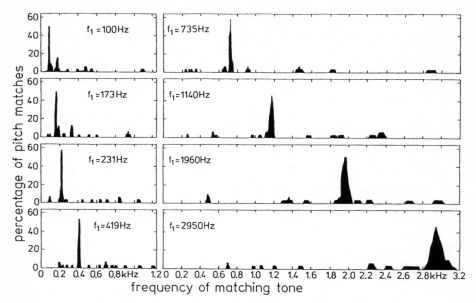

Fig. 5.18. Pitch of peaked ripple noise. Histograms of pitch matches as a function of matching frequency. Different frequencies f_1 of the first peak in the spectrum indicated in the different panels

Figure 5.18 shows histograms of pitch matches for peaked ripple noises of different frequencies f_1 for the first peak. The percentage of the pitch matches is given as a function of the frequency of the pure tone, matched in pitch to the peaked ripple noise. For the production of the histograms, a window with a bandwidth of 0.2 critical bands was shifted along the abscissa. The results displayed suggest that the pitch of peaked ripple noise corresponds to the frequency of the first peak in the spectrum. At low frequencies ($f_1 = 100\,\text{Hz}$ or $173\,\text{Hz}$), some octave ambiguities show up. In addition, closer inspection reveals that the main maximum in the histogram lies somewhat below the frequency f_1 of the first peak in the spectrum. This effect is comparable to the pitch of complex tones as illustrated in Fig. 5.8. Both the pitch at low frequencies and the octave ambiguities can be regarded as an indication that the pitch of peaked ripple noise at low frequencies also represents a virtual pitch, based on the spectral pitches produced by spectral peaks. Its pitch strength is discussed in Sect. 5.7.

While the pitch of the noises described so far can be traced back to a spectral feature, namely a distinct change in the spectrum, broad-band noises with flat spectra can also produce pitch sensations. For example, amplitude-modulated broad-band noise may elicit a faint pitch that corresponds to the modulation frequency. Although the long term spectrum of amplitude-modulated broad-band noise is independent of frequency, the short term spectrum contains some spectral information which may be correlated with the pitch produced. While narrow-band noise and peaked ripple noise may elicit a relative distinct pitch sensation, the pitch produced by amplitude-modulated broad-band noise is very faint, i.e. its pitch strength is very low. This result will be assessed in quantitative terms in Sect. 5.7.

5.6 Acoustic After Image (Zwicker-tone)

An acoustic after effect was discovered in 1964 and later called the "Zwicker-tone". The phenomenon can be described as follows: if sounds with a spectral gap are switched off, a faint tone lasting for several seconds can be heard. Its pitch strength (see Sect. 5.7) corresponds to the pitch strength of a pure tone of same pitch and sensation level. This result is at first astonishing, since after a broad-band noise with a spectral gap is switched off, one might expect a "negative acoustic afterimage" with a "noise quality" similar to a narrow-band noise corresponding to the spectral gap. Contrary to this expectation, however, the afterimage has a clear tonal quality as evidenced by its pitch strength.

The results displayed in Fig. 5.19 illustrate how frequently the Zwicker-tone phenomenon is found. Figure 5.19a and 5.19d show the spectral distributions of band-stop noises with spectral gaps at 1 and at 4 kHz. The distributions in Figs. 5.19b and 5.19e show the probability of hearing a Zwicker tone as a function of the overall level of the noise. There is a clear maximum around medium levels of about 43 dB (density level around 0 dB). At these medium levels, more than 60% and more than 80% of the subjects can perceive a Zwicker tone for gaps in band-stop noise centred at 1 and 4 kHz, respectively. Figures 5.19c and 5.19f show probabilities of the durations of Zwicker tones when a band-stop noise with 43 dB overall level and 10 seconds duration is switched off. For both band-stop noises, the most likely duration is around 2 s, however durations longer than 6 s occur about 5% of the time. Zwicker tones can be produced not only by band-stop noises but also by line spectra transmitted through a band-stop filter, or by line spectra produced by a computer omitting a certain number of adjacent lines. In all these cases, the loudness of the Zwicker tones corresponds to that of a pure tone at about 10 dB SL.

The existence limits of the Zwicker tone are illustrated in Fig. 5.20. The minimum width of the producer's spectral gap depends both on the line spacing and the critical-band rate of the spectral gap. For dense line spacing of 1 Hz that resembles a Gaussian noise in good approximation, a gap, one critical-band wide, is needed near 1 kHz (8.5 Bark), while at a centre frequency of 4 kHz less than 0.5-Bark gap width is sufficient. The dashed lines displayed in Fig. 5.20 fit the measured data well and correspond to the following simple equations:

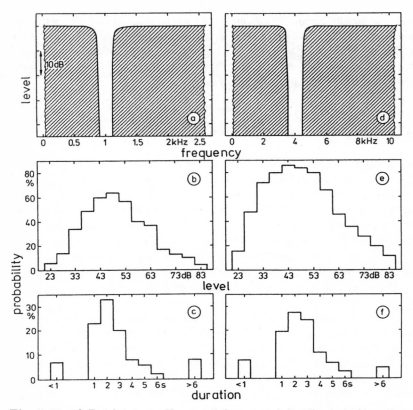

Fig. 5.19a–f. Zwicker tone. *Upper panels:* spectral distribution of band-stop noises used to elicit Zwicker tones. *Middle panels:* probability of the occurence of Zwicker tones as a function of the overall level of the band-stop noise. *Lower panels:* distribution of the duration of Zwicker tones elicited by band-stop noises with 43 dB level and 10 s duration

$$\Delta z/\mathrm{Bark} = 24\,\mathrm{Bark}/z \quad \text{for 200-Hz line spacing}\,, \tag{5.8}$$

$$\Delta z/\mathrm{Bark} = 12\,\mathrm{Bark}/z \quad \text{for 20-Hz line spacing}\,, \tag{5.9}$$

and

$$\Delta z/\mathrm{Bark} = 8\,\mathrm{Bark}/z \quad \text{for 1-Hz line spacing}\,, \tag{5.10}$$

where Δz is the minimal spectral gap width producing Zwicker tones.

These equations can be interpreted as follows: the minimum spectral gap necessary to produce a Zwicker tone, when expressed in critical bands, corresponds to the ratio of a constant number and the critical-band rate of the spectral gap's centre. The magnitude of the constant number depends on the line spacing. Starting from a line spacing of 1 Hz, the width of the spectral gap expressed in critical bands increases by a factor of 1.5 for 20-Hz line

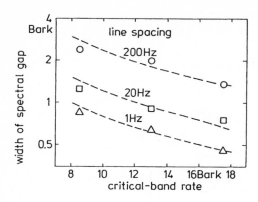

Fig. 5.20. Existence limits of the Zwicker tone. Minimum width of the spectral gap necessary to produce a Zwicker tone as a function of the gap's critical-band rate. Parameter: spacing of spectral lines

spacing and by a factor of 3 for 200-Hz line spacing, in order to delimit the existence region of the Zwicker tone for narrow spectral gaps.

Figure 5.21 shows what happens to the pitch of the Zwicker tone if the spectral gap is widened. The upper panel shows the results for line spectra with 20-Hz line spacing, the lower panel for 200-Hz line spacing (the four symbols belong to four subjects). All spectral gaps are centred at 2 kHz. The comparison of the two upper rows in Fig. 5.21 reveals that with 20-Hz line spacing, a Zwicker tone is produced at the centre of the spectral gap, whereas for 200-Hz line spacing, in accordance with the data given in Fig. 5.20, no Zwicker tone is heard. Figure 5.21 shows that with increasing width of the spectral gap, the pitch of the Zwicker tone shifts from the centre towards values near the lower edge of the gap. The same tendency is found for a centre frequency of 4 kHz.

The results displayed in Fig. 5.21 and those at other centre frequencies can be more uniformly described if, instead of the frequency scale, the critical-band rate scale is used. The critical-band rate of the pure tone matched in pitch to the Zwicker tone can then be given in good approximation by the simple equation

$$z_T = z_L + 1\,\text{Bark} , \tag{5.11}$$

where z_L represents the critical-band rate of the lower edge of the spectral gap and z_T the critical-band rate of a pure tone matched in pitch to the Zwicker tone. Hence, the pitch of the Zwicker tone elicited by line spectra with spectral gaps of different width corresponds to a critical-band rate 1 Bark above the critical-band rate of the lower edge of the spectral gap.

Not only spectral gaps but also spectral enhancements can produce a Zwicker-tone. Some results are illustrated in Fig. 5.22. When switching off a broad-band line spectrum with 1 Hz line spacing and a spectrum level of −2 dB (overall level 41 dB) plus a pure tone with a level of 50 to 80 dB, a Zwicker-tone becomes audible with a pitch *lower* than the pure tone. The data displayed in Fig. 5.22a indicate that the pitch of the Zwicker-tone is hardly influenced by the level of the pure tone producing the spectral enhancement.

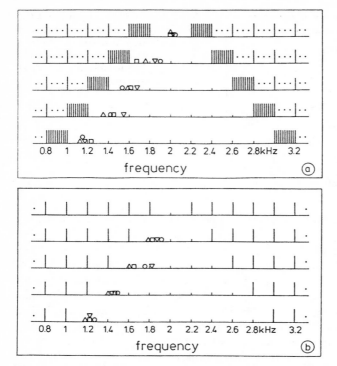

Fig. 5.21a,b. Pitch of the Zwicker tone for different width of the spectral gap. Different symbols belong to results of each of the four subjects

If, on the other hand, the level of the pure tone is kept constant and the level of the broad-band line spectrum increased, according to the data shown in Fig. 5.22b the pitch of the Zwicker-tone clearly increases.

The correlations between the pitch of the Zwicker-tone and the related masking pattern are illustrated in Fig. 5.23. The upper panels show data for a spectral gap, the lower panels for a spectral enhancement. According to results displayed in Fig. 5.23a the pitch of the Zwicker-tone increases with the level of the broad-band sound with spectral gap. As illustrated schematically in Fig. 5.23b with increasing level, the slope of the masking pattern at the lower edge of the gap becomes flatter and hence the minimum of the resulting masking pattern shifts to higher values of the critical band-rate as does the pitch of the Zwicker-tone. Therefore, the pitch of the Zwicker-tone corresponds to the minimum of the masking pattern or to the crossing point of the masking pattern and the hearing threshold (dashed).

Results in Fig. 5.23c show that for spectral enhancements the pitch of the Zwicker-tone is not much affected by the level of a tone added to a broad-band line spectrum. The schematic masking patterns illustrated in Fig. 5.23d nicely predict this behavior if it is assumed that the pitch of the Zwicker-

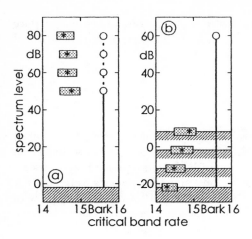

Fig. 5.22a,b. Zwicker-tones elicited by spectral enhancements. *Shaded areas*: Range of subjectively evaluated Zwicker-tones. *Asterisks*: Pitch of the Zwicker-tones derived from masking patterns. *Hatched areas*: Illustration of broadband line spectra. *Circles*: Illustration of pure tones added to the broadband line spectra. (**a**) Spectrum level of broadband line spectrum fixed. (**b**) Level of pure tone fixed

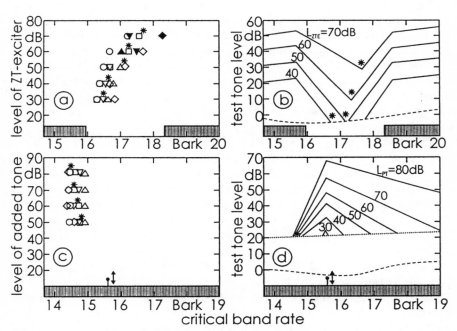

Fig. 5.23a–d. Pitch of the Zwicker-tone and masking patterns. (**a**) Spectral gap: Dependence of the pitch of the Zwicker-tone on the level of the Zwicker-tone exciter. *Symbols*: Subjective pitch matches; *asterisks*: Calculated values. (**b**) Spectral gap: Masking patterns (*solid*), threshold of hearing (*dashed*). (**c**) Spectral enhancement: Dependence of the pitch of the Zwicker-tone on the level L_{PT} of a pure tone added to a broadband line spectrum. *Symbols*: Subjective pitch matches; *asterisks*: Calculated values. (**d**) Spectral enhancement: Masking patterns of pure tones (*solid*) and broadband line spectrum (*dotted*). *Dashed*: Threshold of hearing. *Shaded areas* and *double arrows*: Illustrations of spectral distributions

tone corresponds to the crossing point of the lower slope of the added pure tone's masking pattern (solid) with the masking pattern of the broad-band line spectrum (dotted).

5.7 Pitch Strength

Experiments on pitch sensation generally explore variations along a scale from high to low, normally called pitch. Independent of the pitch, the sensation can also be labelled as faint pitch or strong (distinct) pitch, leading to a scale of pitch strength. For example, a pure tone of 1 kHz elicits a very distinct strong pitch sensation, whereas a high-pass noise with a cut-off frequency of 1 kHz produces an indistinct or faint pitch. Despite these differences in pitch strength, both sounds produce approximately the same pitch.

The pitch strength of sounds can be assessed quantitatively using a procedure of magnitude estimation (see Sect. 1.3). Figure 5.24 shows the spectral distributions for a large variety of sounds for which pitch strength was measured. For sound 10 only, an amplitude-modulated broad-band noise, the time function is indicated instead of the spectrum. The sounds include a selection of pure tones, complex tones, narrow-band noises, low-pass and high-pass noises, comb-filtered noises, AM tones, and AM noises. All sounds within a column elicit approximately the same pitch but differ considerably in pitch strength.

The relative pitch strength of the sounds 1–11 is displayed in Fig. 5.25. Each of the three frequency regions is represented by a separate panel, and in each panel pitch strength was normalized relative to the maximum value in the panel. Figure 5.25 shows that relative pitch strength decreases with increasing sound number in all three frequency regions. The largest pitch strength is produced by a pure tone (sound 1). The pitch strength of complex tones achieves on average at least half the pitch strength of a pure tone. But the pitch strength elicited by different types of noises (sounds 7–11) is generally a factor of 5 to 10 smaller than the pitch strength of a pure tone. The only exception is narrow-band noise (sound 4): in this case, the pitch strength is comparable to the pitch strength of complex tones. As described in Sect. 5.5, high-pass noise with low cut-off frequencies produces no pitch and on average, a pitch strength of 0 is obtained for sound 11.

To summarize, it can be stated that sounds with line spectra generally elicit relatively large pitch strength, whereas sounds with continuous spectra produce only small values of pitch strength. An exception to this rule is sound 4, a narrow-band noise, which has a continuous spectral distribution but elicits a relatively large pitch strength. This result is in line with the data displayed in Fig. 5.16, which demonstrate that pitch matches between narrow-band noise and a pure tone can be performed with the same accuracy as pitch matches between two pure tones.

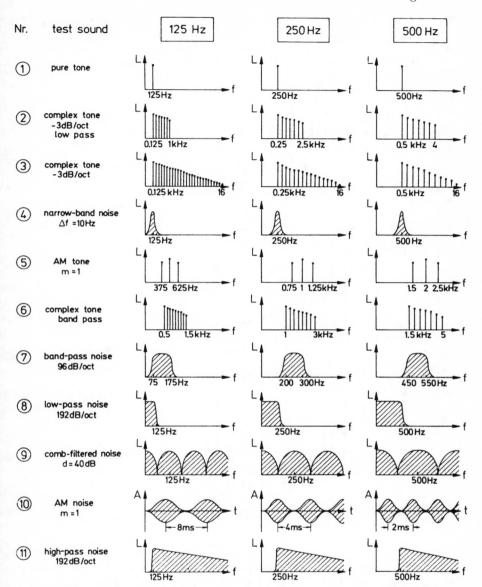

Fig. 5.24. Schematic representation of the sounds used for scaling pitch strength

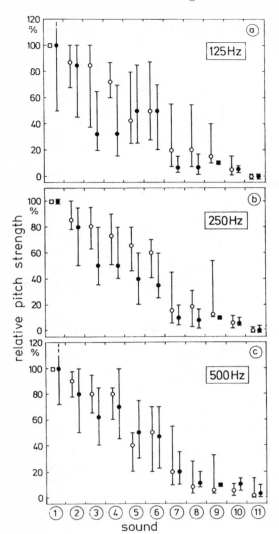

Fig. 5.25a–c. Relative pitch strength of sounds 1–11 of Fig. 5.24. Three different pitch ranges are given in the three panels

With increasing duration the pitch strength of pure tones increases. Results displayed in Fig. 5.26 suggest an almost linear increase of pitch strength with the logarithm of the test-tone duration up to a duration of about 300 ms.

With increasing sound pressure level the pitch strength of pure tones increases. Data displayed in Fig. 5.27 suggest an increase in relative pitch strength by about 10% for each increase in level by 10 dB. Within a level range of 20 to 80 dB pitch strength increases by a factor of about 2.5, whereas the loudness increases by about a factor of 100. Therefore, despite the effect that loudness decreases with tone duration, the decrease in relative pitch

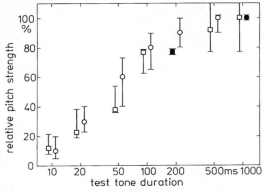

Fig. 5.26. Relative pitch strength of pure tones at 1 kHz with 80 dB sound pressure level as a function of test-tone duration

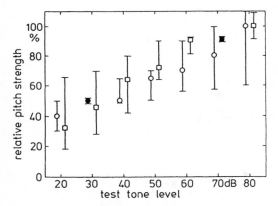

Fig. 5.27. Relative pitch strength of pure tones at 1 kHz with 500 ms duration as a function of test-tone level

strength shown in Fig. 5.26 cannot be explained on the basis of the reduced loudness alone.

As a function of test-tone frequency, pitch strength of pure tones reaches largest values at mid frequencies. Results displayed in Fig. 5.28 illustrate that pitch strength of pure tones at low (125 Hz) and high (8 to 10 kHz) frequencies is about a factor of 3 smaller than pitch strength of tones at mid frequencies around 1.5 kHz.

The dependence of the pitch strength of noise bands at different center frequencies f_c on the bandwidth is illustrated in Fig. 5.29. All data were normalized relative to the pitch strength of a pure tone with frequency f_c.

With increasing bandwidth pitch strength of noise bands decreases. Irrespective of the center frequency at narrow (3.16 Hz) versus large (1000 Hz) bandwidth, large versus small values of pitch strength are obtained. For in-

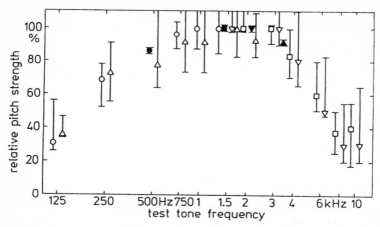

Fig. 5.28. Relative pitch strength of pure tones at 80 dB sound pressure level with 500 ms duration as a function of test-tone frequency

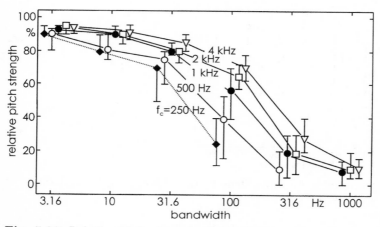

Fig. 5.29. Relative pitch strength of noise bands with center frequency f_c as a function of their bandwidth. Sound pressure level 50 dB

termediate bandwidth (100 Hz) however, a distinct influence of the center frequency on the pitch strength of noise bands shows up. As soon as the bandwidth of noise bands exceeds a critical band (see Chap. 6) only faint pitches with a relative pitch strength of some 20% are perceived.

As described in Sect. 5.3, AM-tones can produce a virtual pitch which roughly corresponds to their modulation frequency. The dependence of relative pitch strength of AM-tones on their carrier frequency is illustrated in Fig. 5.30 for AM-tones with 50 dB sound pressure level and a modulation frequency f_{mod} of 125 Hz (left) or 1 kHz (right). All data are normalized to the

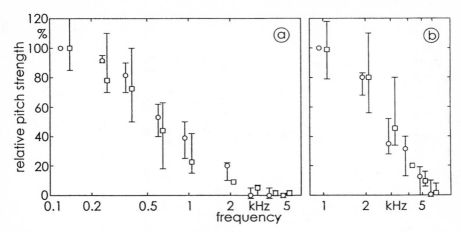

Fig. 5.30a,b. Relative pitch strength of the virtual pitch of AM-tones as a function of their frequency. Sound pressure level 50 dB, modulation frequency 125 Hz (*left panel*), and 1 kHz (*right panel*)

pitch strength of a pure tone at the modulation frequency indicated by the left most symbols in each panel.

For a carrier frequency of 3 times the modulation frequency the relative pitch strength for an AM-tone at 375 Hz with $f_{\mathrm{mod}} = 125$ Hz reaches some 77% whereas the corresponding value for $f_{\mathrm{mod}} = 1$ kHz at 3 kHz amounts to only 41%. For low modulation frequency (125 Hz) negligible values of pitch strength are obtained for a center frequency of 3 kHz which is 24 times as large, whereas for $f_{\mathrm{mod}} = 1$ kHz pitch strength of AM-tones vanishes at 6 kHz, a center frequency only 6 times larger than the modulation frequency.

This behavior nicely corroborates the existence region of virtual pitch described in Sect. 5.3. This reasoning is illustrated by results displayed in Fig. 5.31. The modulation frequency of the AM-tones representing their virtual pitch is given as a function of the frequency of the lowest component, i.e. the lower sideband of the AM-tone. Circled numbers indicate percentage values of relative pitch strength. Starting from values between 80 and 90%, with increasing frequency of AM-tones their pitch strength decreases and approaches 0% just at the border of the existence region of virtual pitch.

Peaked ripple noise, the spectrum of which is displayed in Fig. 5.17, can produce a pitch strength that is relatively large, like that of complex tones. The relative pitch strength of peaked ripple noise is plotted in Fig. 5.32 as a function of the depth of spectral modulation for three different frequencies of the first spectral peak, f_1. Up to a spectral modulation depth of about 10 dB, peaked ripple noise produces no pitch. Nevertheless, small spectral modulations only a few dB deep can lead to the perception of sound colouration. For spectral modulation depths larger than 10 dB, the relative pitch strength

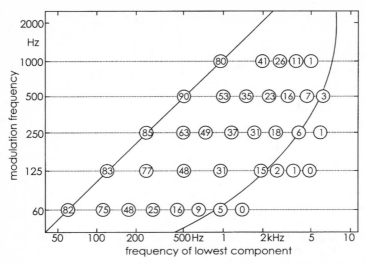

Fig. 5.31. Relative pitch strength of virtual pitch of AM-tones (*circled numbers*) in comparison to the existence region of virtual pitch (*curve*)

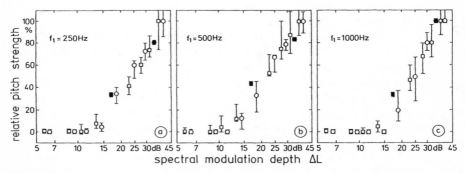

Fig. 5.32a–c. Relative pitch strength of peaked ripple noise as a function of spectral modulation depth. Frequency, f_1, of the first peak in the spectrum is indicated in the three panels

increases almost linearly with the logarithm of spectral modulation depth at all pitch ranges considered.

Pitch strength of peaked ripple noise is almost independent of sound pressure level. Within a level range of 30 to 70 dB, pitch strength decreases only by about 10%.

Figure 5.33 shows the dependence of pitch strength of peaked ripple noise on the frequency of its first spectral peak. The largest pitch strength is produced for peaked ripple noises with narrow peak spacing, i.e. low frequency f_1. With increasing peak spacing, the pitch becomes fainter and therefore pitch strength decreases. This result is related to the fact that the number

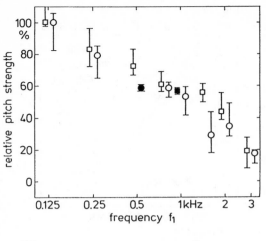

Fig. 5.33. Relative pitch strength of peaked ripple noise as a function of the frequency, f_1, of the first spectral peak

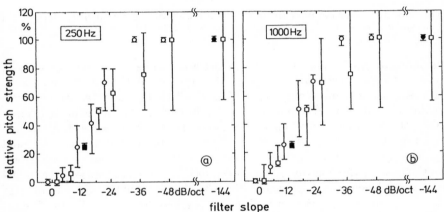

Fig. 5.34a,b. Relative pitch strength of low-pass noise as a function of the filter slope. Cut-off frequency 250 Hz (*left*) and 1000 Hz (*right*)

of spectral peaks contained within the hearing area decreases with increasing frequency f_1.

When comparing the spectra of peaked ripple noise displayed in Fig. 5.17 and of comb-filtered noise displayed in Fig. 5.24 (sound 9), the two sounds show a figure-ground relation: peaked ripple noise shows sharp spectral peaks and broad spectral valleys, whereas comb-filtered noise shows broad spectral peaks and narrow spectral valleys. Because the pitch strength of peaked ripple noise is larger by a factor of more than 5 than that of comb-filtered noise, it can be assumed that large pitch strength is produced by sounds containing narrow bands with steep spectral slopes rather than by narrow gaps.

As shown in Fig. 5.25, low-pass noise (sound 8) with extremely steep spectral slopes produces a pitch strength about 5 to 10 times smaller than the

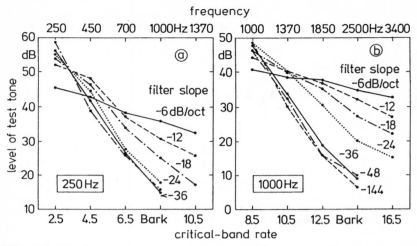

Fig. 5.35a,b. Masking patterns of low-pass noises with different filter slope but a constant loudness of 8 sone. Cut-off frequency 250 Hz (*left*) and 1000 Hz (*right*)

pitch strength of a pure tone, indicating a dependence on the steepness of the filter slope. Figure 5.34 shows relative pitch strength as a function of the filter slope for cut-off frequencies of 250 Hz and 1 kHz. The pitch strength was again normalized to the maximum value in each panel. As expected, pitch strength increases with steeper filter slope. However, filters with extremely steep spectral slopes (-144 dB/octave) are *not* required to reach the maximal value; slopes of -48 dB/octave are sufficient. This result can be understood on the basis of the masking patterns produced by low-pass noises with different spectral slopes as displayed in Fig. 5.35. The level of the just-audible test tone is given as a function of both frequency and critical-band rate. The results indicate that the masking pattern definitely depends on the filter slope for both cut-off frequencies up to a filter slope of -36 dB/octave, whereas the masking patterns are similar for steeper filter slopes. The difference of 10 dB in test-tone level between the illustrations in the left and right panels is due to the fact that a constant loudness of 8 sone was chosen for all low-pass noises.

In Fig. 5.36, relative pitch strength of low-pass noise is plotted as a function of the slope of the respective masking pattern. A comparison of Figs. 5.34 and 5.36 reveals that the results on pitch strength can be more easily described if, instead of the filter slope, the slope of the masking pattern is used as the abscissa. In this case, pitch strength of low-pass noise increases almost linearly with the slope of the corresponding masking pattern. The maximum pitch strength of low-pass noise is reached with a slope of the masking pattern of about 9 dB/Bark. It has to be kept in mind, however, that even the largest relative pitch strength of low-pass noise (100%) amounts to not more than 1/5 of the pitch strength produced by a pure tone.

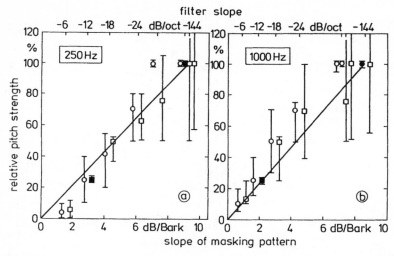

Fig. 5.36a,b. Relative pitch strength of low-pass noise as a function of the slope of the masking pattern. Cut-off frequency 250 Hz (*left*) and 1000 Hz (*right*)

Not only the pitch strength but also the pitch of low-pass noise *per se* shows strong correlations to the respective masking patterns. When increasing the level of low-pass noise from 50 to 80 dB SPL at 250 Hz cutoff frequency, the frequency of a pure tone matched in pitch increases from 260 Hz to 285 Hz. At 1000 Hz cutoff frequency of the low-pass noise the corresponding shift is from 1022 Hz to 1073 Hz. Since with increasing level the upper slope of the masking pattern for low-pass noise becomes flatter, the increase in pitch of low-pass noise with level can be described when assuming that the pitch corresponds to the 3 dB down point of the masking pattern.

The pitch strength of pure tones can be reduced considerably by partial masking sounds. As described in Sect. 5.1.2, with increasing level of the partial masking sound, the magnitude of the pitch shift of pure tones increases. However, at the same time the pitch strength of pure tones decreases. This means that a large value of pitch shift is always correlated with a small pitch strength of the shifted pure tone. In Fig. 5.37, some examples of the relative pitch strength of partially masked pure tones are given. Panel (a) shows the situation for a pure tone partially masked by another pure tone. In order to avoid the detection of difference tones, a low-pass noise masker is added (see inset). Panel (b) gives the results for pure tones partially masked by a narrow-band noise centred at a lower frequency. Panel (c) illustrates a similar condition except that the narrow-band masking noise lies above the partially masked pure tone. Finally, panel (d) shows a pure tone partially masked by a broad-band noise. For all four cases, the relative pitch strength is given as a function of the level of the test tone above masked threshold. Partially masked test tones, which lie only 3 dB above masked threshold, frequently produce

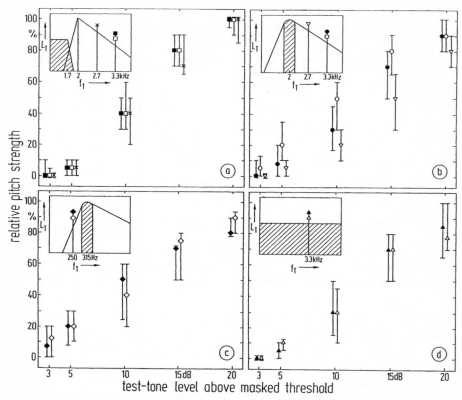

Fig. 5.37a–d. Pitch strength of partially masked pure tones. Relative pitch strength as a function of the test-tone level above masked threshold. Partial masking by (**a**) pure tone at lower frequency, (**b**) narrow-band noise at lower frequency, (**c**) narrow-band noise at higher frequency, and (**d**) broad-band noise (see insets)

very small pitch strength. For test tones 10 dB above masked threshold, the pitch strength reaches almost half the value obtained with an unmasked pure tone. At levels 20 dB above masked threshold, pitch strength is almost equal to the pitch strength of an unaffected pure tone.

For pure tones partially masked by broad-band noise, data similar to those displayed in Fig. 5.37d for 3300 Hz were also obtained at other frequencies between 55 and 8000 Hz.

So far the pitch strength of different sounds separated by pauses has been described. However, continuous variations in parameters of sounds may clearly influence their pitch strength. For example, amplitude modulation of pure tones with low modulation frequencies (4 Hz) can reduce their pitch strength by some 10 to 20%. This decrease in pitch strength holds also true for AM-tones partially masked by broadband noise.

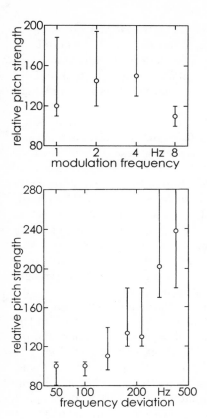

Fig. 5.38. Relative pitch strength of low-pass noise with modulated cutoff frequency as a function of the modulation frequency. Cutoff frequency 1 kHz ±85 Hz, spectrum level 40 dB

Fig. 5.39. Relative pitch strength of low-pass noise with modulated cutoff frequency as a function of frequency deviation. Cutoff frequency 1 kHz, spectrum level 40 dB, modulation frequency 4 Hz

On the other hand, the (small) pitch strength of low-pass noise can be increased by frequency modulation of its cutoff frequency. Figure 5.38 shows data for a low-pass noise with 1 kHz cutoff frequency and a density level of 40 dB (overall level 70 dB) where the cutoff frequency was periodically swept by ±85 Hz leading to a frequency deviation of 170 Hz. The speed of the frequency modulation was varied. For an unmodulated low-pass noise a relative pitch strength of 100 is obtained.

When sweeping the cutoff frequency, the relative pitch strength increases. For a modulation frequency of 4 Hz, the pitch strength of the time-varying low-pass noise is by about a factor of 1.4 larger than the pitch strength of a steady-state low-pass noise.

The increase in pitch strength when modulating the cutoff frequency of low-pass noise depends on the frequency deviation, i.e. the width of the cutoff-frequency sweep. Figure 5.39 shows the dependence of relative pitch strength on frequency deviation for a low-pass noise with 1 kHz cutoff frequency, 40 dB spectrum level, and 4 Hz modulation frequency. Again the unmodulated low-pass noise produces the relative pitch strength 100.

Results displayed in Fig. 5.39 suggest that the pitch strength of low-pass noise is hardly influenced by small (up to ± 40 Hz) sweeps of the cutoff frequency. On the other hand, for large frequency deviations with instantaneous cutoff frequencies varying between 800 and 1200 Hz, pitch strength is by a factor of 2.4 larger than the pitch strength of a stationary low-pass noise with 1 kHz cutoff frequency.

So far the pitch strength of sounds that are easily described in terms of their spectrum has been assessed. In the following, the pitch strength of amplitude-modulated broad-band noise, most easily described by its time function, is discussed. Figure 5.40 shows the relative pitch strength of rectangularly-gated broad-band noise as a function of the ratio of impulse-to-gap duration. The left panel gives the results for a repetition rate of 100 Hz, the right panel for a repetition rate of 400 Hz. The insets in the left panel indicate the temporal envelopes of broad-band noises for impulse-to-gap ratios of 0.1 and 10. In the first case, short impulses are separated by relatively long temporal gaps, and in the second case the situation is reversed. Up to a ratio of impulse duration and gap duration of 0.1, the maximum pitch strength of amplitude-modulated broad-band noise appears. With increasing impulse duration, and consequently decreasing gap duration, the pitch strength decreases. For a ratio larger than about 3, pitch strength drops to zero. In this case, long broad-band noise bursts are followed by extremely short temporal gaps which are no longer resolved by the hearing system (see Sect. 4.4). Hence, these gated broad-band noises sound like continuous broad-band noise and produce no pitch. The maximum pitch strength (100%) of amplitude-modulated broad-band noise is smaller by a factor of 5 to 10, than the pitch strength of a pure tone. This result is in line with the data displayed in Fig. 5.25 for sinusoidally amplitude-modulated broad-band noise (sound 10).

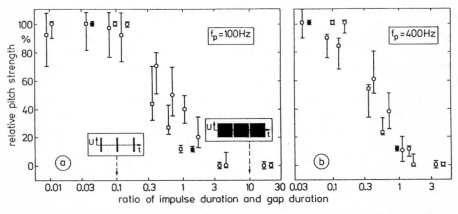

Fig. 5.40a,b. Pitch strength of gated broad-band noise. Relative pitch strength as a function of the ratio of impulse duration and gap duration. Repetition rate, f_{p}, is 100 Hz (*left*) and 400 Hz (*right*)

The pitch strength of rectangularly amplitude-modulated broad-band noise can be expected to decrease with decreasing temporal modulation depth, since for very small modulation depth, almost broad-band noise is produced. This produces no pitch, and hence has no pitch strength. For a ratio of burst-to-gap duration of 0.04, up to a temporal modulation depth of 5 dB (degree of modulation 28%), amplitude-modulated broad-band noises elicit *no* pitch sensation. For larger temporal modulation depth, the pitch strength of amplitude-modulated broad-band noise increases almost linearly. About half of the maximally possible pitch strength is obtained for temporal modulation depths around 20 dB (degree of modulation some 80%), and maximally possible pitch strength is obtained for temporal modulation depths larger than about 30 dB (degrees of modulation in excess of 95%). However, it should be noted again that even with very short bursts of broad-band noise and with large temporal modulation depth, amplitude-modulated broad-band noise produces a pitch strength which reaches only about 15% of the pitch strength produced by a pure tone.

6. Critical Bands and Excitation

The concept of critical bands is introduced in this chapter, methods for determining their characteristics are explained, and the scale of critical-band rate is developed. The definitions of critical-band level and excitation level are given and the three-dimensional excitation level versus critical-band rate versus time pattern is illustrated.

The concept of critical bands was proposed by Fletcher. He assumed that the part of a noise that is effective in masking a test tone is the part of its spectrum lying near the tone. In order to gain not only relative values but also absolute values, the following additional assumption was made: masking is achieved when the power of the tone and the power of that part of the noise spectrum lying near the tone and producing the masking effect are the same; parts of the noise outside the spectrum near the test tone do not contribute to masking. Characteristic frequency bands defined in this way have a bandwidth that produces the same acoustic power in the tone and in the noise spectrum within that band when the tone is just masked. Fletcher's assumptions may be used to estimate the width of characteristic bands, and we shall see later on how these values compare with the critical bandwidths determined by other measurements.

As outlined in Fig. 4.1, white noise produces masked thresholds that are not independent of frequency, although white noise has a frequency-independent density level. Such masked thresholds are only frequency independent up to about 500 Hz but increase for frequencies above 1 kHz with a slope of about 10 dB per decade. The relatively pronounced frequency selectivity of our hearing system has already been described in Sects. 4.1 and 4.2, indicating that it can be assumed that our hearing system processes sounds in relatively narrow frequency bands. If it is assumed that our hearing system produces masked thresholds using a criterion that is frequency independent, the frequency bands we are looking for have to be independent of frequency below 500 Hz; in this range masked threshold is independent of frequency, and so is the density of white noise. Therefore, the critical frequency band should have a constant width. For higher frequencies, masked threshold increases by 10 dB per decade which means that the intensity within the frequency bands in question increases in proportion to frequency. Therefore, the bandwidth of the bands has to increase by a factor of 10 when frequency is increased by the same factor of 10. Assuming, as Fletcher did, that a tone is audible within a noise when the acoustic power of the tone matches the acoustic power of the noise, falling into the critical frequency band where the frequency of the just-masked tone is centred, then the bandwidth in question can be estimated as follows: for frequencies below 500 Hz, masked threshold is 17 dB higher than the density level of the white noise that is masking the tone. Under the assumption of equal acoustic power at threshold for noise and tone within this band, we can calculate the bandwidth as being $10^{17/10}$, i.e. about 50 times larger than 1 Hz. This would lead to a bandwidth of 50 Hz at low frequencies.

However, the assumption that the criterion used by our hearing system to produce masked threshold is independent of the frequency of the tone is incorrect. As will be discussed later, the power of the tone at masked threshold is only about half to a quarter of that of the noise falling into the band in question. Using this additional information, the width of the bands in question, the critical bands, can be estimated quite closely. At low frequencies, critical bands show a constant width of about 100 Hz, while at frequencies above 500 Hz critical bands show a bandwidth which is about 20% of centre frequency, i.e., in this range critical bandwidth increases in proportion to frequency.

In contrast with the estimation of the width of the critical band using the assumption described above, there exist several direct methods for measuring the critical band. These methods and the results obtained will be described in the following sections.

6.1 Methods for the Determination of the Critical Bandwidth

Threshold measurements are the basis of the first method for obtaining critical bandwidths. As with all other methods for direct measurement of the critical band, either the bandwidth or a value directly correlated with bandwidth has to be the variable. In this case, the threshold for a complex of uniformly spaced tones as a function of the number of tones in the complex, where each tone has the same amplitude, was used to estimate the critical bandwidth near 1 kHz.

Figure 6.1 shows the threshold level (in terms of the level of each tone in the complex) as a function of either the number of test tones or the frequency difference between the lowest and the highest tone. The number of test tones is also the parameter differentiated by the symbols in Fig. 6.1. The frequency difference between the tones remains constant at 20 Hz. Using the tracking procedure, threshold is measured for a pure tone at 920 Hz and occurs at a sound pressure level of +3 dB. Another tone of the same level is added at a frequency of 940 Hz (20 Hz higher) and the threshold of the two-tone complex is again measured. It is found at a level of 0 dB for each tone of the

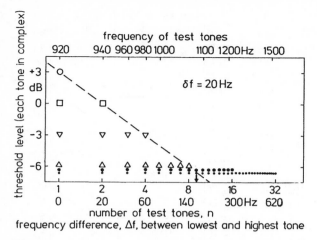

Fig. 6.1. Threshold in quiet as a function of the number of test tones (different symbols) equally spaced at a frequency distance of 20 Hz. The level of the constituent equal-amplitude test tones is given as a function of the number of test tones or the frequency difference between lowest and highest test tone. Frequency of test tones is also given on the upper scale. The arrow indicates the transition point where the critical bandwidth is estimated

complex. We proceed further in this fashion by adding two more tones at 960 and 980 Hz and measure threshold again. The threshold lies at a level of -3 dB for each tone. For eight tones, the threshold is found at -6 dB SPL of each tone. This means that threshold expressed in level per tone decreases with the increasing number of tones, as expected. However, beyond a certain number of tones, no further decrease occurs, as Fig. 6.1 indicates for the sets of 16 and 32 tones used. With a certain number of individual tones, in our case about nine, the decrement in level produced by adding tones stops. The transition point is marked in the figure by an arrow and is used as a measure of the critical bandwidth.

It is interesting to note that in the range between one and eight tones in the complex, the threshold level decreases 3 dB per doubling of the number of tones. This means that, at threshold, the overall sound pressure level of the complex remains constant regardless of the number of tones. This rule holds only up to about nine tones. Beyond this number, threshold expressed as level of each tone in the complex no longer decreases with increasing number of tones, i.e. overall sound pressure level increases. This means that the threshold in quiet is determined in our hearing system by the sound intensity of the total complex, as long as the components of this complex fall within a certain bandwidth. Parts outside that bandwidth do not contribute to threshold in quiet. This bandwidth can be calculated from the number of tones and the distance per tone, and leads in our case to about 160 Hz according to $(9-1) \cdot 20\,\text{Hz} = 160\,\text{Hz}$.

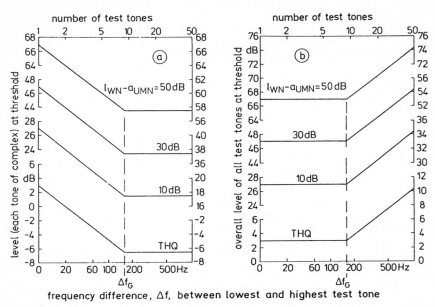

frequency difference, Δf, between lowest and highest test tone

Fig. 6.2a,b. Two different plots of the results of Fig. 6.1. The left part has the same ordinate as is used in Fig. 6.1, i.e. the level of the constituent test tones. The number of tones and the frequency difference between lowest and highest test tone are again on the abscissa. In addition to threshold in quiet, masking results produced with uniform masking noise of a given density level are also indicated. Critical bandwidth, Δf_G, separates the two regions within which the results follow different rules. The right panel shows the same abscissa but the ordinate is now the overall level of the test tones. This leads to a horizontal curve up to the critical band. Beyond that, the results indicate rising functions. The data of both panels were produced at a centre frequency of 1 kHz

This experiment was started at threshold in quiet. It can be performed in a meanigful way only in a frequency range in which threshold in quiet is independent of frequency. This is rarely the case and arises only in the frequency range between 500 Hz and 2 kHz. However, as described in Fig. 4.2, uniform masking noise has the advantage of producing masked thresholds independent of frequency. If it could be shown that the effect described in Fig. 6.1 takes place not only at threshold in quiet but also at the threshold produced by uniform masking noise, then it would be possible to measure this effect over the whole frequency range of hearing. Results of measurements comparable to those outlined in Fig. 6.1 for tones near 920 Hz and for different levels of uniform masking noise are outlined in Fig. 6.2a. The number of equally-spaced tones of equal amplitude is given in the upper abscissa, the bandwidth resulting from the number of tones is shown as the lower abscissa. The noise density level at low frequencies is given as the parameter with masked threshold. The results clearly show that the two rules described above, i.e. the decreasing threshold expressed in level of a single tone for small Δf and the independence of threshold for larger Δf's remain, although the threshold in quiet is raised to the threshold masked by uniform masking noise. In order to characterize the rule for a small number of tones, i.e. smaller frequency separation Δf, the data are plotted in Fig. 6.2b in a different form. The abscissa is the same, but the ordinate is now the overall sound pressure level of the complex of tones. It therefore becomes clear that the overall sound pressure level remains constant at threshold in quiet (or at masked threshold) for bandwidths smaller than the critical band. The overall sound pressure increases for bandwidths above the critical bandwidth, indicating that the components outside the critical band do not contribute either to the threshold in quiet or to the masked threshold. The sound intensity falling into one critical band is responsible for threshold in quiet as well as for masking.

Using uniform masking noise, it is possible to produce meaningful data at all frequencies so that the critical bandwidth can be measured as a function of frequency. Further, instead of using a number of tones of equal amplitude, noises can be used in the same way. In this case, uniform masking noise is used as masker and the threshold of an additional noise is measured as a function of the bandwidth of this additional noise. The results show the same effect: the threshold of this noise masked by uniform masking noise is independent of bandwidth for noise narrower than the critical bandwidth but increases – just like the data outlined in Fig. 6.2b – for bandwidths larger than the critical band.

Masking in frequency gaps is the second method used in determining critical bandwidth. A relatively simple combination of masker and test sound involves the use of two tones of equal level as masker, and a narrow-band noise as the test sound. The threshold of the test sound masked by the two tones is measured as a function of the frequency separation between the two masker tones with the narrow-band noise centred between them. The bandwidth of the test sound has to be small in relation to the expected critical bandwidth. The inset in Fig. 6.3 shows the frequency composition, and Fig. 6.3 shows the data that are produced by two 50-dB tone maskers

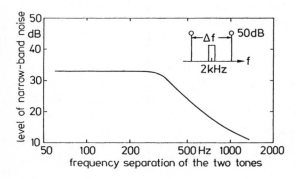

Fig. 6.3. The threshold of a narrow-band noise centred between two masking tones of equal level (indicated in the inset) as a function of the frequency separation of the two tones

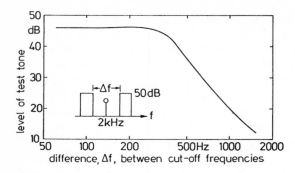

Fig. 6.4. Threshold of a test tone masked by two band-pass noises (indicated in the inset) as a function of the frequency difference between the cut-off frequencies of the noises

and a narrow-band noise centered at 2 kHz. The threshold of the narrow-band noise masked by the two tones is plotted as a function of the frequency separation between the two tones. For narrow frequency separations, Δf, the masked threshold remains independent of Δf. Beyond a certain Δf, called the critical bandwidth, threshold decreases. The crossing point between the horizontal part and the decaying part is taken as the critical bandwidth. The location of the critical transition point in this experiment appears to be invariant with level, at least for a masker level up to about 50 dB. At higher levels, the masking audiogram shows a basic asymmetry that influences the effect we are looking for. Often, the knee point is still indicated even though the decaying part is distorted by effects that are related to the ear's own nonlinearity. Measurements at medium and small levels can easily be produced for different centre frequencies so that critical bandwidth can be measured as a function of frequency using this method.

A similar method with interchanged stimuli is indicated in the inset of Fig. 6.4. In this procedure two bands of noise, a lower one with an upper cut-off frequency and an upper one with a lower cut-off frequency, are presented together as the masker. The difference between the upper and the lower cut-off frequency of the two noises, Δf, varies and the threshold of a tone centred geometrically within the gap is measured as a function of Δf. Results of such measurements are outlined in Fig. 6.4 for a centre frequency of 2 kHz and a level of 50 dB for each of the 200 Hz wide flanking noises. Masked threshold, in this case that of the tone, remains constant for small Δf but decreases for Δf larger than a critical value. This critical value, the critical bandwidth, can be obtained from Fig. 6.3 as well as from Fig. 6.4 and is 300 Hz for a centre frequency of 2 kHz.

The third method for determining critical bandwidth is based on the detectability of phase changes. Three-component complexes can be changed from representing amplitude-modulated tones to quasi-frequency modulated tones by changing one component by 180°. When the amplitude of a tone is modulated sinusoidally, the result is the original tone (the carrier) and a sideband equally spaced in frequency to either side. The spacing between carrier and sidebands corresponds to the rate of modulation (modulation fre-

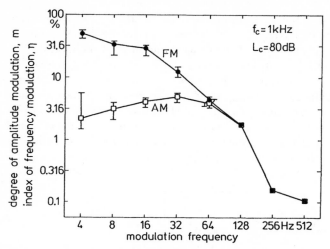

Fig. 6.5. The medians and interquartile ranges for just-noticeable degree of amplitude modulation (AM) and just-noticeable index of frequency modulation (FM) of a 1-kHz tone at 80 dB SPL, as a function of modulation frequency. Note that the thresholds for the two kinds of modulation coincide at modulation frequencies above 64 Hz

quency). When the frequency of a tone is modulated sinusoidally, the same thing happens. For a small modulation index, i.e. the ratio between the frequency deviation and the modulation frequency, a carrier and two sidebands are again produced. Significant sidebands beyond the first pair are produced only for modulation indices larger than 0.3. The difference between AM and FM with modulation indices less than 0.3 is due to phase: relative to the phases of the components produced by AM, one of the sidebands is 180° out of phase with FM. In other words, in first approximation, AM becomes FM if the phase of one of the sidebands is reversed. The overall bandwidth of the three components for AM or FM is given by twice the frequency of modulation. Sensitivity for the just-detectable amount of modulation can be measured either as the degree of AM or as the modulation index of FM. If differences occur, the difference must be related to the sensitivity of our hearing system for phase changes (in this case reversing of phase of one component of the complex). The results of such measurements, indicated in Fig. 6.5, show that the just-detectable degree of modulation (AM) is smaller than the just-detectable modulation index (FM) at low frequencies of modulation. In other words, in order to be heard as a modulation, the amplitude of the sidebands must be greater in FM than in AM. As the rate of modulation increases, however, and the sidebands are spread further apart, a point is reached beyond which the just-detectable modulation is the same for both FM and AM. At and beyond this point, the phase of the side bands no longer makes any difference to our hearing capabilities.

The data given in Fig. 6.5 were obtained at an 80-dB level of the carrier and a centre frequency of 1 kHz. The results and the distinguishing point of the two ranges become even clearer if the logarithmic ratio of the modulation index and the degree of modulation is plotted. This is done in Fig. 6.6 where this ratio is plotted as a function of $2f_{\mathrm{mod}}$, the overall spacing of the three components. The spacing $\Delta f = 2f_{\mathrm{mod}}$ at which the decreasing part reaches 0 dB is a measure for the critical bandwidth, 150 Hz for the 1-kHz case plotted in Fig. 6.6. Again, these measurements can be carried out for various carrier frequencies and lead to a determination of the critical bandwidth as a function of centre frequency, in this case the carrier frequency.

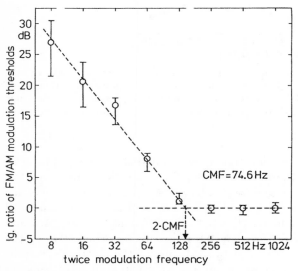

Fig. 6.6. Replot of the data given in Fig. 6.5 on an ordinate scale showing the logarithm of the ratio between the just-noticeable frequency modulation and the just-noticeable amplitude modulation (medians and interquartile ranges are given). This way, two almost straight lines approximate the results and indicate, at the crossing point, the critical modulation frequency (CMF), which is equal to half the critical bandwidth. Because of this, twice the modulation frequency is used on the abscissa

A fourth method for determining the critical bandwidth is that of loudness measurement as a function of bandwidth for constant sound pressure level. Although loudness measurements will be discussed in much more detail in Chap. 8, a typical result is given here. In Fig. 6.7, the subjectively measured loudness of a noise is plotted as a function of its bandwidth, with the overall sound pressure level of the noise kept constant. The result shows that loudness is constant as long as the bandwidth of the band-pass noise is smaller than a critical value, in this case 300 Hz at a centre frequency of

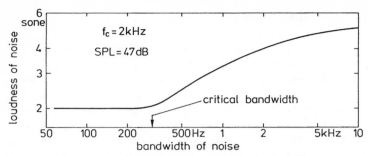

Fig. 6.7. Loudness of band-pass noise centred at 2 kHz with an overall SPL of 47 dB as a function of its bandwidth

2 kHz, which corresponds to the critical bandwidth at that frequency. Beyond that bandwidth, loudness increases by up to a factor of three for very large bandwidths. At that point, the loudness of a broad-band noise is reached. The important condition for these measurements is the fact that the overall sound pressure level remains constant, i.e., the density of the sound intensity has to be reduced as the bandwidth of the noise is increased. In this case the critical bandwidth is determined directly by measuring loudness as a function of bandwidth and searching for the knee point that devides the two ranges. Many measurements, some with tone and some with noises and all as a function of the overall spacing, have been performed for various centre frequencies so that the critical bandwidth could be estimated as a function of centre frequency.

A fifth method stems from binaural hearing. The localization of short impulses is used as an indication of the development of critical bands. Just-noticeable delay between the envelope of two tone bursts, somewhat different in frequency and each presented to one ear, is measured as a function of the frequency separation of the tone bursts. The hearing system is quite sensitive to envelope delay as long as the two tone bursts are of high frequency and of the same or of nearly the same frequency. The sensitivity decreases drastically as soon as the frequency difference between the two tone bursts becomes larger than the critical bandwidth. The results obtained in this way lead to critical frequency distances that are very close to those measured with the four methods mentioned above, at least where they can be measured.

Since all the methods, except that using Fletcher's assumption, lead to similar values of the critical bandwidth, it seems reasonable to accept the latter estimates and conclude that the equal power assumption is wrong. It will be shown in Chap. 7 that threshold is reached when the signal power is one half (at low frequencies) to one quarter (at higher frequencies) that of the masker. Introducing these ratios, masking of a tone by noise (Fletcher's method) leads to the same critical bandwidth as found using the five methods described.

6.2 Critical-Band Rate Scale

Data from many subjects have been collected to produce a reasonable estimation of the width of the critical band. The discussion of this width at low frequencies confirmed that the frequency response of the transducer must be considered to produce meaningful data for critical-band estimation. Below 200 Hz, it seems that the method using the detectability of phase effects when switching from FM to AM is the most reliable one. Although the lowest critical bandwidth in the audible frequency region may be very close to 80 Hz, it is attractive to add the inaudible range from 0 Hz to 20 Hz to that critical band, and to assume that the lowest critical band ranges from 0 Hz to 100 Hz. Using this approximation, Fig. 6.8 shows the average of our data using five methods, more than 50 subjects, and levels between threshold in quiet and about 90 dB. Although there is a small tendency for the critical band to increase somewhat for levels above about 70 dB, the curve given in Fig. 6.8 represents a good approximation for critical bandwidth as a function of frequency. The critical bandwidth remains near 100 Hz up to a frequency of about 500 Hz. Above that frequency, the critical bandwidth increases a little slower than in proportion to frequency and for frequencies above about 3 kHz a little faster. It is useful to assume constant bandwidth of 100 Hz up to a centre frequency of 500 Hz, and a relative bandwidth of 20% for centre frequencies above 500 Hz. More exact values are given in Table 6.1, which gives the lower and upper limit of the critical bands if they are accumulated in such a way that the upper cut-off frequency of the lower critical band is identical to the lower cut-off frequency of the next higher critical band.

The critical-band concept is important for describing hearing sensations. It is used in so many models and hypotheses that a unit was defined leading to the so-called critical-band rate scale. This scale is based on the fact that our hearing system analyses a broad spectrum into parts that correspond

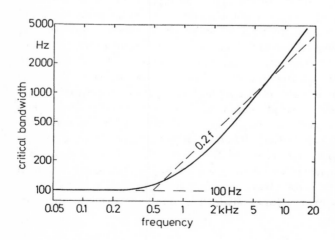

Fig. 6.8. Critical bandwidth as a function of frequency. Approximations for low and high frequency ranges are indicated by broken lines

Table 6.1. Critical-band rate z, lower (f_l) and upper (f_u) frequency limit of critical bandwidths, Δf_G, centred at f_c

z	f_l, f_u	f_c	z	Δf_G	z	f_l, f_u	f_c	z	Δf_G
Bark	Hz	Hz	Bark	Hz	Bark	Hz	Hz	Bark	Hz
0	0				12	1720			
		50	0.5	100			1850	12.5	280
1	100				13	2000			
		150	1.5	100			2150	13.5	320
2	200				14	2320			
		250	2.5	100			2500	14.5	380
3	300				15	2700			
		350	3.5	100			2900	15.5	450
4	400				16	3150			
		450	4.5	110			3400	16.5	550
5	510				17	3700			
		570	5.5	120			4000	17.5	700
6	630				18	4400			
		700	6.5	140			4800	18.5	900
7	770				19	5300			
		840	7.5	150			5800	19.5	1100
8	920				20	6400			
		1000	8.5	160			7000	20.5	1300
9	1080				21	7700			
		1170	9.5	190			8500	21.5	1800
10	1270				22	9500			
		1370	10.5	210			10500	22.5	2500
11	1480				23	12000			
		1600	11.5	240			13500	23.5	3500
12	1720				24	15500			
		1850	12.5	280					

to critical bands. Adding one critical band to the next in such a way that the upper limit of the lower critical band corresponds to the lower limit of the next higher critical band, leads to the scale of critical-band rate. If the critical bands are added up this way, then a certain frequency corresponds to each crossing point (see Table 6.1). The procedure is illustrated in Fig. 6.9. The first critical band spans the range from 0 to 100 Hz, the second from 100 to 200 Hz, the third from 200 to 300 Hz and so on up to 500 Hz where, of course, the frequency range of each critical band increases. Plotting the ordinal number of each critical band lined up as a function of frequency produces a series of dots plotted in Fig. 6.9. It can be seen that the audible frequency range to 16 kHz can be subdivided into 24 abutting critical bands. The series of dots does not mean that critical bands exist only between two neighbouring dots; rather, they should be thought of as able to be shifted continuously along a scale produced by a curve through the dots. The scale produced in this way is called critical-band rate. It grows from 0 to 24 and

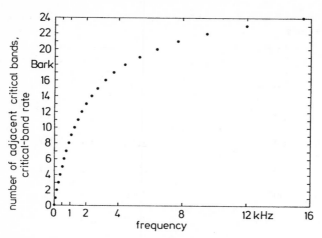

Fig. 6.9. The numeral associated with the sequence of adjacent critical bands, a value that is equal to the critical-band rate, is plotted as a function of frequency. Both coordinates are linear

has the unit "Bark" (in memory of Barkhausen, a scientist who introduced the "phon", a value describing loudness level for which the critical band plays an important role). The relation between critical-band rate, z, and frequency, f, is important for understanding many characteristics of the human ear.

The critical-band rate is closely related to several other scales that describe characteristics of the hearing system. For example, both the just-noticeable increment in frequency and the threshold for frequency modulation are closely related to critical bandwidth. Although this relation is discussed later in Chap. 7, the frequency dependence of just-noticeable frequency variations may be compared with that of critical bandwidth. Furthermore, critical bandwidth seems to bear a relation to the function relating frequency to ratio pitch, and to the function relating frequency to the position of maximal stimulation on the basilar membrane. For comparing critical bandwidth, just-noticeable variations in frequency and the position of maximal stimulation on the basilar membrane, it is convenient to advance in constant step sizes (0.2 mm) along the basilar membrane, and to plot the increment in frequency, Δf, as a function of frequency corresponding to each point. Near the helicotrema, i.e. at low frequencies, the step of 0.2 mm leads to a frequency increment of about 15 to 20 Hz. At high frequencies near the oval window, however, a step size of 0.2 mm produces a frequency increment, Δf, of about 500 Hz.

The relationship produced in this way is shown in Fig. 6.10 as a dashed line which indicates the frequency increment, Δf, as a function of frequency for a step size of 0.2 mm along the basilar membrane. The other two curves shown as solid lines in Fig. 6.10 represent the critical bandwidth, Δf_G, and the difference limen, $2\Delta f$, for frequency modulation, both as a function of

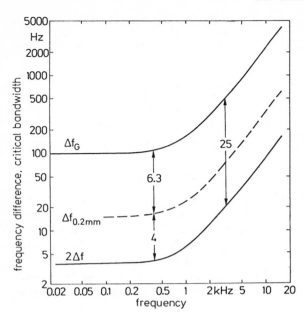

Fig. 6.10. Critical bandwidth, Δf_G, and just-noticeable frequency variation, $2\Delta f$, are given as solid lines as a function of frequency. The broken line indicates the frequency variation necessary to shift the maximum of displacement on the basilar membrane 0.2 mm. The double arrows indicate the factor by which the curves are displaced from each other

frequency. The shape of all three curves is very similar, and one can be reproduced from the other by a parallel shift upwards or downwards. Because the abscissa and ordinate are given in logarithmic scales, such a parallel shift corresponds to multiplication by a certain factor. These factors are indicated by double arrows in the figure, and the just-noticeable modulation in frequency, $2\Delta f$, is about 25 times smaller than the critical band. The frequency shift, Δf, which corresponds to a change of 0.2 mm of the place of maximal stimulation along the basilar membrane, is about 4 times larger than $2\Delta f$ but a factor of 6.3 times smaller than the width of the critical band. This means that a frequency shift of a sinusoidal tone by $2\Delta f$ produces a shift of constant distance along the basilar membrane of about 0.05 mm. The shift is independent of the frequency of the tone and constant along the basilar membrane (the small difference between frequency and frequency pitch as described in Sect. 5.1.2 is ignored in this discussion because it amounts only to a small percentage). The width of the critical band corresponds to a distance along the basilar membrane of about 1.3 mm. Assuming that the abutting haircells have a distance of about 9 μm along the whole length of the 32 mm basilar membrane, the total number of 3600 haircells in one row from helicotrema to oval window is achieved. Taking into account the discussion in Chap. 5, the fact that both the summation of critical bandwidths and the summation of just-noticeable steps measured by frequency modulation produce the same function as that relating the pitch of pure tones to their frequency, leads to the interesting relationship indicated in Table 6.2.

Table 6.2. Relationship between distance of critical-band rate (left column), distance along basilar membrane (second column), number of abutting just-audible pitch steps (third column), difference in ratio pitch (fourth column) and equivalent number of abutting haircells

24 Bark $\doteq$	32 mm $\doteq$	640 steps $\doteq$	2400 mel $\doteq$	3600 haircells
1 Bark $\doteq$	1.3 mm $\doteq$	27 steps $\doteq$	100 mel $\doteq$	150 haircells
0.7 Bark $\doteq$	1 mm $\doteq$	20 steps $\doteq$	75 mel $\doteq$	110 haircells
0.04 Bark $\doteq$	50 μm $\doteq$	1 step $\doteq$	3.8 mel $\doteq$	5.6 haircells
0.01 Bark $\doteq$	13 μm $\doteq$	0.26 steps $\doteq$	1 mel $\doteq$	1.5 haircells
0.007 Bark $\doteq$	9 μm $\doteq$	0.18 steps $\doteq$	0.7 mel $\doteq$	1 haircell

These relations can also be drawn on scales. Six of them are shown in Fig. 6.11. The upper part shows the inner ear and basilar membrane (hatched) unwound so that its total length is visible. It starts at the helicotrema, where low frequencies are located, and becomes smaller until it reaches the oval window, where high frequencies are located. The total length of the basilar membrane is 32 mm, indicated in the second linear scale. The third scale gives the number of just-noticeable steps measured by frequency modulation that can be achieved in proceeding from helicotrema to the oval window. All together, 640 steps based on just-noticeable frequency modulation can be added one after the other. The ratio pitch of tones is given on the fourth scale. It grows from 0 to 2400 mel on a linear scale. The critical-band rate plotted from 0 to 24 Bark is on the fifth scale, again plotted linearly. The final and bottom scale is that of the frequency. It has a nonlinear subdivision; up to about 500 Hz the scale is almost linear, but for frequencies above 500 Hz, the scale is almost logarithmically subdivided.

From the scales plotted in Fig. 6.11 an important fact becomes clear: the frequency scale, a physical scale, is not very useful in describing effects produced in the inner ear; over the whole length of the basilar membrane neither a linear nor a logarithmic scale will serve. In contrast to frequency,

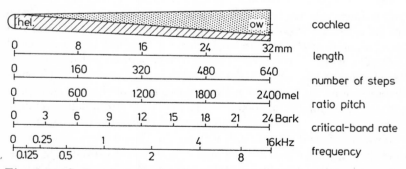

Fig. 6.11. Scales of pitch-related sensations transformed to the length of the unwound cochlea. Note that the scales of length, number of steps, ratio pitch and critical-band rate are linear scales but that the frequency scale is not

all other values, such as the number of frequency steps, the ratio pitch of tones and the critical-band rate, can be plotted on linear scales along the basilar membrane. Therefore, it seems reasonable to use a frequency to place transformation as early as possible when discussing characteristics of the hearing system, or when elaborating models to describe these characteristics. Either the critical-band rate scale or the ratio-pitch scale are much more useful than the frequency scale. In many cases, an early transformation of frequency into critical-band rate is sufficient to describe the effects taking place along the basilar membrane in a simple and unique way.

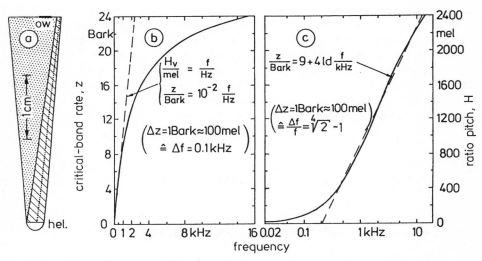

Fig. 6.12a–c. The scale of the unwound cochlea (**a**), on the ordinate the critical-band rate, and the ratio pitch on linear scale are shown as the function of the frequency on linear scale (**b**) and on logarithmic scale (**c**). Useful approximations are indicated by the broken lines and their equations

The relationship between frequency on one hand, and length of the basilar membrane or critical-band rate or ratio pitch of tones on the other is important. This relationship is outlined in Fig. 6.12 using different frequency scales, one divided linearly and the other logarithmically. Sometimes approximations may be useful, especially if only the low frequency or the high frequency ranges are considered. These approximations, shown as broken straight lines in the figures, are also given numerically. On the left of Fig. 6.12, the uncoiled inner ear, including the basilar membrane from helicotrema to oval window, is shown. The dotted line drawn along the centre of the basilar membrane may be assumed to be the row of the inner haircells. Part (b) shows frequency on linear abscissa scale, and critical-band rate as the ordinate, also on linear scale. Ratio pitch is given as the ordinate on the right. Because critical bandwidth at low frequencies is 100 Hz, and because frequency and ratio pitch are linearly correlated at low frequencies with the factor of proportionality being unity, the approximation given in part (b) for low frequencies becomes evident: 1 Bark is equal to 100 mel. It may again be mentioned that the pitch shifts discussed in Chap. 5, which remain mostly in the range of a few percent, have been ignored. The approximation of direct proportionality shown as the broken line in part (b) indicates the range within which the ratio pitch in mel is equal to the frequency in Hz. This is the range that governs harmony in music. The critical-band rate in Bark is proportional too, but 100 times smaller than frequency (Hz) in this range. An increment of the critical-band rate of 1 Bark corresponds to a change in ratio pitch of 100 mel.

A logarithmic frequency scale is used in part (c) as the abscissa. The straight broken line indicates that a logarithmic relation between critical-band rate and frequency is a very useful approximation for frequencies above 500 Hz. This approximation leads to the relation between increments in Bark (or increments of 100 mel) to a relative frequency change of about 20%.

In many cases an analytic expression is useful to describe the dependence of critical-band rate (and of critical bandwidth) on frequency over the whole auditory frequency range. The following two expressions have proven useful:

$$z/\mathrm{Bark} = 13 \arctan (0.76 f /\mathrm{kHz}) + 3.5 \arctan (f/7.5\,\mathrm{kHz})^2 , \tag{6.1}$$

and

$$\Delta f_\mathrm{G}/\mathrm{Hz} = 25 + 75 \left[1 + 1.4(f/\mathrm{kHz})^2\right]^{0.69} . \tag{6.2}$$

6.3 Critical-Band Level and Excitation Level

The frequency selectivity of our hearing system can be approximated by sub-dividing the intensity of the sound into parts that fall into critical bands. Such an approximation leads to the notion of critical-band intensities. If instead of an infinitely steep slope of the hypothetical critical-band filters, the actual slope produced in our hearing system is considered, then such a procedure leads to an intermediate value called excitation. Mostly, these values are not used as linear values but as logarithmic values similar to sound pressure level. The critical-band level and the excitation level are the corresponding values that play an important role in many models as intermediate values. The critical-band intensity, I_G, can be calculated by the follwing equation that takes into account the frequency dependence of critical bandwidth:

$$I_\mathrm{G}(f) = \int_{f-0.5\Delta f_\mathrm{G}(f)}^{f+0.5\Delta f_\mathrm{G}(f)} \frac{dI}{df} df . \tag{6.3}$$

We have already seen that the critical-band rate is useful in describing the characteristics of our hearing system. Because critical-band rate, z, is a definite function of frequency, (6.3) can also be expressed in critical-band rates:

$$I_G(z) = \int_{z-0.5\,\text{Bark}}^{z+0.5\,\text{Bark}} \frac{dI}{dz} dz \; . \tag{6.4}$$

In logarithmic expressions, and using $I_0 = 10^{-12}\,\text{W/m}^2$ as reference value, the critical-band level, L_G, is defined as

$$L_G = 10 \cdot \log \frac{I_G}{I_0}\,\text{dB} \; . \tag{6.5}$$

Critical-band intensity can be seen as that part of the overall unweighted sound intensity that falls within a frequency window that has the width of a critical band. The transformation of frequency into critical-band rate transfers the frequency-dependent window width into a window width of 1 Bark, independent of critical-band rate. This window of 1-Bark width can be continuously shifted along the critical-band scale. Consequently, a critical-band wide narrow-band noise produces a critical-band intensity which is a function of critical-band rate, and which shows the form of a triangle with a base width of 2 Bark. A sinusoidal tone, however, produces a function with a rectangular shape and the width of 1 Bark.

The intermediate values such as excitation or excitation level, however, represent a much better approximation to the frequency selectivity of our hearing system. The upper and lower slopes of thresholds for sinusoidal tones masked by narrow-band noises are used to construct the excitation level versus critical-band rate pattern. The so-called main excitation corresponds in this transformation to the maximum value of the critical-band level. The slope excitation corresponds to the subjectively measured slopes of masked thresholds. In most cases, the excitation level defined as L_E and given by the equation,

$$L_E = 10 \log\left(\frac{E}{E_0}\right)\,\text{dB} \; , \tag{6.6}$$

is used.

The excitation level can be constructed most simply from the critical-band level as a function of critical-band rate, by calculating first the critical-band level in the range of the main excitation. There, the excitation level is identical to the critical-band level. In cases of an abrupt change of the intensity density as a function of the critical-band rate, as for low-pass noise or sinusoidal tones, the maximum value of the critical-band level corresponds to the excitation level. Starting from this point or from the centre of this range, the slopes of the excitation level are added. These slopes are defined by shifting the slopes obtained for masked threshold level upwards in such a way that the slopes of the excitation levels fit the main excitation levels already available. This means that masked thresholds are shifted upwards by the value of the masking index, which is the difference between the critical-band level and the masked threshold in the region of the main excitation.

Threshold in quiet is also interpreted as a masked threshold produced by internal noise. This internal noise is frequency independent at medium and high frequencies, but increases strongly towards low frequencies and is responsible for the rise in threshold in quiet at low frequencies. Heart beats and spontaneous activity of muscles are typical noise sources which produce an acoustical stimulation of our hearing system especially at very low frequencies. The effect of these noises can be increased by closing the outer ear canal. In this case, the sound pressure level can be measured using a probe microphone and appears larger in relation to the values measured with an open ear canal.

When searching for better approximations of the internal activity produced by external stimuli in our hearing system, the frequency response of the transmission factor relating intensity measured in the free field to that being active internally has to be taken into account. The form of our head, the size of the outer ear, the length of the ear canal, and the transfer characteristics of the middle ear are the reasons for the frequency dependence of this transformation factor. It is normally introduced as a logarithmic value by the corresponding attenuation a_0. In the case of exact calculations of the excitation level based on the critical-band rate level, this transformation attenuation a_0 has to be taken into account. A typical example is the calculation of loudness for which a_0 plays an important role.

For didactical reasons a_0 is often neglected, as for example in Fig. 6.13, where the construction of the excitation level is outlined for three different types of sounds. The left side of the figure introduces all the details of the construction of the excitation level for a narrow-band noise centred at 2 kHz (hatched) and for white noise. On the right side, the construction is outlined for 11 harmonics with a fundamental frequency of 500 Hz. The upper drawing on each side represents the intensity density, dI/df, or the intensity, I, as a function of frequency. The next drawing indicates the transformation of frequency, f, into critical-band rate, z. This is only a change in the scaling of the abscissa from linear frequency into linear critical-band rate. The third drawing shows the critical-band intensity related to the reference value I_0 as a function of the critical-band rate. White noise shows a rise of I_G/I_0 above 5 Bark because critical bandwidth increases above 500 Hz (5 Bark). Narrow-band noise, one critical band wide, is indicated in this drawing by a triangular shape with a base of 2 Bark. This is so because a window of 1-Bark width starts to collect intensity if moved from low to high z-values when its centre is 0.5 Bark left of the limit of the critical-band wide noise. The peak of the triangularly shaped area corresponds to the total intensity, which is also reached by the white noise at exactly this value. This coincidence indicates that the width of the narrow-band noise is exactly one critical band wide. The tones, evenly separated along the frequency scale, are transferred to increasingly narrowly separated tones along the critical-band rate. The third drawing shows rectangles indicating the critical-band intensity. These rectangles are separated from each other in cases where the partial tones have a distance of more than 1 Bark. For smaller distances, as in higher fre-

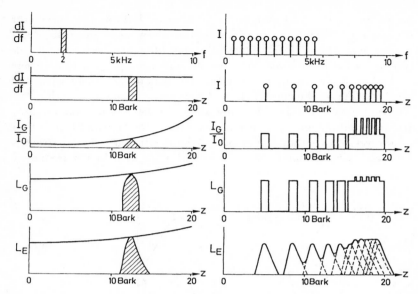

Fig. 6.13. Development of the excitation level versus critical-band rate patterns from intensity versus frequency patterns. The first step transforms frequency in critical-band rate as the abscissa (*2nd panel*). From this, the critical-band intensities are calculated (*3rd panel*) and transformed into logarithmic values (*4th panel*). Finally, using the shape of masked thresholds the excitation patterns are constructed (*5th panel*). The examples indicated apply to white noise (*solid line*) and narrow-band noise (*hatched area*) on the left, and to an 11-tone complex in the right part (see text for details)

quency regions, additional rectangles higher by a factor of two are produced. The transformation from relative critical-band intensity to critical-band level (step four) is only a transformation into logarithmic values. The triangularly-shaped area is transformed in this way into an area corresponding to the form of a gothic window. The double height of the rectangles for sinusoidal tones at high frequencies corresponds to an enlargement of 3 dB. The lowest drawings represent the excitation level as a function of critical-band rate we have been searching for.

In the case of main excitations, (at critical-band rates corresponding to the centre frequency of the critical-band wide noise or to the frequency of the tones) critical-band level and excitation level are identical. Because broadband noise, like white noise produces only main excitations, the excitation level and critical-band level of white noise are indistinguishable. For critical-band wide band-pass noise, however, only one single value, the maximum value, is identical. Starting from this point, the slope of masked threshold towards lower frequencies and the slope towards higher frequencies have to be added in moving from critical-band level to excitation level. The form of these slopes outlined as a function of critical-band rate with centre frequency as the parameter in Fig. 6.14 and with critical-band level as the parameter in Fig. 6.15, will be discussed later. The bottom drawing in Fig. 6.13 (right side) for sinusoidal tones clearly illustrates the difference between critical-band level and excitation level. Main excitations are only existent in the centre of the top of the rectangles. The lowest harmonic is clearly separated from the others. There, the construction of the excitation level by adding the slope excitation corresponding to masking slopes to the critical-band levels at the centre, becomes obvious. Performing the same operations with the other partials, however, shows that the slopes of the excitations overlap towards lower and higher critical-band values more and more with higher frequency range.

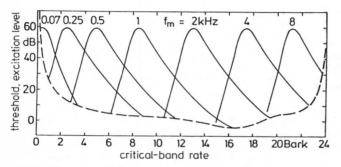

Fig. 6.14. Excitation level versus critical-band rate pattern for narrow-band noises of given centre frequency and a sound pressure level of 60 dB. Note that a_0 is ignored. The broken line indicates the threshold in quiet

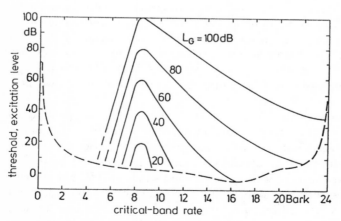

Fig. 6.15. Excitation level versus critical-band rate pattern for a critical-band wide noise with a centre frequency of 1 kHz and the critical-band levels as indicated. a_0 is ignored. The broken line indicates threshold in quiet

Although it is not yet known exactly how the slope excitations add up, a reasonable approximation seems to be to round the excitations of the slopes, especially where a deep valley is produced. For equal excitation level of the lower and upper slopes, this means that the minimum is enhanced by 3 dB. Experimental results have shown that this enhancement may even be larger (see Sect. 4.5.1). In most practical cases, one excitation is dominant so that the other excitation can be ignored.

A comparison between the top drawings in Fig. 6.13 with the bottom drawings indicates how the characteristics of the hearing system, expressed by critical-band rate and excitation level, transform the physical values into intermediate values more meaningful for the development of models for psychoacoustical perception.

The form of masked thresholds as a function of critical-band rate plays an important role in constructing excitation patterns. It is relatively difficult to exactly measure thresholds of tones masked by tones. Therefore, the thresholds of pure tones masked by narrow-band noises are used to construct the excitation level versus critical-band rate patterns. Thresholds obtained with narrow-band noise maskers having the width of a critical band, have been measured for several centre frequencies. Because the dependence of the masking levels on critical-band rate are important in constructing slope excitations, the excitation level versus critical-band rate pattern produced by narrow-band noises for levels of 60 dB and for seven different centre frequencies are plotted as a function of critical-band rate in Fig. 6.14. Threshold in quiet is drawn as the dashed line. A comparison of the slope excitations of the seven narrow-band noises indicates that the lower slope (towards smaller critical-band rates) remains constant independent of centre frequency. The steepness of this slope is about 27 dB/Bark. The upper slopes (towards larger critical-band rates) of the excitations are steeper for low-frequency narrow-band noises. Above about 200-Hz centre frequency, however, the upper slopes are again identical. This means that in most cases the shapes of the lower and upper slope can be produced by simply shifting the pattern for $f_c = 1$ kHz along the abscissa. Threshold in quiet produces a limitation at both low and high frequencies.

A comparison between the excitation level versus critical-band rate pattern outlined in Fig. 6.14, with the corresponding masking level versus frequency pattern drawn in Fig. 4.3, indicates the clear advantage of the excitation level versus critical-band rate patterns; they are produced from one another merely by a shift in the horizontal direction.

The upper slope of the masking pattern is level dependent. This kind of nonlinearity can be taken into account most easily using the excitation level versus critical-band rate patterns produced by narrow-band noises of different levels. Figure 6.15 shows such patterns for a centre frequency of 1 kHz. While the pattern looks almost symmetrical for narrow-band levels below 40 dB, it becomes more and more asymmetrical at higher levels. The steepness of the slope towards lower critical-band rates remains level independent at about 27 dB per Bark. The slope towards higher critical-band rates shows the nonlinear effect of flattening with increasing level. For a narrow-band level of 100 dB, this slope is only 5 dB per Bark. The patterns outlined in Fig. 6.15 also hold for narrow-band noises of centre frequencies different from 1 kHz; the corresponding excitation level versus critical-band rate patterns can be produced by a horizontal shift of the pattern with the appropriate level towards lower or higher critical-band rates. In shifting the patterns, it is necessary to be aware of the limitation produced by the threshold in quiet.

Another important difference between main excitation and slope excitation is worth noting: main excitation is related to frequency by the frequency to critical-band rate relationship, while slope excitation is related to the main frequency of the main excitation producing the slope excitation. This fact is evident in excitation patterns produced by low-frequency high-level tones (masking-period patterns). The excitation produced at a critical-band rate as high as 13 Bark (2 kHz) can "vibrate" with a frequency as low as 20 Hz.

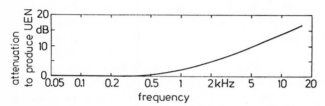

Fig. 6.16. Frequency dependence of the attenuation of a filter that, connected to a white-noise generator, produces uniform exciting noise (UEN)

In some cases, it may be of interest to have a noise that produces the same intensity in each critical band. Such a uniform exciting noise (not identical to uniform masking noise) can be produced from white noise by using a filter with an attenuation as outlined in Fig. 6.16. The frequency response of the attenuation a_{UEN} corresponds to the increment of critical bandwidth as a function of frequency and corresponds to the following equation:

$$a_{\mathrm{UEN}} = 10 \log \left[\Delta f_{\mathrm{G}}(f)/100 \,\mathrm{Hz} \right] \mathrm{dB} . \tag{6.7}$$

Uniform exciting noise produces an excitation level as a function of critical-band rate that, apart from the attenuation a_0 of the hearing system, is independent of critical-band rate.

The difference between uniform exciting noise and uniform masking noise arises because the masking index, a_v, is different at low and high frequencies. Because this effect strongly influences the frequency difference of the critical bandwidth as defined by Fletcher, it may be discussed here in some detail. Let us start with uniform exciting noise with a critical-band level $L_{\mathrm{G}} = 40\,\mathrm{dB}$. The overall level of this noise, which includes 24 critical bands, is therefore

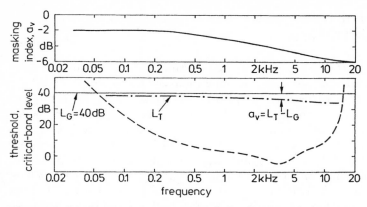

Fig. 6.17. Masking index on an expanded ordinate scale (*upper part*). Critical-band level of uniform exciting noise (*solid line*), masked threshold produced by this noise (*broken-dotted line*) and threshold in quiet (*broken line*) as a function of frequency (*lower part*). Critical-band level of the uniform exciting noise is 40 dB

$(40 + 10 \cdot \log 24)\,\mathrm{dB} = 54\,\mathrm{dB}$. This uniform exciting noise will be used as masker. The threshold of sinusoidal tones masked by this noise is indicated in the lower part of Fig. 6.17 as L_T, the level of the sinusoidal tone necessary to be just-audible. The difference between the critical-band level, L_G, and masked threshold, L_T, represents the masking index

$$a_v = L_\mathrm{T} - L_\mathrm{G} \, . \tag{6.8}$$

This masking index amounts to about -2 dB at low frequencies and decreases at high frequencies to -6 dB. The masking index, a_v, is outlined in the upper diagram of Fig. 6.17 on an enlarged scale. The value of -6 dB for the masking index is the value that would be expexted from the model describing just-noticeable differences in Sect. 7.5.1. This value, however, is reached only at high frequencies where the critical bandwidth is so large that the fluctuation of the noise is not audible. Parts of the noise that are assumed to correspond to the frequency selectivity of our hearing system as being critical-band wide, are effective in that frequency range in a way similar to stationary sounds without fluctuations. At low frequencies, however, the critical-band width is only 100 Hz; at this bandwidth, strong fluctuations of the noise falling into the critical band are produced. These fluctuations reduce the sensitivity of our hearing system for the test tones. Consequently, the masking index, a_v, is increased to -2 dB. The dependence of critical bandwidth on frequency is the reason that the masking index, a_v, depends on frequency and that the density level of uniform exciting noise does not show the same frequency dependence as the density level of uniform masking noise.

6.4 Excitation Level versus Critical-Band Rate versus Time Pattern

The excitation level depends not only on critical-band rate but also on time. Speech sounds, for example, contain strong temporal variations produced not only by the plosives themselves but also by the pauses necessary before the plosives. In the same way as we used the masking level versus frequency patterns for transformations into excitation level versus critical-band rate patterns, masking level versus time patterns can be used. As explained in Sect. 4.4.3, postmasking plays an important role. It can last up to 200 ms, but the decay depends on the duration of the masker. Consequently, sets of data representing masking level versus time patterns for postmasking have to be used in a way similar to masking level versus critical-band rate patterns to produce excitation level versus critical-band rate patterns. It may be argued that this is a relatively complicated procedure. The resulting excitation level versus critical-band rate versus time pattern, however, contains the information used by the hearing system to recognize and understand speech. This is only one example of the use of this very fundamental pattern.

To illustrate this effect, the word "electroacoustics" is recorded and outlined as excitation versus critical-band rate versus time pattern in such a way that frequency selectivity as well as temporal masking are taken into account. The temporal effects can be recognized in Fig. 6.18 by the asymmetric shape

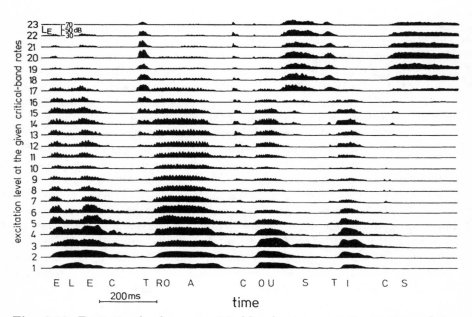

Fig. 6.18. Excitation level versus critical-band rate versus time pattern of the spoken word "electroacoustics". The excitation level is indicated for 23 discrete critical-band rates from 1 to 23 Bark

of excitation as a function of time produced by plosives. Actually, the pattern shown in Fig. 6.18 should contain 640 single excitation-time patterns, since this is the number of pitch variations we can differentiate. However, because of the critical band's filter slope, many adjacent bands out of the 640 will contain very similar information. Therefore, the use of only 24 channels is a significant and helpful approximation. The size of the excitation level can be approximated using the scale that is given on the left side at the top near the patterns for the critical-band rate between 22 and 23 Bark. From the details available in this excitation level versus critical-band rate versus time pattern, only the temporal structure produced by the vowels will be discussed. The formants of the different vowels are clearly visible in the temporally changing pattern of levels across critical-band rate. Besides, the temporal structure of voicing related to the fundamental frequency of 100 Hz (male speaker) also shows up. This means that such a relatively low fundamental frequency produces a strong ripple in the excitation level versus time pattern. Such a ripple may lead to roughness, while more gradual changes in the envelope produced by syllables may elicit fluctuation strength. Both of these sensations are discussed subsequently. Such excitation level versus critical-band rate versus time patterns are intermediate values from which the creation of other psychoacoustical effects or sensations, such as just-noticeable sound variations, loudness as a function of bandwidth or of time, subjective duration, roughness, spectral and virtual pitch, sharpness or fluctuation strength, can be described. Even for automatic speech recognition, the excitation level versus critical-band rate versus time pattern is useful preprocessed information.

7. Just-Noticeable Sound Changes

Two different kinds of sound changes are discussed in this chapter. One is the variation that may be compared to variation in water level: there is always some water but the level varies as a function of time. In acoustics, modulations are typical changes of the sort we call variations. The other kind of change is that of differences. One apple may be different from another apple. In this case, we compare one piece with another piece. In acoustics, this means that we compare one sound with another sound presented after a pause. Because these two kinds of changes may activate different processing features in our hearing system, the first by direct and quick comparison and the second by activating and introducing memory, it is necessary to differentiate strictly between the two kinds of changes. Just-noticeable variations are useful in producing scales of sensations related to position, for example pitch through the frequency-location transformation as discussed in Chap. 5. However, both just-noticeable variations and just-noticeable differences are important as the "stones" on which the "house of sensations" is built.

7.1 Just-Noticeable Changes in Amplitude

The sensation of loudness is an intensity sensation. For such sensations, it is not possible to construct a scale for the magnitude of the sensation from the just-noticeable changes in intensity by adding up the just-noticeable variations between two levels. Nevertheless, just-noticeable level variations as well as just-noticeable level differences play an important role. It seems that amplitude- or level-processing in hearing is based on an element, the size of which is about 1 dB.

7.1.1 Threshold of Amplitude Variation

Abruptly changing the sound pressure or the sound pressure level of a sinusoidal tone leads to a sensation that may contain not only an audible change in level but also the sensation of a click at the time of the abrupt change. The abrupt change produces a short-time spectrum that is relatively broad, the reason for the audible click. To avoid clicks, just-noticeable amplitude variations are often measured by amplitude modulation. The corresponding sound-pressure level differences, ΔL, can be calculated from the degree of modulation, m, using the following equation:

$$\Delta L = 10 \log \left(I_{\max}/I_{\min} \right) \text{dB} = 20 \log \left[(1+m)/(1-m) \right] \text{dB} . \qquad (7.1)$$

This relation can be approximated for $m < 0.3$ by

$$\Delta L = 20 \log e(2m + 2/3 m^3 + \ldots) \text{dB} \approx 20 \log \left(e \cdot 2m \right) \text{dB} \approx 17.5 m \text{ dB.} \quad (7.2)$$

The ordinate scales on the left and right of Fig. 7.1 characterize this relationship. The left side shows the degree of modulation, m, the right side shows the corresponding level difference, ΔL.

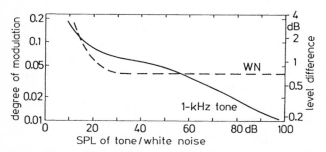

Fig. 7.1. Just-noticeable degree of amplitude modulation (*left scale*) and corresponding level variation (*right scale*) of a 1-kHz tone and of white noise (WN) as a function of sound pressure level (the frequency of modulation is 4 Hz)

Figure 7.1 shows the just-noticeable degree of sinusoidal amplitude modulation for a 1-kHz tone and for white noise as a function of the level for $f_{\mathrm{mod}} = 4 \,\text{Hz}$. The solid line, which holds for the 1-kHz tone, shows that for low levels large degrees of modulation in the range of 20% are necessary to become just audible. At levels of about 40 dB, a degree of modulation of 6% becomes just noticeable. For even higher levels, this just-noticeable degree of amplitude modulation decreases further and reaches, for sound pressure levels of 100 dB, values of about 1%. This dependence on level holds not only for the 1-kHz tone but also for most other pure tones of different frequencies, if instead of the sound pressure level the loudness level (see Chap. 8.1) is used as the abscissa.

The results obtained for white noise are somewhat different and are shown in Fig. 7.1 as a broken line. Again, at very low levels, a relatively large degree of modulation near 20% is necessary to be just audible. The threshold modulation decreases relatively quickly and reaches a value of about 4% at 30 dB. This value does not change with the level up to 100 dB. In searching for a reason of the difference between just-noticeable amplitude modulation for tones and for noises, it is necessary to realize that broad-band noises produce only main excitations. Narrow-band sounds like tones, in contrast, produce not only one main excitation but also slope excitations, the steepness of which is level dependent. It may be that this difference is one of the reasons for the difference in behaviour shown in Fig. 7.1. The two scales used as ordinate in that figure indicate that a degree of modulation of 6% corresponds to a change in level of 1 dB. This is a characteristic value that will occur repeatedly in studying the ear's information processing, and it is interesting to realize that the just-noticeable degree of amplitude modulation for the 1-kHz tone seems to have a tendency to stabilize with increasing level at about 6% (corresponding to 1 dB) but falls off for levels above 50 dB and continues to decrease.

The dependence of the just-noticeable degree of amplitude modulation on modulation frequency is shown in Fig. 7.2. The two solid curves hold for 1-kHz tones at levels of 40 and 80 dB. As can be seen in Fig. 7.2, the ear is most sensitive for amplitude modulation in the modulation-frequency range between 2 and 5 Hz. From very low modulation frequencies, the just-noticeable degree of amplitude modulation decreases a little, reaches a minimum near 4 Hz, increases for higher modulation frequencies up to about 60–70 Hz and finally decreases drastically further on. The increasing part between about 5 and 50 Hz can be approximated by assuming that the just-noticeable degree of amplitude modulation increases with the square root of the modulation frequency. This holds also for carrier frequencies other than 1 kHz, but the maximum reached depends on carrier frequency in such a way that the maximum is reached for lower carrier frequencies at lower modulation frequency. For higher carrier frequencies, the maximum is shifted towards higher modulation frequencies, so that for a carrier frequency of 8 kHz the maximum is reached at $f_{\mathrm{mod}} = 400$ Hz. The roll-off above the maximum is due to the audibility of sidebands so that instead of listening for a kind of modulation or roughness, one listens for additional tones as the modulation frequency is increased. The frequency selectivity of the ear, discussed in Sect. 4.3 on

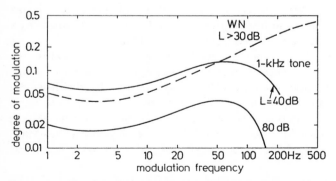

Fig. 7.2. Just-noticeable degree of amplitude modulation as a function of modulation frequency for 1-kHz tones of given level (*solid curves*) and for white noise (*broken line*)

masking, is the reason for this different criterion being used. Sidebands cannot be heard when using broad-band noise, because the long time spectrum of white noise remains constant, even though high-frequency amplitude modulation is used. Consequently for noise, the dependence of the just-audible degree of amplitude modulation on modulation frequency is not affected by the audibility of sidebands.

For low modulation frequencies, the broken curve in Fig. 7.2, which represents the data for white noise, runs parallel to that for tones, but the increment corresponding to the square root of the modulation frequency continues towards higher modulation frequencies up to 500 Hz. At such high modulation frequencies and corresponding high degrees of modulation of about 0.4, another criterion is used to discriminate between modulated and unmodulated noise. The strong amplitude modulation increases the intensity averaged over some hundred milliseconds, and this change in overall intensity or overall sound pressure level is used by the subject to hear the difference. The fluctuations produced by the amplitude modulation, however, remain inaudible at such large frequencies of modulation.

The differences in Fig. 7.1 between the data for white noise and for pure tone lead to the question of whether the different spectral widths or different amplitude distributions of the two sounds are the reason for the difference. The slopes of the excitation, not available for broad-band noise, can be created by decreasing the bandwidth of the noise with band-pass filters. Upon narrowing the bandwidth remarkably, the spectral distribution of a noise becomes comparable to that of a sinusoidal tone. It has to be realized, however, that the narrow-band noise still has a Gaussian distribution of amplitudes, even though the bandwidth may only be a few Hertz. This Gaussian distribution of amplitudes leads to a statistical modulation of the amplitude of narrow-band noise, so that it sounds almost like a sinusoidal tone with statistically-modulated amplitude. In order to measure this effect over a large range of bandwidths, it is most effective to use a centre frequency of 8 kHz for the band-pass noise. At 8 kHz, the critical bandwidth is almost 2 kHz so that the dependence of just-audible modulation on bandwidth can be measured up to that large value without exceeding one critical band.

The dependence of the just-noticeable degree of the 4-Hz square-wave amplitude modulation at a centre frequency of 8 kHz, is indicated by a solid line in Fig. 7.3 as a function of bandwidth. At very narrow bandwidths, the just-noticeable degree of amplitude modulation is near 40%, i.e. quite large and not comparable to that produced by a sinusoidal tone. With increasing bandwidth, the just-noticeable degree of amplitude modulation decreases and reaches the value of about 6% at a bandwidth of 2 kHz (about one critical band).

The critical bandwidth is reached near the end of the solid line. As the bandwidth is further increased up to white noise, all the 24 critical bands finally contribute to the sensation. This range is indicated by the broken line.

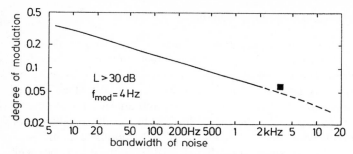

Fig. 7.3. Just-noticeable degree of amplitude modulation of a bandpass-noise as a function of bandwidth. Centre frequency of the noises taken from white noise is 8 kHz; square-wave amplitude modulation is used. The square indicates the level variation of 1 dB at the largest critical band

Increasing the bandwidth beyond the critical band leads to a further decrease, so that for a bandwidth of practically white noise, a degree of modulation of about 3% is reached. This value is somewhat lower than indicated in Fig. 7.1. It should be remembered, however, that the data outlined in Fig. 7.3 are produced using a square-wave amplitude modulation. This was done to compare data produced by masking white noise with white noise.

The filled square shown in Fig. 7.3 belongs to the broadest possible critical band (3.5 kHz) at the upper end of the audible frequency range, and to a sound level change of 1 dB corresponding to a sinusoidal 4-Hz amplitude modulation of 6%. This level change of 1 dB plays a dominant role in the model for just-noticeable sound variations of slow rate.

For spectral widths of the band-pass noise larger than a critical band, the just-noticeable degree of amplitude modulation decreases. This decrease results from the cooperation of several critical bands. Whether this cooperation can be accounted for by adding up the available variations in a coherent fashion across critical bands, or by some other strategy used by the ear remains unclear. However, all psychoacoustical data relevant to this question show a decrease in the just-noticeable amplitude modulation, with an increase in the number of critical bands involved. Thus the curve shown in Fig. 7.3 is shown as a continuous curve, which includes several critical bands at the end with the large bandwidths.

The most important variable in this context seems to be bandwidth. For very small bandwidths, the statistical fluctuation of the amplitude of the narrow-band noise becomes dominantly audible, and thereby strongly disturbs the subjects who are trying to hear periodical amplitude modulations. This is why the just-noticeable degree of amplitude modulation is much larger for narrow bands of noise than for sinusoidal tones, even though the spectral configurations are very similar in the two cases.

7.1.2 Just-Noticeable Level Differences

Although the data for just-noticeable level variations or just-noticeable level differences depend somewhat on the measurement technique used, the values determined for variations are always larger than those produced for just-noticeable differences. A typical example is given in Fig. 7.4. Two sets of data with a 1-kHz tone, one for level changes (left) and another for frequency changes (right), are shown as a function of level. The results produced for level variations (measured as just-noticeable amplitude modulation and indicated by the open circles) decrease from almost 2 dB to about 0.7 dB between 30 and 70 dB sound pressure level. The results for just-noticeable level differences (dots) decrease as a function of level from about 0.7 dB to about 0.3 dB. There is a factor of about 2.5 between the two curves. However, the dependence of just-noticeable amplitude modulation on level remains similar to that for the just-noticeable level difference. The data produced for just-noticeable frequency modulation (open circles in the right part in Fig. 7.4) show no dependence on level. The same is true for the just-noticeable frequency difference (dots) although the data are about a factor of 3 lower than for frequency modulation. The interquartile ranges indicate very clearly that the two sets of data do not overlap, i.e. just-noticeable frequency modulation and just-noticeable frequency difference not only lead to different results but may also be produced by different signal processing in our hearing system.

A typical characteristic of just-noticeable amplitude modulation of sinusoidal tones is the level dependence. Almost the same level dependence is measured for just-noticeable differences in level, as indicated in Fig. 7.5. For low sound pressure levels below 20 dB, JNDL increases greatly towards threshold, but decreases from about 0.4 dB at 40 dB to about 0.2 dB at 100 dB sound pressure level. The decrement seems to be not quite as strong as for amplitude modulation, but the characteristic is similar to the data shown by the solid line in Fig. 7.1. This characteristic is almost independent of fre-

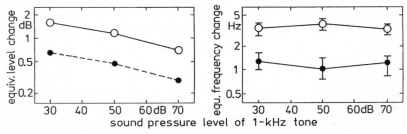

Fig. 7.4. Just-audible equivalent level changes (*left*) and frequency changes (*right*) as a function of the sound pressure level of a 1-kHz tone. The open circles connected by solid lines indicate data produced by sound variations (amplitude modulation and frequency modulation). The data indicated by dots connected with broken lines are for just-noticeable level and just-noticeable frequency differences. Medians and interquartile ranges for six subjects are given

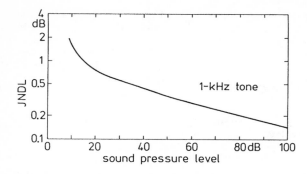

Fig. 7.5. Just-noticeable level difference of a 1-kHz tone as a function of sound pressure level

quency if, instead of the sound pressure level, the level above threshold (or, even better, the loudness level) is used as the abscissa. In this case, the level dependence of JNDL at both low and high frequencies remains almost the same.

For measuring just-noticeable differences, a pause is needed between the sounds to be compared. For pause durations ranging between 0.1 and 2 s, the results are independent of the duration of the pause. The data given in Fig. 7.4 and 7.5 were obtained with a pause of 200 ms, which is still within the range of independence but short enough to make it a very easy task for the subject. Besides the pause separating the two sounds to be compared, the duration of the sounds influences the size of the just-noticeable difference in level. With the level difference measured at 200 ms used as a reference, the dependence of just-noticeable level differences on tone-burst duration is indicated in Fig. 7.6. JNDL increases by almost a factor of 5 for a reduction from 200 to 2 ms. This corresponds to a slope of about −6 dB per decade within this range. For a duration longer than 200 ms the JNDL does not decrease much.

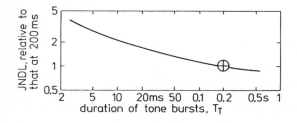

Fig. 7.6. Just-noticeable level difference of a 1-kHz tone related to the difference observed for a duration of 200 ms as a function of the duration of tone bursts

Changing from tones to broad-band noises affects the dependence of JNDL on sound pressure level. As indicated in Fig. 7.7, the level dependence disappears almost completely for white noise in the sound pressure level range above 40 dB. If white noise is limited by a low-pass filter so that the upper slope of masking with its level dependence can be seen, the level dependence becomes noticeable again, as indicated by open circles connected by broken

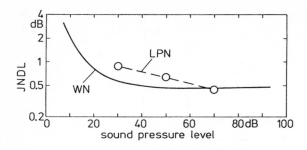

Fig. 7.7. Just-noticeable level difference of a white noise and of a low-pass noise with a cut-off frequency of 1 kHz as a function of sound pressure level

lines. The size of JNDL around 60 dB sound pressure level is about 0.5 dB for most broad-band noises.

7.2 Just-Noticeable Changes in Frequency

In Chaps. 5 and 6, it was argued that the frequency-to-place transformation of the inner ear is the fundamental factor in pitch sensation. Low frequencies stimulate the sensory cells in the organ of Corti near the helicotrema, while high frequencies stimulate cells near the oval window. The frequency-to-place transformation suggests that the sensation of pitch belongs to the category of sensations of position, so that it is possible to construct the sensation function of pitch from just-noticeable changes in frequency. From this point of view, just-noticeable variations in frequency are more important than those of amplitude or level. Because just-noticeable variations in frequency, i.e. in the stimulus, lead to constant values of the corresponding steps in pitch, i.e. in sensation, we are able to construct a relationship between frequency and pitch by integrating just-noticeable variations. This way, a pitch function very similar to that constructed from data of pitch doubling or halving can be calculated from JNVs.

7.2.1 Threshold for Frequency Variation

An abrupt change in frequency is in most cases correlated with an audible click so that, as with the measurement of just-noticeable variations in amplitude, just-noticeable variations in frequency are measured using sinusoidal frequency modulation. In system theory, the deviation in frequency, Δf, is defined as the change in frequency between the unmodulated frequency, f, and the maximum frequency in one direction. Using this definition of Δf, it is important to note that the frequency changes between $f - \Delta f$ and $f + \Delta f$. Thus the value for the total variation in frequency is $2\Delta f$.

Our hearing system is most sensitive for sinusoidal frequency modulations at frequencies of modulation in the neighbourhood of 4 Hz. Therefore, we shall

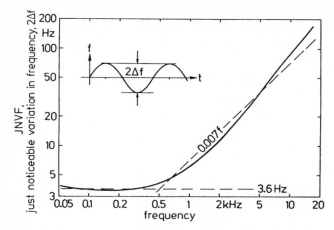

Fig. 7.8. Just-noticeable frequency modulation as a function of frequency for sinusoidal frequency modulation at a modulation frequency of 4 Hz. Broken lines indicate useful approximations. Note that the total variation is $2\Delta f$ as marked in the inset

concentrate first on data related to that modulation frequency. The just-noticeable value, $2\Delta f$, is given in Fig. 7.8 as a function of carrier frequency. The parameters are the frequency of modulation (4 Hz) and the loudness level of the frequency-modulated tones (60 phon). At low frequencies, $2\Delta f$ is approximately constant and, averaged over many subjects, has a value of about 3.6 Hz. Above about 500 Hz, $2\Delta f$ increases nearly in proportion to frequency; in this range, $2\Delta f$ is approximately $0.007f$. This means that a change in frequency of about 0.7% is just noticeable in this frequency range. At low frequencies, the relative just-noticeable change increases and has a value of 3.6% at 100 Hz. This means that at 50 Hz, $2\Delta f$ corresponds to a semi-tone in music; our hearing system is relatively insensitive for changes in the frequency of sinusoidal tones in this low-frequency range. Musical tones, however, are rarely sinusoidal tones; they comprise many harmonic components and the frequency changes of these high-frequency harmonics can be more easily detected than frequency changes of the fundamental. Thus we usually listen to higher frequency harmonics when tuning a musical instrument. The just-noticeable frequency changes of 0.7% at medium and high frequencies are surprisingly small; our hearing system is very sensitive to frequency variations in this range.

The strong correlation between the critical band and the value of $2\Delta f$ was already shown in Fig. 6.10 and discussed there. Assuming that pitch belongs to the sensations of position, it is possible to construct the relationship between pitch and frequency by using the fact that a just-noticeable increment in frequency leads to a constant increment in sensation, independent of frequency. Figure 7.9 shows the number $n_{2\Delta f}$ of the many adjacent steps

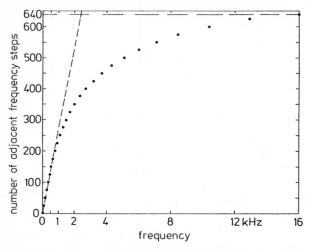

Fig. 7.9. Number of abutting frequency steps based on just-noticeable frequency variations, consecutively arranged up to a certain frequency which is used as the abscissa. Note that 25 steps are summarized from dot to dot. The broken line indicates the approximation of proportionality at low frequencies

in frequency that fit between 0 Hz and frequency f, which is used as the abscissa. Because the just-noticeable steps in frequency are very small, only every 25th step is shown as a point in Fig. 7.9. Starting from zero, the number $n_{2\Delta f}$ at first increases in proportion to frequency. Above about 500 Hz, the function illustrated by the dots starts to deviate from the proportionality, which is shown in Fig. 7.9 as a broken line. For higher frequencies, the number $n_{2\Delta f}$ increases less than proportionally; the function illustrated by the dots actually seems to increase logarithmically. This can be seen by the fact that an increment of one octave in frequency leads to a constant number of about 100 steps. Altogether, it is possible to add 640 adjacent steps in the frequency range up to 16 kHz. This frequency resolution is very high. Remembering that about 3600 inner haircells extend in a row between the helicotrema and the oval window with a distance of 9 μm from one to the next, we can estimate that one step in frequency corresponds to a shift of the excitation over a distance of 6 inner haircells. A similar function was already discussed in Chap. 6, where the relationship between critical-band rate and frequency was based on the integration of critical bandwidths over frequency.

The just-noticeable value of $2\Delta f$ depends on the frequency of modulation. This dependence is outlined in Fig. 7.10 and illustrates again that our hearing system is most sensitive to frequencies of modulation around 4 Hz. Figure 7.10 holds for a frequency of 1 kHz and a level of 60 dB. The minimum value of $2\Delta f$ is about 6 Hz, corresponding to the data given in Fig. 7.8. For modulation frequencies between 10 and 50 Hz, $2\Delta f$ increases markedly with a slope that corresponds to the square root of modulation frequency. This increment

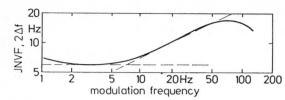

Fig. 7.10. Just-noticeable frequency modulation as a function of modulation frequency (centre frequency 1 kHz). The broken lines indicate useful approximations

ends sooner for low carrier frequencies than for higher carrier frequencies. For a carrier frequency of 8 kHz, the increment continues up to modulation frequencies of about 300 Hz, but for a carrier frequency of 1 kHz (Fig. 7.10) the end of the increment is reached at a modulation frequency of about 70 Hz. This effect is due to the frequency selectivity of our hearing system. Modulation in frequency produces sidebands similar to amplitude modulation. The narrow spacing of the many sidebands for a low frequency of modulation is much wider for high frequencies of modulation. The sidebands are spaced at integer multiples of f_{mod} from the carrier frequency and become audible separately at high frequencies of modulation. In this case, the subject no longer listens for a change in frequency but for additional sidebands. At very low frequencies of modulation, the increment of $2\Delta f$ seems to be due to a limited memory. We do not remember the pitch of a tone very precisely after several seconds have elapsed. Consequently, the value $2\Delta f$ increases towards very low frequencies of modulation.

The dependence of the value $2\Delta f$ on the loudness level of the frequency-modulated tone is relatively small. A reduction of loudness levels from 100 to 30 phon increases the step size by a factor of not more than about 1.5. Near threshold in quiet, however, the increase in $2\Delta f$ is pronounced.

7.2.2 Just-Noticeable Frequency Differences

Although the dependence of just-noticeable frequency differences on frequency and on sound pressure level are similar to those of just-noticeable frequency modulations, the absolute values are smaller by a factor of three. The direction of this difference is astonishing: our hearing system is more sensitive to frequency changes if the task is to recognize differences rather than to recognize modulations. The pause between the two sounds to be compared does not reduce the sensitivity, on the contrary, it increases it! Displacing the data given in Fig. 7.8 down by a factor of three produces a reasonable approximation to the results at the two asymptotes: i.e. at frequencies below 500 Hz, we are able to differentiate between two tone bursts with a frequency difference of only about 1 Hz; above 500 Hz, this value increases in proportion to frequency and is approximately $0.002f$.

It should be noted that just-noticeable differences in frequency and the corresponding results from frequency modulation are sometimes mixed up in the literature. This leads to confusion that ought to be avoided. A clear separation of the two kinds of data sets is helpful; it may even be necessary to separate the data further on the basis of the method used.

Just-noticeable differences in frequency are level dependent only below sensation levels of about 25 dB. Below this value, the just-noticeable difference rises with decreasing level so that the just-noticeable difference in frequency is about 5 times larger at a sensation level of 5 dB than at 25 dB. The data discussed so far correspond to durations greater than 200 ms for the two tone bursts to be compared. This corresponds to quasi-steady-state condition. For burst durations shorter than 200 ms, the just-noticeable frequency difference increases.

Figure 7.11 shows the just-noticeable relative frequency difference between test and comparison tone obtained by a yes/no-procedure plotted as a function of the test tone duration. The dots connected by solid lines in each panel represent medians and interquartile ranges of 8 subjects; the shaded areas include *all* individual results.

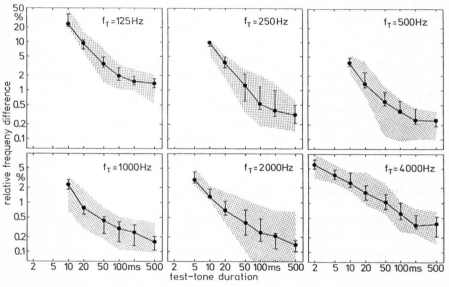

Fig. 7.11. Frequency discrimination of pure tones at short durations. Just-noticeable relative frequency difference as a function of test tone duration. 6 different test tone frequencies are indicated in the different panels

The data in Fig. 7.11 show that the just-noticeable relative frequency difference increases with decreasing test tone duration. The magnitude of this increase, however, depends on frequency.

This dependence is considerably reduced if, instead of the frequency scale, the critical bandrate scale is used. At long durations (around 500 ms) the critical bandrate difference of about 0.01 Bark represents the just-noticeable difference, whereas at a duration of 10 ms the JNDF amounts on average to 0.2 Bark.

The just-noticeable differences in cut-off frequencies can be measured using noises instead of tones. Data produced for cut-off frequencies near 1 kHz (low-pass on the left, high-pass on the right) are given in Fig. 7.12 as a function of the critical-band level at 1 kHz. It is interesting to realize that the just-noticeable difference in cut-off frequency increases with increasing level for low-pass noise (see left part of Fig. 7.12), but that just-noticeable difference in cut-off frequency for high-pass noise remains independent of level (see right part of Fig. 7.12). A comparison of the data produced for just-noticeable differences (filled circles) with the data produced using modulation of the cut-off frequency (open circles) indicates a difference by a factor of about three. This is consistent with the data for tones. It shows again the astonishing effect that just-noticeable differences create smaller values, even though the sounds to be compared are separated by a pause.

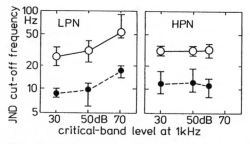

Fig. 7.12. Just-noticeable change of the cut-off frequency of low-pass noise (*left*) and high-pass noise (*right*) as a function of the critical-band level of these noises at 1 kHz. The open circles connected with solid lines are for just-noticeable variations (frequency modulation). The dots connected with broken lines correspond to just-noticeable differences of the cut-off frequencies (i.e. measured for noises presented as bursts separated by a pause)

7.3 Just-Noticeable Phase Differences

Phase differences become audible if the spectrum consists of three tones. Variation of phases may also be detected by a change in rhythm, a sensation discussed in Chap. 13. Only data of just-noticeable phase differences are discussed here, and sounds composed of three tones of equal frequency separation will be used. In this case, changes in phase can produce a change from amplitude modulation to quasi-frequency modulation. If the phase of one of these tones is changed by 90°, then the effect discussed in Sect. 6.1 appears. The experimental results discussed there, and the experimental results produced for just-noticeable phase changes, lead to the conclusion that changes

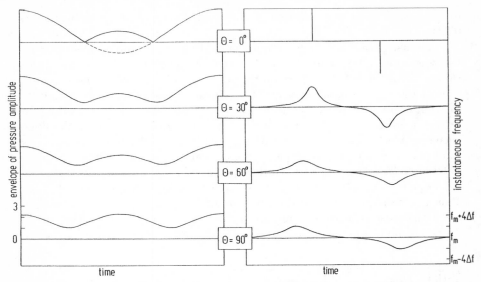

Fig. 7.13. Envelope of the sound pressure amplitude (*left part*) and instantaneous frequency (*right part*) of a three-tone complex as a function of time within its period

of the phase of one of the three components corresponding to the following equation

$$p(t) = \quad p_0[a_1 \cos(2\pi(f_m - \Delta f)t + \varphi_1)$$
$$+ a_2 \cos(2\pi f_m t + \varphi_2) + a_3 \cos(2\pi(f_m + \Delta f)t + \varphi_3)] , \quad (7.3)$$

become audible if a value, defined as the effective phase angle by the equation

$$|\theta| = |\varphi_2 - (\varphi_1 + \varphi_3)/2| , \qquad (7.4)$$

exceeds a certain value. To illustrate the effect of such a phase change, Fig. 7.13 shows the envelope (left side) of the resulting time function, $p(t)$, and the instantaneous frequency, $f(t)$, (right side) of a complex composed of three tones of equal amplitude and equal frequency distance, Δf. For $\theta = 0$, the top panel shows the envelope to be an overmodulated amplitude modulation. The corresponding instantaneous frequency shows periods of constant frequency and Dirac impulses of opposite signs at the zero-crossings of the envelope. For $\theta = 30°$, the envelope shows no zero-crossing and the instantaneous frequency consists of deviations that are relatively large. For $\theta = 60°$ and $90°$, the envelope of the sound pressure function becomes more symmetrical, and shows two equal variations within one period. The corresponding changes in instantaneous frequency again show deviations in both the positive and negative direction that reach only half the value of the peak obtained for $\theta = 30°$. Figure 7.13 indicates that two effects are associated with phase changes, i.e. variations in the envelope of the sound pressure's temporal pattern and variations of the instantaneous frequency.

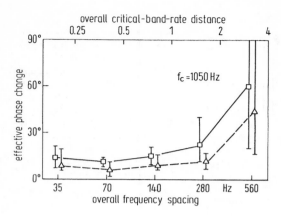

Fig. 7.14. Just-noticeable effective phase change as a function of the frequency separation of the sidebands (*lower abscissa*) and of the overall critical-band-rate distance (*upper abscissa*). The solid line and the broken line correspond to data with different reference phase (0° : *solid*; 90° : *broken*)

There are a number of variables that can influence the just-noticeable phase difference, one being level difference amongst components. When two sidetones equally spaced in frequency from the centre tones are used, the results of psychoacoustical measurements indicate that our hearing system is most sensitive when all three tones have almost equal amplitude. The effect of overall sound pressure level on the just-noticeable phase difference seems to be small in relation to the dependence on the centre frequency. The effect of the frequency separation between the sidetones and the effect of the refence phase (parameter) is outlined in Fig. 7.14. The just-noticeable difference $\Delta\theta$ of the effective phase angle is shown as the ordinate. Besides the frequency separation between the sidetones and the centre frequency, the overall critical-band-rate distance is given as the abscissa (top). The data given in Fig. 7.14 correspond to a centre frequency of 1050 Hz. Three effects are clear: our hearing system is most sensitive to phase differences when the frequency separation is small. For frequency separations larger than about 200 Hz, i.e., larger than about one critical band, the just-noticeable effective phase difference increases drastically. The data indicate that we are most sensitive for the situation in which the reference phase is 90° (broken line). For a reference phase of 0°, the sensitivity is decreased by about a factor of two in the frequency separation range in which we are most sensitive. The data outlined in Fig. 7.14 were obtained with a level difference of 6 dB between the centretone and the two sidetones or sidebands.

For the same level difference and for a reference phase of 0°, Fig. 7.15 shows just-noticeable phase difference as a function of the frequency separation between the sidebands and the centretone, with the frequency of the centretone changing from panel to panel. Data measured in a sound absorbing room (SAR) are compared with data measured in a reverberation room (RR) at the indicated distances. Two effects are clear: firstly, we are most sensitive to phase changes at small values of frequency separation and secondly, the sensitive range is limited by the critical bandwidth, independent of

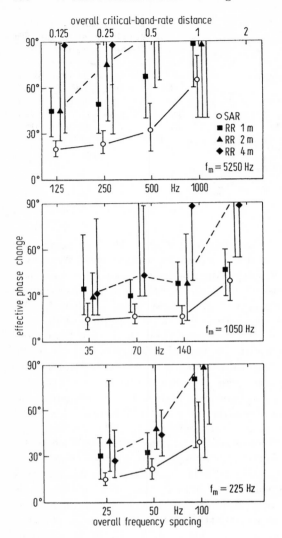

Fig. 7.15. Just-noticeable effective phase change as a function of the frequency spacing (*lower scale*) or the critical band rate distance (*upper scale*). The data correspond to results produced in the sound absorbing room (SAR) or in the reverberation room (RR) at a distance of 1, 2, and 4 m from the source. The centre frequency of the complex is indicated in each panel

centre frequency. The just-noticeable effective phase difference increases for a centre frequency of 5 kHz at a separation near 500 Hz, while for a centre frequency of 225 Hz this increase starts near 50 Hz. In all three cases, the overall frequency distance of the constituent tones transformed to the corresponding critical-band rate relates closely to 0.5 Bark.

The just-noticeable phase differences reach values near 10° in the most sensitive condition, i.e., levels of 70 dB, frequencies around 1 kHz, reference phase near 90°, and small frequency separations. This value increases by about a factor of two when the reference phase angle is changed from 90° to

$0°$. All the data can only be measured using either earphones or loudspeakers in an anechoic chamber. As soon as the sounds are presented in a normal living room or in a concert hall, the just-noticeable phase changes increase by a factor of about three.

A model describing these effects was established by Fleischer; it accounts for his extensive data collection measured psychoacoustically. The model makes use of the hearing system's characteristics and starts with the excitation level versus critical-band rate pattern. The maximal changes in level are picked up, weighted as a function of centre frequency and time, and added before they are transferred to a threshold detector. A comparison between measured and calculated data indicates that it is the difference in the envelope of sound pressure that most often leads to the just-noticeable phase difference. Only in the neighbourhood of a reference phase of $90°$, the change of the instantaneous frequency seems to be picked up by our hearing system. It is interesting to note that our hearing system shows reduced sensitivity for phase changes in normal surroundings, as in a living room with loudspeaker transmission. The first incoming wave which plays a dominant role in localizing the sound source, seems to play a much less, almost negligible, role for just-noticeable phase changes. In the latter case, the steady-state condition reached about 50 to 100 ms after the onset of the sound, seems to be much more important for perceiving phases changes. Under normal room conditions, the influence of the electroacoustic system is negligible relative to the influence of the room with its resonances and strong influences on the time function of the envelope. This makes it understandable that under normal living room conditions, phase distortions are rarely audible and that consequently the phase characteristics of the amplifier and loudspeakers may play a secondary role.

7.4 Influence of Partial Masking on Just-Noticeable Changes

Often when listening to music or speech we cannot listen in the quiet. Additional tones or noises influence just-noticeable sound changes. Such influences must occur because the threshold in quiet is shifted upwards to become a masked threshold. The important question is whether additional sound that produces partial masking influences just-noticeable sound changes only in the neighbourhood of the masked threshold or also at higher levels. This question can be answered if just-noticeable modulations are measured as a function of level under noisy conditions.

For amplitude modulations, the results are indicated in Fig. 7.16. The just-noticeable degree of modulation is shown as a function of level for the condition in which the sinusoidal tone is presented, either in quiet or during the simultaneous presentation of a broad-band masking noise with levels of

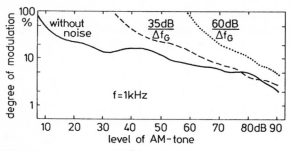

Fig. 7.16. Just-noticeable degree of amplitude modulation of a 1-kHz tone as a function of its level. The solid line corresponds to data produced without background noise, while the broken and dotted lines are produced with additional uniform exciting noise at the given level per critical band

35 or 60 dB in each critical band. The frequency of the tone was 1 kHz, and the modulation frequency remained at 4 Hz. The masking sound shifts the threshold in quiet to a masked threshold at 32 and 57 dB for masking noise with levels of 35 and 60 dB per critical band, respectively. Above this threshold the modulated tone can be heard. Just-noticeable degree of modulation decreases quickly for levels near threshold from 100% down to about 4% at levels 25 dB above masked threshold. At this level of the sinusoidal tone, the just-noticeable degree of amplitude modulation measured for a tone in quiet is almost the same. Similar data are found for a level of 60 dB per critical band. In this situation, the just-noticeable degree of amplitude modulation almost reaches that measured under unmasked conditions at levels between 80 and 90 dB. It seems that values that correspond to those measured in quiet are reached about 20 to 30 dB above masked threshold.

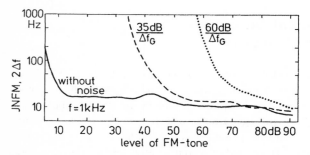

Fig. 7.17. Just-noticeable frequency modulation of a 1-kHz tone as a function of the level of the tone. Solid line corresponds to data produced without background noise. Broken and dotted lines indicate data produced with additional uniform exciting noise at the given level per critical band

Corresponding data for frequency modulation are presentend in Fig. 7.17 which shows the just-noticeable frequency deviation, $2\Delta f$, as a function of the level of the modulated tone. The curve with the parameter "without noise" has already been discussed. For partial masking, the just-noticeable frequency modulation decreases quickly from high values down to a value that is near that measured in quiet without a masker. Additional noise shifts the masked threshold towards higher levels but, relative to the masked threshold behaviour, shows similar behaviour for masker levels of both 35 and 60 dB per critical band. Similar to amplitude modulation, just-noticeable frequency modulation is not greatly influenced by masking noise if levels of the modulated tone 20 to 30 dB above masked threshold are used. Similar measurements have been performed for centre frequencies of 250 and 4000 Hz with similar results.

Just-noticeable modulations were also measured after narrowing the broad-band masking noise to critical bandwidth, and keeping the same critical-band level used to obtain the results illustrated in Figs. 7.16 and 7.17. The data are similar to those with a broad-band masker when the centre frequency of the noise and the frequency of the modulated tone coincide. With frequency modulation, however, some side effects do appear, understandably, because a frequency deviation of ± 500 Hz masked, for example, by a critical-band wide noise at 1 kHz, shifts the frequency into a range where no masker occurs. This means that large frequency modulation becomes audible even though the tone at 1 kHz is inaudible because of the narrow-band masking.

The influence of partial masking can be summarized as follows: each masker produces an excitation pattern that is described for broad-band noises by main excitations, and for narrow-band noises by slope excitations and main excitations. For signal levels more than 20 dB above this defined excitation pattern of the masker, the masker scarcely influences the thresholds for modulation. This means that about 20 dB above masked threshold and higher, the just-noticeable degree of amplitude modulation and the just-noticeable frequency deviation remain almost the same, regardless of whether the masker is present. This is not obvious, but represents the basis of the fact that we are able to communicate with speech even though our communication may take place in noisy surroundings.

Few data are available for just-noticeable differences determined with additional background noise. In most cases, only white noise has been used as background. The data available indicate that just-noticeable difference in frequency shows large values in the neighbourhood of the masked threshold. However, 10 dB above threshold, the just-noticeable difference in frequency is only a factor of two larger than without masking. About 20 dB above masked threshold, just-noticeable differences in frequency reach almost the same value as measured without masking. The data available for just-noticeable difference in level lead to similar results. In this case, not only broad-band noise but also low-pass and high-pass noises have been used as additional maskers. The data available for the latter conditions show relatively large scatter. There is general agreement, however, that JNDL is higher by about a factor of 5 close to masked threshold, and seems to reach values corresponding to those measured under unmasked conditions for levels of the test tone of more than 20 dB above masked threshold. It can therefore be summarized that just-noticeable changes (variations as well as just-noticeable differences) are larger by about a factor of 10 close to masked threshold and reach almost normal conditions for levels of test tones about 20 dB above masked threshold.

7.5 Models of Just-Noticeable Changes

It is clearly nessary to differentiate between variations and differences in the development of models for just-noticeable changes.

7.5.1 Model for Just-Noticeable Variations

The amplitude resolution of communication systems is one of their most important characteristics, but it is not easy to measure this characteristic in our hearing system. Using pure tones, we have problems with the nonlinearities of our hearing system. Using narrow-band noise, the self-modulation of the noises distorts the effect that we are searching for. Only at very high centre frequencies can a narrow-band noise be produced that sounds like a steady sound, even though it is limited to a critical bandwidth. As outlined in Fig. 7.3, the just-noticeable degree of modulation is found to be near 6% in this case. This correlates with a threshold factor

$$s = \Delta I/I = 0.25 \ , \tag{7.5}$$

and a corresponding logarithmic threshold factor

$$\Delta L_S = 10 \log \left(1 + \Delta I/I\right) \mathrm{dB} \approx 1 \, \mathrm{dB} \ . \tag{7.6}$$

The second important factor in such systems is the frequency resolution. We have seen that the critical bands provide a good first approximation for frequency resolution. Actually, slopes of critical-band filters are not infinitely steep but finite, and can be approximated by the steepness of 27 dB/Bark towards low critical-band rates (low frequencies). Towards larger critical-band rates (higher frequencies), the slope depends on level and decreases with increasing level from about 27 dB/Bark to 5 dB/Bark over the 60-dB range above 40 dB. In total, the frequency selectivity of our hearing system can be approximated by the excitation level versus critical-band rate pattern.

The model for just-noticeable slow sound variations makes use of this pattern. It exchanges the logarithmic threshold factor, ΔL_S, for the corresponding excitation level variation ΔL_{ES}. In this way a relatively simple model for just-noticeable slow sound variations can be given: the threshold of variation of the sound is reached when the excitation level produced by the sound, L_E, changes somewhere along the critical-band rate scale by more than $\Delta L_{ES} = 1\,\mathrm{dB}$.

This model assumes that the excitation level is picked up along the critical-band rate scale by many detectors that act independently from each other. In case of excitation-level variations that are in phase along the critical-band rate scale, i.e. coherent, there may be an addition of probabilities or other laws (not clear so far) that may influence the detectability somewhat. These effects, however, are side effects that are only expected in rare cases, and the assumption of the independently acting channels is satisfactory in most cases.

For band-pass noise, we have an audible self-modulation of the sound. This self-modulation is far above threshold, which means that detectors that might signal the modulation are already responding. Such a situation also occurs when a tone burst is compared with another tone burst of different amplitude or frequency and the tone bursts are separated by a pause; the subject has to say whether there is an audible difference between the two bursts. In this case, there is also a strong change in amplitude or level that leads to the fact that the detectors have already responded. This brings us to the extended model described in the next paragraph. Let us first restrict the model to just noticeable amplitude and frequency variations.

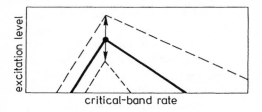

Fig. 7.18. Schematic drawing of the change of excitation level versus critical-band rate pattern as a consequence of amplitude modulation of a pure tone

The excitation level versus critical-band rate pattern for a tone modulated in amplitude is outlined schematically in Fig. 7.18. The solid curve indicates the unmodulated condition, while the two broken lines indicate the extreme values in an exaggerated fashion. Main excitation (the peak) and lower slope excitation can be usefully approximated as moving up and down in parallel. The upper slope, however, changes with level according to the level dependence of the upper slope as the figure shows, i.e. *not* in parallel. An increment of 1 dB may produce an increment of from 2 to 5 dB on the upper slope (Fig. 4.9 gives this relation in quantitative form). This means that threshold for modulation is achieved when the upper slope – the most sensitive one in this case – varies by more than 1 dB somewhere along its extent. The corresponding variation of the main excitation is then two to five times smaller than 1 dB. In other words, a modulation corresponding to less than a quarter of one dB may be enough to produce an excitation-level change at the upper slope that, at 1 dB, corresponds to threshold. Therefore, the just-noticeable degree of modulation is smaller than 6% (i.e. smaller than 1 dB) at high levels; indeed, it can be almost as small as 1% (0.2 dB). Thus it is understandable why amplitude modulation of pure tones at high levels is so much more easily audible than at levels near 40 dB; the nonlinear rise of the upper slope documented in Fig. 4.9 is the reason.

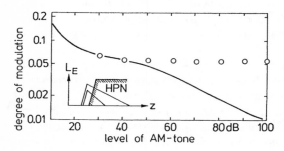

Fig. 7.19. Just-noticeable degree of amplitude modulation of sinusoidal tones without background noise (*solid curve*) and with additional high-pass noise (*open circles*) as a function of the sound pressure levels of the tones. Excitation level versus critical-band rate patterns are indicated schematically in the inset

The model for just-noticeable level variations can be checked by adding a simultaneous high-pass noise to the amplitude-modulated sinusoidal tone. This is indicated in the inset of Fig. 7.19. For levels of the high-pass noise adjusted in such a way that the upper slope of the excitation level versus critical-band rate pattern for the sinusoidal tone is masked, these upper slopes can be prevented from contributing to the audibility of the modulation of the sinusoidal tone. The sinusoidal tone itself remains clearly audible. However, the just-noticeable degree of amplitude modulation measured under such conditions remains almost constant as a function of level. The open circles in Fig. 7.19 clearly show this effect. The just-noticeable degree of amplitude modulation measured under these conditions is more than a factor of four larger at high levels near 90 or 100 dB than without additional high-pass noise. Using the high-pass noise to eliminate the efficiency of the upper slope of the excitation level versus critical-band rate pattern, leads to a dependence of the just-noticeable degree of amplitude modulation on level that corresponds almost exactly to that measured with broad-band noise. This is to be expected in cases in which the ear is unable to use the information presented at the upper slope and indicates that the model predicts the results correctly.

There is another way to check the model: listening to an 80-dB, 1-kHz tone that is modulated only little above its threshold amplitude modulation, one hears that the amplitude modulation leads to a fluctuation of the sensation that corresponds only to a higher pitch. That part of the sensation that corresponds to the pitch of the tone itself remains constant and does not fluctuate. The model was created in this way, i.e., by sitting in the

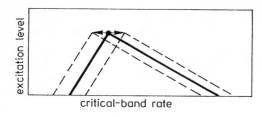

Fig. 7.20. Schematic drawing of the change of the excitation level versus critical-band rate pattern as a consequence of frequency modulation of a pure tone

booth, listening as much as possible in detail and thinking about the detail's meaning.

Frequency modulation of a sinusoidal tone leads to an excitation level versus critical-band rate pattern that can be approximated by a triangular-like distribution that moves to the left and right along the critical-band rate scale. Figure 7.20 shows such a pattern, where the solid distribution is that of an unmodulated tone and the broken curves correspond to the patterns created by the extreme frequency shifts of the modulation. Assuming that the detectors along the critical-band rate scale are only able to react to level changes, it becomes clear that we have to search for the location along the critical-band rate scale that first reaches an excitation-level change of 1 dB as the frequency variation is increased. Figure 7.20 indicates that this would be the position of the lower slope of the excitation level versus critical-band rate pattern. There, the slope is steepest and reaches 27 dB per Bark. The upper slope of the pattern is flatter and produces a 1 dB level increment much later. The excitation level increment of 1 dB is reached at the lower slope when $2\Delta f \times 27$ dB/Bark = 1 dB. Because 1 Bark corresponds to the critical bandwidth, we can equally well express the steepness of the lower slope by 27 dB/Δf_G. The just-noticeable variation in frequency, $2\Delta f$, of sinusoidal tones can therefore be calculated to be

$$2\Delta f = \Delta f_G/27 . \tag{7.7}$$

Since the lower slope is independent of critical-band rate, the model leads to direct proportionality between just-noticeable frequency variation, $2\Delta f$, and the width of the critical band. The factor between the two should be 27, the constant lower slope in dB/Δf_G. This value corresponds quite well to the data outlined in Fig. 6.10. There, a factor of 25 is marked between the two values. Our model therefore assumes that the high sensitivity of our hearing system for variations in frequency is not established by the use of specialized systems that react to frequency changes. Rather, the steep slopes of the frequency selectivity of our hearing system are the reason for the high sensitivity to frequency changes. This sensitivity is based on our hearing system's ability to indicate changes in excitation level that are of the size of only 1 dB. It seems that our hearing system is using this sensitivity for level changes together with its frequency selectivity to reach the high sensitivity for frequency variations.

Using the model, we may be able to find a situation in which the upper slopes of excitation level versus critical-band rate pattern may also be used by our hearing system for detecting frequency changes. In order to establish this, the effect of the lower slope has to be drastically reduced. This can be done most easily by using low-pass noise in which the cut-off frequency is varied periodically and was realized here for a cut-off frequency of 1 kHz. The inset of Fig. 7.21 indicates the excitation level versus critical-band rate patterns produced by such low-pass noises of different level. In the region of small critical-band rates, only main excitations are produced; their level is independent of critical-band rate up to the cut-off frequency. One effect has to be considered in using noises: the quantitative value of the just-noticeable modulation of a low-pass noise's cut-off frequency is much larger in comparison to the frequency modulation of a sinusoidal tone, because of the statistical self-modulation of the noise.

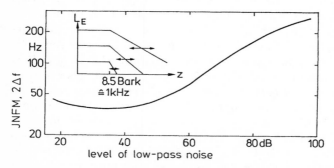

Fig. 7.21. Just-noticeable modulation of the cut-off frequency (at 1 kHz) of low-pass noise as a function of the level of the low-pass noise. The change of the excitation level versus critical-band rate pattern with increasing level is shown schematically in the inset

As indicated by the inset of Fig. 7.21, the upper slopes of the excitation patterns become flatter with increasing level of the low-pass noise. Because the only variation of excitation level – besides the fluctuation of the noise – is that of the upper slope, the deviation of the cut-off frequency has to become larger for increasing level, in order to reach the criterion corresponding to the just-noticeable increment in excitation level. The just-noticeable frequency modulation, $2\Delta f$, of the cut-off frequency of the low-pass noise was measured as a function of the level of the low-pass noise. The curve outlined in Fig. 7.21 indicates a result that may at first be unexpected, but that is predicted by the model. Although the level increases, the sensitivity of our hearing system to changes of the cut-off frequency of the low-pass noise decreases, i.e., the value $2\Delta f$ increases. This means that our hearing system becomes – in contrast to our expectation – less sensitive to cut-off frequency variations at higher levels. Normally, we think that we can hear better if we increase the level. This unexpected behaviour is understandable in terms of the model. The upper slope becomes flatter and flatter with increasing level; it decreases from about 30 dB/Bark at low level to about 5 dB/Bark at high levels near 100 dB. This decrement of the slope corresponds to the rise in $2\Delta f$ from about 40 Hz at 30 dB to about 240 Hz with the low-pass noise at a level of 100 dB.

The model can be used for threshold in quiet and masked thresholds. At low frequencies it is assumed that the subject produces his/her own noise as a consequence of the blood supply in the inner ear and of vibrations produced by the heart and muscles. We adapt to this noise because it is always there and therefore we don't usually hear it, particularly in our normal surroundings. This kind of internal noise is especially strong at low frequencies. At medium and high frequencies there may be noise produced in the neural processing. The effect of low-frequency noise can be measured quantitatively by closing up the outer ear canal comparing the noise measured with a probe microphone in this closed condition, with the same measurement made in the open condition. The noise increases and so does the threshold in quiet which – under the closed condition – may no longer be called threshold in quiet because the internal noise is audible although it is faint.

Thresholds masked by noises can be calculated easily by using main excitation and slope excitation levels. Because the model is based on masking patterns that are transformed to excitation level patterns, it can also be used in the other direction. This may be discussed quantitatively for white-noise stimuli which produce main excitations. The self-modulation produced by the noise in a high-frequency critical band becomes almost inaudible. Therefore, the just-noticeable excitation level increment becomes 1 dB and the corresponding threshold factor 0.25. The corresponding masking index is −6 dB. For a test-tone frequency of 5 kHz, we may calculate the masked threshold using the model. The width of the critical band reaches $\Delta f_\mathrm{G} = 1000$ Hz at 5 kHz. For an assumed density level of $l_\mathrm{WN} = 40$ dB, the level of the noise that falls within the critical band at 5 kHz becomes

$$L_\mathrm{G} = l_\mathrm{WN} + (10\log 1000)\,\mathrm{dB} = 70\,\mathrm{dB}\ . \tag{7.8}$$

Because the masking index, a_v, corresponds to the difference between test-tone level and critical-band level, i.e. $L_\mathrm{T} = L_\mathrm{G} + a_v$, the level of the test tone at masked threshold becomes $L_\mathrm{T} = 64$ dB. This value corresponds to the data given in Fig. 4.1.

The calculation of the threshold of sinusoidal tones masked by white noise can be generalized as follows:

$$L_\mathrm{T} = l_\mathrm{WN} + 10\log\left[\Delta f_\mathrm{G}(f)/\mathrm{Hz}\right]\mathrm{dB} + a_v(f)\ . \tag{7.9}$$

The masking index, a_v, has negative values, and changes from −6 dB at high frequencies to −2 dB at low frequencies. This has to be taken into account if the threshold of a sinusoidal tone at 500 Hz, masked by white noise of the same density used above, is calculated. There, critical bandwidth is 100 Hz and the masking index is $-2\,\text{dB}$. This leads to the level of the 500-Hz tone at threshold

$$L_\text{T} = 40\,\text{dB} + (10\,\log 100)\,\text{dB} - 2\,\text{dB} = 58\,\text{dB} \ . \tag{7.10}$$

This calculated value also corresponds to the data given in Fig. 4.1. Thresholds masked by low-pass or high-pass noises can be calculated in the same way. Besides main excitations, slope excitations also have to be used. Moreover, the level L_T of a tone at threshold can be calculated for each critical-band rate, z, using the following equations for the masking index $a_v(z)$:

$$a_v(z) = [10\,\lg s(z)]\,\text{dB} \tag{7.11}$$

with $s(z)$ being the threshold factor correlated with the logarithmic threshold factor $\Delta L_\text{S}(z)$ by the equation

$$s(z) = 10^{\Delta L_\text{S}(z)/10\,\text{dB}} - 1 \ . \tag{7.12}$$

Because of the different self modulation of critical band wide noise at different critical band rates, the logarithmic threshold factor $\Delta L_\text{S}(z)$ depends on the critical band rate z as follows:

$$\Delta L_\text{S}(z) = (2.2 - 0.05z/\text{Bark})\,\text{dB} \ . \tag{7.13}$$

The use of excitation level versus critical-band rate patterns in the model is an example of the usefulness of these patterns. It seems to be one of the most basic patterns that our hearing system uses and appears to be crucial in many psychoacoustical effects and sensations.

7.5.2 Model for Just-Noticeable Differences

Memories for pitch and loudness have to be included in the process of modelling just-noticeable differences in frequency and amplitude. Therefore, the model for just-noticeable differences has to include models for pitch and loudness. Consequently, it becomes more complicated, although it contains submodels for pitch and loudness that have already been discussed, or will in the next chapter. Figure 7.22 shows the whole model. The incoming sound produces the excitation level versus critical-band rate pattern that is used to produce pitch and pitch strength, and loudness and specific loudness. The corresponding data are transferred to the pitch memory and the loudness memory. These values of the first of the two sounds to be used in measuring just-noticeable differences, are stored and can be compared with the incoming values of the second sound. The just-noticeable difference in frequency depends strongly on pitch strength and increases with decreasing pitch strength. Therefore, as a first approximation, it is assumed that the just-noticeable difference in frequency can be expressed in an increment along the critical-band rate scale that is independent of frequency, but which increases reciprocally with the pitch strength of the sound in question (pitch strength was discussed in Chap. 5).

Just-noticeable differences in amplitude, JNDL, are received as just-noticeable differences in loudness. It seems that it is not the absolute increment of loudness that is responsible for JNDL, but a relative increment. The just-noticeable increment in specific loudness at a certain critical-band rate, or the sum of up to 3 neighbouring specific loudnesses, are the values that seem to be related to the total loudnesses of the two sounds being compared. These partial loudnesses cannot be defined exactly with the data so far available, but seem to play an important role if JNDL is measured under partial masking. The lowest block in Fig. 7.22 indicates the production of just-noticeable variations, in particular modulations, discussed in former sections.

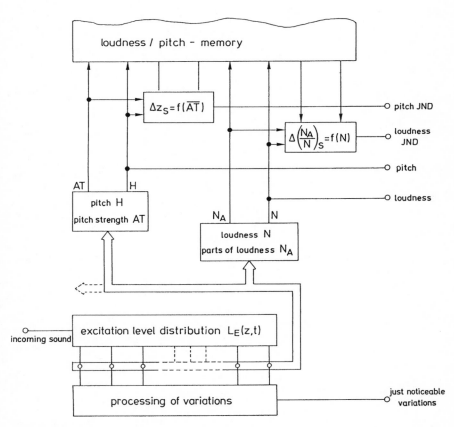

Fig. 7.22. Block diagram representing a model for just-noticeable differences in pitch and loudness and for just-noticeable variations (the lowest box, simplified)

8. Loudness

Loudness belongs to the category of intensity sensations. The stimulus-sensation relation *cannot* be constructed from the just-noticeable intensity variations directly, but has to be obtained from results of other types of measurement such as magnitude estimation. In addition to loudness, loudness level is also important. This is not only a sensation value but belongs somewhere between sensation and physical values. Besides loudness in quiet, we often hear the loudness of partially masked sounds. This loudness occurs when a masking sound is heard in addition to the sound in question. The remaining loudness ranges between a loudness of "zero", which corresponds to the masked threshold, and the loudness of the partially masked sound is mostly much smaller than the loudness range available for unmasked sound. Partial masking can appear not only with simultaneously presented maskers but also with temporary shifted maskers. Thus the effects of partially masked loudness are both spectral and temporal.

8.1 Loudness Level

Loudness comparisons can lead to more precise results than magnitude estimations. For this reason the loudness level measure was created to characterize the loudness sensation of any sound. It was introduced in the twenties by Barkhausen, the researcher whose name was shortened to create a unit for critical-band rate, the Bark. Loudness level of a sound is the sound pressure level of a 1-kHz tone in a plane wave and frontal incident that is as loud as the sound; its unit is "phon". Loudness level can be measured for any sound, but best known are the loudness levels for different frequencies of pure tones. Lines which connect points of equal loudness in the hearing area are often called equal-loudness contours. They have been measured in several laboratories and hold for durations longer than 500 ms. Equal-loudness contours for pure tones are shown in Fig. 8.1. Because of the definition, all curves have to go through the sound pressure level at 1 kHz that has the same value in dB as the parameter of the curve in phon: the equal-loudness contour for 40 phon has to go through 40 dB at 1 kHz. Threshold in quiet, where the limit of loudness sensation is reached, is also an equal-loudness contour. Be-

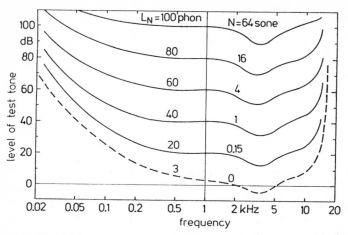

Fig. 8.1. Equal-loudness contours for pure tones in a free sound field. The parameter is expressed in loudness level, L_N, and loudness, N

cause threshold in quiet corresponds to 3 dB at 1 kHz and not to 0 dB, this equal-loudness contour is indicated by 3 phon.

The equal-loudness contours outlined in Fig. 8.1 have a shape for low loudness levels such as 20 phon that is almost parallel to the threshold in quiet. This is especially true for frequencies above about 200 Hz and, in that frequency range, also holds for larger loudness levels. At low frequencies, however, equal-loudness contours become shallower with higher levels. A 50-Hz tone of 50 dB sound pressure level reaches a loudness level of about 20 phon, but a 50-Hz tone of 110 dB sound pressure level reaches 100 phon. The difference between the numbers in phon and the numbers in dB is 30 at low levels but only 10 at high levels. The most sensitive area of threshold in quiet, the frequency range between 2 and 5 kHz, corresponds to a dip in all equal-loudness contours. For high levels, this dip seems to become even deeper, i.e. tones in that frequency range are even louder than expected from a parallel shift of threshold in quiet.

Equal-loudness contours are normally drawn for a frontally-incident plane sound field. However, in many cases the sound field is not a plane sound field but similar to what is known as a diffuse sound field, in which the sound comes from all directions. Our hearing system is not equally sensitive to sounds coming from different directions and the direction dependence depends also on frequency. For this reason, the equal-loudness contours for a plane sound field and contours for a diffuse sound field are different. This difference can be most easily expressed as the attenuation, a_D, that is necessary to produce equal loudness in the plane field and the diffuse field. Figure 8.2 shows the dependence of this attenuation a_D on frequency. At low frequencies, this attenuation is negligible because our hearing system acts as

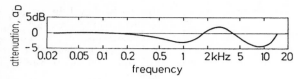

Fig. 8.2. Attenuation, a_D, necessary to produce the same equal loudness of a pure tone in a diffuse and in a free sound field as a function of the pure tone's frequency

an omnidirectional receiver. At 1 kHz, the attenuation is −3 dB. This means that the sound pressure level of the 1-kHz tone in a diffuse field must be 3 dB smaller than the sound pressure level of the 1-kHz tone in a plane sound field, to produce equal loudness. The measurements leading to the results expressed as a_D are not performed with sinusoidal tones but with narrow-band noises. Therefore, the frequency plotted as the abscissa in Fig. 8.2 may be seen as the centre frequency of a narrow-band noise. Towards higher frequencies, the attenuation, a_D, increases again and reaches about 2 dB at 2.5 kHz. For even higher frequencies, however, a_D decreases. Using the data plotted in Fig. 8.2 together with those plotted in Fig. 8.1, the equal-loudness contours of the diffuse sound field can easily be constructed. Equal-loudness contours for sinusoidal tones or narrow-band noises indicate the interesting dependence of loudness on frequency. However, loudness depends on many more variables such as bandwidth, frequency content, and duration. Consequently, it is too simple to approximate loudness level by a single weighting such as the A-weighted sound pressure level. Because A-weighting has a frequency dependence that corresponds to that of the equal-loudness contours at low levels, A-weighted sound pressure level approximates loudness level only for sinusoidal tones or narrow-band noises at lower levels. Therefore, dB(A)-values of noises or complex tones, or combinations of both, are misleading when used as indications of subjectively perceived loudness.

8.2 Loudness Function

The sensation that corresponds most closely to the sound intensity of the stimulus is loudness. The sensation-stimulus relation of loudness can be measured by answering the question how much louder (or softer) a sound is heard relative to a standard sound. This can be achieved by either searching for a ratio by changing the stimulus, or by judging a ratio of two sensations produced by two given stimuli. In electroacoustics and psychoacoustics, the 1-kHz tone is the most common standard sound. Instead of sound intensity, the sound intensity level is usually given. In the free-field condition, this value corresponds to the sound pressure level. The level of 40 dB of a 1-kHz tone was proposed to give the reference for loudness sensation, i.e. 1 sone.

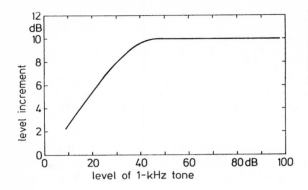

Fig. 8.3. Level increment (or decrement) necessary to produce a doubling (or halving) of the loudness of a 1-kHz tone as a function of its level

For loudness evaluations, the simplest ratio is doubling and halving. In this case, the subject searches for the level increment that leads to a sensation that is twice as loud as that of the starting level. The average of many measurements of this kind indicates that the level of the 1-kHz tone in a plane field has to increase by 10 dB in order to enlarge the sensation of loudness by a factor of two. For example, the sound pressure level of 40 dB has to be increased to 50 dB in order to double the loudness, which then corresponds to 2 sone. In order to plot the loudness function over the whole level range, experiments of halving and doubling the loudness at different levels have to be performed. The data outlined in Fig. 8.3 show the level increment needed to produce a sensation that would be called twice as loud as the starting level. A decrement of the same size would correspond to halving the loudness. The results show that this level increment or decrement is almost independent of the 1-kHz level for values larger than 40 dB. This means that the exponent of the loudness function, which in this range corresponds to a power law, can be calculated as follows: a 10 dB increment produces an increment in loudness of a factor of 2, which in logarithmic values would be equivalent to an increment of 3 dB. Therefore, the exponent of the power law expressing the loudness function of the 1-kHz tone for levels above 40 dB is 3/10, i.e. 0.3. At sound pressure levels below 40 dB, the level difference needed for loudness doubling and halving becomes smaller. It reaches 5 dB at levels of 20 dB and only 2 dB at levels near 10 dB SPL. This means that the loudness function becomes much steeper at these low levels.

Using as the reference point the loudness of a 40-dB 1-kHz tone, corresponding to a loudness of 1 sone, the loudness function can be calculated. It is shown in Fig. 8.4 by the solid line. The abscissa is the level of the 1-kHz tone, the ordinate the loudness on a logarithmic scale. The exponent of the power law that corresponds to a straight line is given by the steepness of the straight line that occurs for levels above about 30 dB. At levels below 30 dB of the 1-kHz tone, the power function is no longer usable as an approximation. The difference between the broken line (corresponding to the power law) and the solid line shows the disagreement. At levels below 10 dB, the loudness

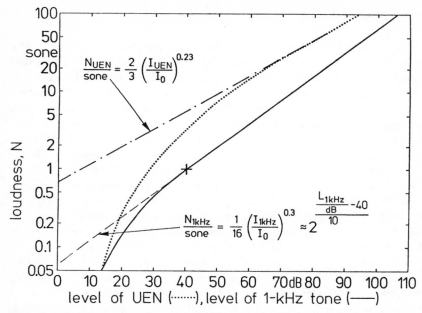

Fig. 8.4. Loudness function of a 1-kHz tone (*solid line*) and of uniform-exciting noise (*dotted*); loudness is given as a function of the sound pressure level. Approximations using power laws are indicated as broken and as dashed-dotted lines together with their corresponding equations

of the 1-kHz tone decreases drastically and reaches zero at an SPL of 3 dB. This "zero" corresponds on a logarithmic scale to a value at minus infinity. In other words, the loudness function must reach a vertical asymptote for the threshold in quiet that corresponds to a level of 3 dB.

The loudness function is normally given for the 1-kHz tone. However, it can also be plotted for other frequencies using equal-loudness contours. Because loudness level measured in phon corresponds to the sound pressure level of the equally loud 1-kHz tone, the equal-loudness contours plotted in Fig. 8.1 can be marked – besides the loudness level in phon – also with the corresponding loudness. Because an increment of 10 phon corresponds to a doubling of loudness, an increment of 20 phon corresponds to an increment in loudness by a factor of 4. Therefore, the parameter of 60 phon can be transformed to a loudness of 4 sone, 80 phon to a loudness of 16 sone, 100 phon to a loudness of 64 sone, and so on. At loudness levels below 40 phon, the loudness decreases more quickly. This leads to the fact that the loudness corresponding to 20 phon is more than a factor of 4, indeed a factor of 6.6 lower than the loudness corresponding to 40 phon. The loudness value corresponding to 20 phon is 0.15 sone. As discussed above, the threshold in quiet, which corresponds to the 3-phon equal-loudness contour, must correspond by definition to the 0-sone curve.

8.3 Spectral Effects

The spectral distribution of a sound can be either narrow or broad. The narrowest sound with respect to bandwidth, is a sinusoidal tone. The sound being most broad and at the same time most equally distributed – not with respect to physical values but with respect to the characteristics of the human ear – is uniform exciting noise. This noise produces the same intensity for each critical band. It indicates the end of a scale that represents a sinusoidal tone at one end and the uniform exciting noise at the other; different spectral bandwidths fall in between. If loudness depends on spectral effects, the uniform-exciting noise ought to produce the biggest effect.

When measuring loudness level of an object sound by a method of adjustment, one type of experiment involves adjusting the object sound until it is as loud as the standard sound. However, a second experiment is needed to remove the bias from the experimental results. In this experiment, the standard sound has to be adjusted to be equally as loud as an object sound of fixed level. Using uniform exciting noise as the object sound, the results of the two experiments are outlined in Fig. 8.5. In this figure – in contrast to all other figures displayed in this chapter – both the medians and interquartile ranges of the standard loudness level and object loudness level are given. Inspection of these data shows that varying the object sound leads to somewhat different results from varying the standard sound. This means that object loudness level and standard loudness level are different. It is the interpolated loudness level that represents the average results from the two kinds of measurements.

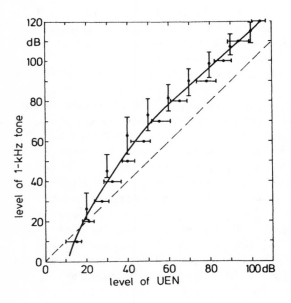

Fig. 8.5. Level of a 1-kHz tone necessary to sound as loud as a uniform exciting noise (UEN), the level of which is the abscissa. Medians and interquartile ranges are given for the two possible variables (variation of the 1-kHz tone and variation of the uniform exciting noise), using the method of adjustment. The solid line corresponds to the interpolated loudness level, the broken line corresponds to the relationship "equal level = equal loudness"

Inspecting the two data sets more carefully, it becomes clear that there is a systematic difference between the results. It seems that at least for sounds that are not extremely loud, i.e. have an SPL smaller than about 80 dB, the subjects have the tendency to set the variable sound louder than expected. This seems to be reasonable for faint and intermediate loud sounds. In this case, the subjects want to have the sound in a loudness range that corresponds to good audibility, i.e. fairly loud. In consequence, we see vertical bars related to the standard loudness level *above* the interpolated loudness level, indicated as the solid line. The object loudness level is shown by the horizontal bars and is located to the right of the solid line, again at higher levels. The solid line indicates the interpolated loudness level, i.e. the loudness level that would be measured directly using a method of constant stimuli instead of the method of adjustment. This is the curve that is usually used when referring to loudness level.

A comparison between the loudness level, i.e. the level of the equally loud 1-kHz tone, and the level of the uniform exciting noise makes it clear that uniform exciting noise is much louder than a 1-kHz tone at the same sound pressure level. The broken line in Fig. 8.5 represents equal level of the two sounds. For low levels, the two sounds are almost equally loud at equal level; for very faint levels the broad-band noise is even less loud than the 1-kHz tone. Above about 20 dB, however, the uniform exciting noise produces a larger loudness than the 1-kHz tone of the same level. A uniform exciting noise of 40 dB is as loud as a 1-kHz tone of 55 dB. This difference is largest near 60 dB of the uniform exiting noise, where the 1-kHz tone of equal loudness has to have an SPL of 78 dB, i.e. 18 dB more. For larger levels, the difference becomes somewhat smaller. However, even for an SPL of 100 dB for the uniform exciting noise, it still has a 15 dB lower level than the equally-loud 1-kHz tone.

Because uniform exciting noise is an important sound characterized by its extremely large bandwidth, it is of interest to know its loudness function. Using the loudness function of the 1-kHz tone as one set of data, and the relationship outlined in Fig. 8.5 as another set of data, it is possible to construct the loudness function for uniform exciting noise. The result of that procedure is shown in Fig. 8.4 by the dotted line, using the same reference point as for the 1-kHz tone, namely that 1 sone corresponds to the loudness produced by a 40-dB, 1-kHz tone. This loudness of 1 sone is already reached for uniform exciting noise at a level of about 30 dB. The loudness of the uniform exciting noise rises somewhat more steeply with level than the loudness of the 1-kHz tone, at least for levels of the uniform exciting noise below about 60 dB. Above this level, the loudness function can be approximated by the dotted-dashed straight line in Fig. 8.4. This means that again a power-law relationship exists; its exponent however, is smaller than that for the loudness function of the 1-kHz tone. It has a value of only 0.23 with the consequence that the two loudness functions, that of the 1-kHz tone, and that of uniform exciting

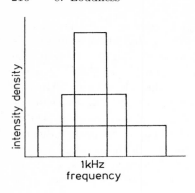

Fig. 8.6. Intensity density of band-pass noises producing the same overall intensity for different bandwidths

noise come closer together at higher levels. It is interesting to note that the loudness of a 60-dB uniform exciting noise is about 3.5 times larger than the loudness of the 1-kHz tone of the same level. This distinct effect plays an important role in judging the loudness of noises. Because A-weighted sound pressure level is relatively close to the overall sound pressure level of broadband noises, it becomes evident that A-weighted sound pressure level creates misleading values when used as indication for the loudness of broad-band noises.

The 1-kHz tone and the uniform exciting noise differ especially in bandwidth. It therefore seems reasonable to measure loudness as a function of bandwidth. Doing so, it is necessary to take care of an effect that is outlined in Fig. 8.6. The intensity density, dI/df, also called spectral density, is frequency independent for white noise which is normally created by a noise generator. Adding a band-pass filter in sequence with the noise generator, and varying the bandwidth of the filter, leads to sounds with unchanged intensity density. Their total intensity, however, changes in direct proportion to bandwidth. Because we already know that loudness depends strongly on sound level, we have to take care to keep the sound pressure level of the noise constant when the bandwidth is changed. This means that the total sound intensity, given by the area below the curves in Fig. 8.6, has to be kept constant. This is only possible by reducing the intensity density in the same way as the bandwidth is increased, as demonstrated in Fig. 8.6 at a centre frequency of 1 kHz.

The results of loudness comparisons between band-pass noise and 1-kHz tones are given in Fig. 8.7. The level of the equally-loud 1-kHz tone is used as the ordinate, and the bandwidth of the band-pass noise at a centre frequency of 1 kHz is the abscissa. The parameter in the diagram is the level of the band-pass noise that is kept constant along each curve. The data show that band-pass noise of small bandwidths is as loud as the 1-kHz tone of the same level. This simple relationship holds only up to a certain bandwidth. This bandwidth at 1 kHz is about 160 Hz, and corresponds to the critical bandwidth. For bandwidths above the critical bandwidth, the loudness of

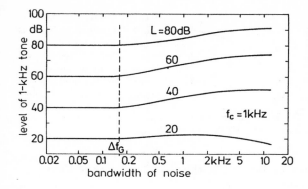

the band-pass noise increases, i.e. the level of the equally loud 1-kHz tone given as the ordinate increases. Only for very low levels of the noise, such as 20 dB, the increment of the curve is very small; it even changes to a decrement at larger bandwidth at this low level. At high levels of the band-pass noise, such as 80 dB, the increment in loudness above the critical bandwidth is not as pronounced and reaches only 11 dB in contrast to 15 dB at medium levels such as 60 dB. This means that a white noise with a bandwidth of 16 kHz and a sound pressure level of 60 dB produces a loudness level of 75 dB. This value is somewhat smaller than that found for uniform exciting noise, but indicates that the loudness of the white noise is almost three times that of the 1-kHz tone of equal sound pressure level.

The data of Fig. 8.7 indicate clearly that the dependence of loudness on the bandwidth of band-pass noise follows two different rules: one rule indicates independence of bandwidth at small bandwidths, and only holds up to a characteristic bandwidth. The other rule holds for bandwidths beyond that characteristic bandwidth, and indicates that loudness increases with increasing bandwidth. The characteristic bandwidth that separates these two regions from each other is the critical bandwidth which amounts to 160 Hz at a centre frequency of 1 kHz. The results of comparable measurements for different centre frequencies show that the characteristic bandwidth agrees with the critical bandwidth measured using other methods. This means that critical bandwidth plays a very important role in loudness sensation. Actually, results of measurements such as those described in Fig. 8.7 for different centre frequencies, have been used to define critical bandwidth.

Another way to study the dependence of loudness on bandwidth is to measure the loudness of two tones of equal level as a function of their frequency separation, Δf. The results of such measurements, again at a centre frequency of 1 kHz, are plotted in Fig. 8.8. The level of each tone was 60 dB. The figure shows the level of a 1-kHz tone judged to be equal in loudness to the two-tone complex. For small frequency separations up to about 10 Hz, the hearing system is able to follow the beats between the two tones that lead to a loudness sensation corresponding to the peak value reached within

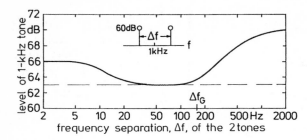

Fig. 8.8. Level of a 1-kHz tone judged as loud as a two-tone complex, each tone with 60 dB SPL and centred at 1 kHz, as a function of the frequency separation Δf of the two tones. The broken line indicates the overall level of the two tones

the beating. The peak corresponds to twice the sound pressure of the single tone, and leads to a loudness level of 66 phon. This means that for narrow frequency separations, the loudness level is determined by the peak value of the sound pressure reached within the beating period. In order to estimate the loudness level of such a two-tone complex, we have to add the sound pressures of the two tones.

For larger frequency distances between the two tones, i.e. in the range of 20 Hz up to about 160 Hz, the equally loud tone is constant in level and matches at about 63 dB. This corresponds to an addition of the sound intensities. In other words, in that region of frequency separation in which the ear no longer takes into account the temporal variation of the envelope of the sound pressure, the sound intensity is the value that creates loudness. For a frequency separation larger than about 160 Hz, the level of the equally loud 1-kHz tone, i.e. the loudness level, increases strongly. This increment shows a continuous rise that leads, at a frequency separation of 2000 Hz (1 kHz is the geometric mean of f_1 at 400 Hz and f_2 at 2400 Hz), to an increment of 10 dB relative to the single 1-kHz tone. Because an increment of 10 dB of the 1-kHz tone corresponds to a doubling of loudness, we can represent the development of loudness for large frequency distances of the two tones by assuming that the loudness of the complex is the sum of the loudnesses of each tone. The transition of frequency separations between intensity addition at intermediate frequency separation, and loudness addition at larger frequency separations, leads again to the critical bandwidth. Similar dependencies have been measured for other centre frequencies. In other words, critical bandwidth plays an important role not only for the loudness of noises of different bandwidth, but also for the loudness of two-tone complexes as a function of the frequency separation of the tones.

Similar results can be achieved by using three instead of two tones. In this case, it is convenient to use amplitude or frequency modulation. Because frequency modulation can be described by three lines only for relatively small modulation indices, it is better to use three lines that produce amplitude modulation, and to put the carrier for 90° out-of-phase to produce quasi-frequency modulation (QFM). It is meaningful, in this case, not to use the frequency distance between two adjacent lines, but to use the total bandwidth of this three-component sound, which is equivalent to $2f_{\mathrm{mod}}$, i.e. twice the

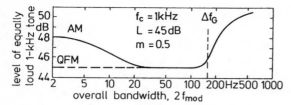

Fig. 8.9. Level of a 1-kHz tone judged as loud as a 50% amplitude-modulated 1-kHz tone of 45 dB SPL as a function of the overall bandwidth, i.e. twice the modulation frequency. The horizontal broken line indicates data produced for quasi-frequency-modulated tones. The vertical broken line indicates the critical bandwidth

modulation frequency. The results outlined in Fig. 8.9 correspond to a centre frequency (equivalent to the carrier frequency) of 1 kHz, to a sound pressure level of the carrier of 45 dB and to a degree of amplitude modulation of 0.5. For amplitude modulation and low modulation frequency, we again see an increment of the level of the equally loud 1-kHz tone (ordinate) by about 3 dB, corresponding approximately to the peak value of the sound pressure reached in this case.

For modulation frequencies larger than 10 Hz ($2f_{mod} = 20$ Hz), a minimum which corresponds to the sound pressure level increases quickly when $2f_{mod}$ exceeds 160 Hz, i.e. when f_{mod} exceeds 80 Hz. For modulation frequencies of 200 Hz, which means a total bandwidth of 400 Hz, a loudness level of 50 phon is reached. This result again indicates the two regions in which different laws are pursued by our hearing system; the dividing line is indicated even more drastically when quasi-frequency modulation (QFM) is used. In this case, the peak value produced by the three-tone complex does not increase much, and therefore at low frequencies of modulation the equally loud 1-kHz tone shows a level that corresponds to the level of the unmodulated tone. This way, the curve shows two clear-cut regions. One region is characterized by loudness level of the three-tone complex, being independent of modulation frequency up to the value where $2f_{mod}$ reaches the critical bandwidth. In the upper range, loudness increases with increasing modulation frequency. The difference between amplitude modulation (AM) and quasi-frequency modulation (QFM) holds only for conditions in which the ear is able to distinguish between the two different time functions of the envelopes. In this case, AM produces larger peak values and therefore larger loudness sensations than QFM. For modulation frequencies larger than about 10 Hz, the difference in phase of one component of the three-tone complex does not change loudness sensation. The strong influence of the critical bandwidth, however, is clearly seen when the loudness level of a three-tone complex produced by modulation is plotted as a function of modulation frequency.

8.4 Spectrally Partial Masked Loudness

A white noise with a density level of 30 dB shifts the threshold in quiet to a masked threshold in such a way that a 1-kHz tone of 50 dB sound pressure level is just audible in the noise. This means that the loudness of this 1-kHz tone, which in unmasked conditions would be 2 sone, decreases to 0 sone. When the level of the 1-kHz tone is increased to 80 dB, the loudness of the tone is close to the loudness produced in unmasked conditions. This means that the masking sound not only produces a shift of the threshold in quiet to masked threshold, but also produces a masked loudness function that has to be steeper than the unmasked loudness curve. A so-called partial masked loudness curve is produced. Figure 8.10 shows the effect of pink noise with a level of 40 and 60 dB per 1/3 octave band on the loudness of a 1-kHz tone, which is measured as a function of its level. The asymptote towards low loudness is reached by definition at masked threshold, which occurs under this condition near 37 and 57 dB. Above masked threshold, loudness rises very steeply and, at levels about 20 to 30 dB above masked threshold, almost reaches the unmasked loudness. The broken line in Fig. 8.10 corresponds to the unmasked 1-kHz loudness function. In the case of the softer masker, the difference between partially masked and unmasked loudness function shrinks at the 70-dB sound pressure level of the 1-kHz tone to only about 10%. For even higher levels of the 1-kHz tone, the difference becomes unmeasurably small. This means that partial masking produces a loudness function that is comparable to that described by audiologists as recruitment. The effect illustrated in Fig. 8.10 seems reasonable and understandable from what we have learned in everyday conditions where communication very often takes place under partially masked conditions.

Another effect is surprising but also important for understanding the development of loudness. In order to demonstrate this spectrally partial masked

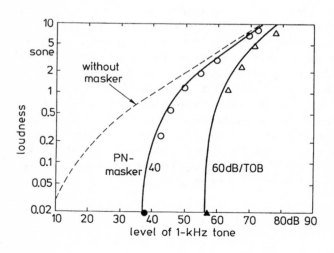

Fig. 8.10. Loudness of a 1-kHz tone as a function of its level. The broken line indicates the unmasked condition while the solid lines and symbols show the loudness of the tone, with an additional masking pink noise having a level of 40 dB and 60 dB per 1/3 octave band

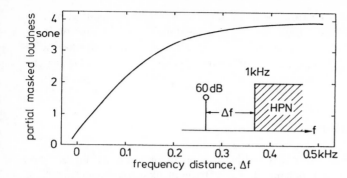

Fig. 8.11. Partially masked loudness of a sinusoidal tone of 60-dB sound pressure level. The frequency separation Δf between the tone and the cut-off frequency of the masking high-pass noise (indicated in the schematic drawing of the inset) is the abscissa

loudness effect, a high-pass noise with a cut-off frequency of 1 kHz outlined in the inset of Fig. 8.11 is used. This high-pass noise has a sound pressure level of 65 dB in each critical band, and would mask a 1.1-kHz tone if the tone and noise were presented simultaneously. In order to start with an almost unmasked condition, a 500-Hz tone is used instead of a 1-kHz tone. In this case, the distance Δf (see inset of Fig. 8.11) amounts to 500 Hz. Alternating with the simultaneously presented 500-Hz tone plus the high-pass noise, a 1-kHz tone is presented without additional noise. This tone is adjusted in level so that it sounds as loud as the 500-Hz tone. This way, the loudness of the partially masked 60-dB tone at 500 Hz can be obtained. The distance Δf between the 60-dB tone and the cut-off frequency is then changed in steps to smaller values. The data indicate that the high-pass noise influences the loudness of the 60-dB tone, even though the tone and the noise have different spectral content.

For large Δf, the influence of the high-pass noise is negligible. For a frequency distance of only 100 Hz, however, the loudness of the 60-dB, 900-Hz tone is reduced to only half of the loudness of the 900-Hz tone presented in quiet. This means that the high-pass noise reduces the loudness of the tone. Figure 8.11 shows the equivalent loudness of the partially masked 60-dB tone, as a function of the frequency separation Δf between the tone and the cut-off frequency of the high-pass noise. For large Δf, the partial masking is very small. It increases with decreasing frequency separation in such a way that the loudness of the partially masked 60-dB tone becomes very small for a frequency separation zero, i.e. when the frequency of the tone is equal to the cut-off frequency of the high-pass noise. Because a high-pass filter with a very steep slope is used to produce the high-pass noise, the effect represents a characteristic of our hearing system and is not determined by the slope of the high-pass noise. The effect is psychoacoustically important. It indicates that the loudness of a pure tone that encompasses a very small spectral width is not developed only at a certain location along the frequency scale, but in a larger area along the critical-band rate: high-pass noise decreases the loudness of the tone even though the tone and the partially masking high-pass noise are spectrally separated. The results of such experiments are the key to the model for loudness.

8.5 Temporal Effects

Most natural sounds are not steady but strongly time dependent; the loudness of such sounds is also a function of time. Typical examples are speech or music, and also quite a few technical noises that sound impulsive or rhythmic where the loudness cannot be described by steady-state condition. Therefore, the loudness of single tone bursts as a function of the duration, and the loudness of sequences of sound bursts as a function of their repetition rate are of interest.

The loudness of a tone burst decreases for durations smaller than about 100 ms. For longer durations, the loudness is almost independent of duration. Figure 8.12 shows this effect with a 2-kHz tone burst which, when presented in steady-state condition at a level of 57 dB, would produce a loudness of 4 sone. A frequency of 2 kHz, i.e. higher than the standard frequency of 1 kHz, was chosen in order to be able to measure at shorter durations without producing additional bandwidth effects. The loudness of this tone is plotted as a function of its duration. The effect of duration occurs for bandwidths larger than critical bandwidth, 300 Hz in this case. Accordingly, durations as short as 3 ms can be used. In order to exclude spectral effects even further, the tone bursts are produced with Gaussian-shaped rise and fall times of 1.5 ms. The curve shown in Fig. 8.12 indicates that loudness remains constant for a duration larger than about 100 ms. Loudness decreases with shorter durations and is reduced by a factor of 2, which means a loudness reduction to 2 sone, at a duration of about 10 ms. The curve continues towards shorter durations in a

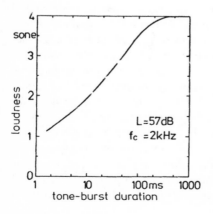

Fig. 8.12. Loudness of a burst extracted form a 2-kHz tone with an SPL of 57 dB as a function of tone-burst duration

similar fashion. Reducing the duration by another factor of 10, the loudness again decreases by a factor of about two and would reach, at a duration of 1 ms, a loudness of 1 sone. At this short duration, however, the rise/fall time has to be reduced and the measurement becomes meaningless because of the greatly enlarged spectral width.

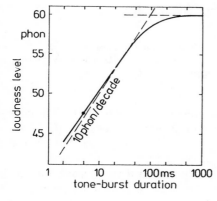

Fig. 8.13. Loudness level of a burst extracted from a 2-kHz tone with 57 dB SPL as a function of burst duration. Broken lines indicate useful approximations

A factor of two in loudness corresponds to a loudness level difference of 10 phon. Therefore, it may be interesting to see the results of Fig. 8.12 with a loudness level ordinate. This transformation leads to Fig. 8.13, which indicates that loudness level decreases by 10 phon if the duration is reduced by a factor of 10. The broken rising line indicates this approximation, which approaches the steady-state condition (the horizontal) at a duration of 100 ms. Similar results have been produced at several other frequencies and lead to the same dependence, which can be expressed most easily for durations below 100 ms by the following approximation: the loudness level increases for 10 phon for each 10-fold rise of the duration of the tone burst. Above 100 ms, the loudness level is approximately independent of duration.

The loudness level of a tone burst of 5 ms duration is marked in Fig. 8.13 with a dot. The corresponding loudness level is 47.5 phon. Starting with this single 2-kHz tone burst of 5 ms duration, the dependence of loudness level on repetition rate is measured. Figure 8.14 indicates the measured dependence as a solid curve. It starts at a very low repetition rate of 1 Hz with a loudness level equivalent to that of a single burst, i.e. 47.5 phon. With increasing repetition rate, the abscissa in Fig. 8.14, the loudness level initially remains constant. For repetition rates higher than about 5 Hz, the loudness increases. The repetition rate can be further increased up to the final value of 200 Hz. There, the tone-burst sequence merges into a steady-state tone and produces, as expected, a loudness level of 60 phon – the loudness level for the steady-state condition (very long duration) shown in Fig. 8.13. Meaningful approximations are plotted in Fig. 8.14 as broken straight lines, and corre-

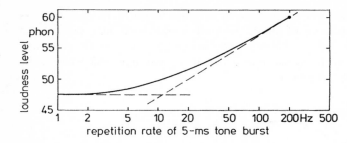

Fig. 8.14. Loudness level of a tone burst lasting 5 ms extracted from a 57-dB, 2-kHz tone as a function of repetition rate. At a repetition rate of 200 Hz, the steady-state tone is reached. Broken lines indicate useful approximations

spond to those outlined in Fig. 8.13. Up to a repetition rate of 10 Hz, loudness level remains approximately constant. For repetition rates above 10 Hz, loudness level increases by 10 phon if the repetition rate is increased by a factor of 10. Similar results have been obtained at other test-tone frequencies.

The repetition rate can be increased above 200 Hz if the duration of the single tone burst is shortened below 5 ms. This can be done only for higher test-tone frequencies. In such cases, the rise of the loudness level continues further and indicates that the approximation, which is not perfect in Fig. 8.14, becomes better at higher frequencies. More important, however, is that the solid curve of Fig. 8.14 is very similar at high test-tone frequencies.

8.6 Temporally Partial Masked Loudness

Besides the spectral effects which influence partial masking that were described in Sect. 8.4, there also exist temporal effects which influence partially masked loudness. Such effects at threshold have already been discussed as pre- and post-masking. The effect of pre-masking on loudness is quite interesting and can be measured if a second loud sound reduces the loudness of a preceding short sound when separated from the second by only some 10 ms. An example of such an effect is indicated in Fig. 8.15. The inset shows the temporal configuration of the two sounds. A 5-ms, 60-dB, 2-kHz tone burst preceeds a uniform exciting noise burst with a critical-band level of 65 dB. The distance between the test-tone burst and the uniform exciting noise burst is indicated by Δt. The loudness of the test-tone burst is measured by comparing it with the loudness of a 2-kHz tone burst, also of 5-ms duration, which is *not* followed by the uniform exciting noise. The solid curve of Fig. 8.15 shows the results of such loudness comparisons by indicating the loudness of the temporally partial-masked tone burst as a function of the gap duration between the tone burst and the following uniform exciting noise. The loudness of the 5-ms, 2-kHz test-tone burst in unmasked condition reaches about 1.9

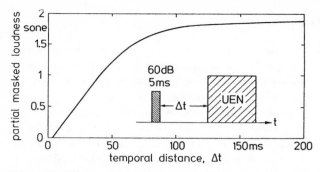

Fig. 8.15. Temporal partially masked loudness of a 5-ms, 60-dB, 2-kHz tone that is presented before the onset of a uniform-exciting noise as a function of the temporal distance indicated in the inset

sone. This value is also obtained with large values of Δt near 200 ms. If the distance Δt is reduced to 100 ms, the loudness of the test-tone burst decreases only a little. For shorter gap durations, however, the loudness of the test-tone burst decreases drastically and reaches zero near $\Delta t = 5$ ms; at this value of Δt, the test-tone burst is totally masked by the following uniform exciting noise. For a gap duration of 40 ms, the loudness of the test-tone burst is reduced by about a factor of two. Decreasing the gap further to about 20 ms reduces the loudness again by another factor of two.

Similar effects can be measured at other test-tone frequencies and with different sounds. The effect can be measured most easily for sounds which have different characteristics in timbre. Tones and noises are very useful for this reason. However, complex tones, or tones composed of many harmonics can also be used, instead of pure tones. The effect is relatively independent of the type of test sound.

Temporally partial masked loudness does not, of course, mean that we are able to listen into the future. Rather, the effect indicates that our hearing system needs some time to develop the sensation of loudness. The short tone burst produces a sensation that is not fully developed within the 5 ms of its presentation, but needs much more time. If, during that time, a strong uniform-exciting noise is presented, the development of the loudness of the short tone burst is interrupted. This means that the full development of loudness of the tone burst is partially reduced by the following noise. The development of sensations follows a rule well known in many other systems: the cause and its consequences are often non-simultaneous!

8.7 Model of Loudness

In previous sections, we have described several effects that are useful in developing a model of loudness. Two tones of equal level with a frequency separation greater than the critical bandwidth produce a loudness which is larger than the loudness of a single tone with a frequency midway between that of the two tones, and with a level corresponding to the total intensity of the two tones. With increasing frequency separation of the two tones, loudness increases. This means that loudness is not produced from separate spectral components, but rather the two components influence each other, especially if their frequency separation is small. Only for quite large frequency separation, where the two single tones do not influence each other, does a loudness value occur which corresponds to the addition of the loudnesses of each tone.

Because critical bandwidth plays an important role in loudness, the loudness on the slopes of the frequency selectivity characteristic of our hearing system may also play an important role. The excitation level versus critical-band rate pattern represents the pattern that describes not only the influence of critical bandwidth, but also that of the slopes. Therefore, it seems reasonable to use the excitation level versus critical-band rate pattern as a basis from which the loudness of the complex may be constructed. Because we have seen that the two tones influence each other in the creation of the total loudness, even though they are spectrally separated, it may be useful to treat total loudness as an integral of a value that we have to find, but which can be drawn as a function of critical-band rate. If such an integral leads to the loudness that is given in sone, the value we require has to have the unit of sone/Bark. In this case, total loudness, given as the integral of this value over critical-band rate, will lead to the unit of sone.

As usual, we call the value we require "specific loudness" with the symbol N'. Loudness, N, is then the integral of specific loudness over critical-band rate (see Fig. 8.16 lower right) or in mathematical expression:

$$N = \int_0^{24\,\text{Bark}} N' \, dz \,, \tag{8.1}$$

where the integral is taken over all critical-band rates. Two examples may show how this total loudness is created. In order to make the situation less complicated at the beginning, we start with the steady-state condition and use a uniform-exciting noise as the first example, and a narrow-band noise of critical bandwidth at 1 kHz as the second. Figure 8.16 shows how loudness in these two cases is determined. The two left-hand diagrams give the critical-band level as a function of critical-band rate. The upper part shows the condition for uniform exciting noise, which produces frequency-independent critical-band rate levels (for didactical reasons, the value a_0 is ignored). The level in each of the 24 abutting critical bands is 50 dB, leading to an overall sound pressure level of 50 dB+$(10 \times \log 24)$ dB = 64 dB. In the lower panel,

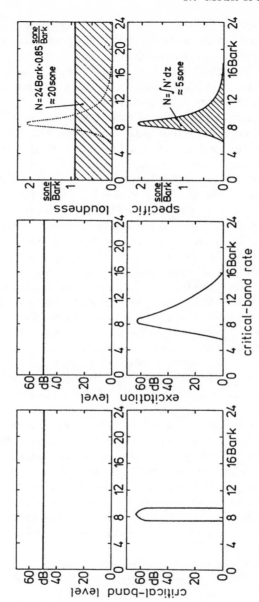

Fig. 8.16. Development of total loudness for uniform exciting noise and for critical-band wide noise centered at 1 kHz: the left part shows the critical-band levels, the centre part the excitation level and the right part the specific loudness, each as a function of critical-band rate. The upper row indicates distributions for uniform exciting noise, the lower row those for narrow-band noise. The hatched areas correspond to the total loudnesses. The area of total loudness of the narrow-band noise is indicated in the upper right drawing as a dotted curve for comparison

we concentrate the intensity of the uniform exciting noise into one critical band at 1 kHz, corresponding to 8.5 Bark. This allows a comparison between uniform exciting noise and critical-band wide noise at 1 kHz, with equal sound pressure level. The critical-band level L_G of such an ideal narrow-band noise assumes, according to (8.5), the shape of a gothic window, the peak value of which lies at 8.5 Bark with at a level of 64 dB.

The transformation from critical-band rate level into the excitation level versus critical-band rate pattern is the next step, which is indicated by the two middle diagrams of Fig. 8.16. For uniform exciting noise, only main excitations occur, and hence we have the same figure as on the left. For the narrow band noise, however, we have to introduce the slopes of excitation, which are quite steep towards lower critical-band rates corresponding to lower frequencies but flatter towards higher critical-band rates. This is indicated by the excitation level in the lower middle diagram of Fig. 8.16. The main excitation level is identical to the critical-band rate level, but we also see the slopes of excitations. In Sect. 8.3, the loudness of uniform exciting noise is discussed. From Fig. 8.4, we see that uniform exciting noise of 64 dB creates a loudness of 20 sone. In the same figure we can also see that the loudness of the 64-dB, 1-kHz tone is only 5 sone. Because uniform exciting noise (ignoring a_0) produces a specific loudness independent of critical-band rate, we expect a rectangularly-shaped specific loudness versus critical-band rate area that corresponds to total loudness. Calculating backwards, we have to assume that the specific loudness is 0.85 sone/Bark in order to produce 20 sone of total loudness from this stimulus (see Fig. 8.16, top right). Similar calculations can be performed using the dependence of the loudness of uniform exciting noise as a function of its level, as indicated in Fig. 8.4. This dependence can be approximated for large levels of the uniform exciting noise by relating its intensity to the loudness produced, using a power law with an exponent of 0.23. The distribution of specific loudness becomes independent of critical-band rate when a uniform exciting noise is used. Therefore, the exponent measured for uniform exciting noise is directly related to the exponent we expect for specific loudness. Towards low levels of uniform exciting noise, the slope of the loudness curve becomes steeper. This effect is due to the influence of threshold in quiet, which has to be taken into account for calculations in the lower level region.

The distribution of specific loudness as a function of critical-band rate is not so simple for the narrow-band noise. It is plotted in Fig. 8.16 in the lower right diagram. The area below the specific loudness pattern again corresponds to total loudness. This area, however, is concentrated not only in that region in which the narrow-band noise is physically present, but also shows slope-specific loudnesses towards lower, and especially towards higher critical-band rates. The dashed area indicates clearly the large amount that contributes to the total loudness in the area of critical-band rates above 9.5 Bark. This distribution of the specific loudness corresponds to the distribution seen in the excitation level versus critical-band rate pattern. Corresponding to the main excitations, we see main specific loudnesses, and corresponding to slope excitations we see slope-specific loudnesses. The transformation from excitation level into specific loudness has to be known in order to transform excitation level versus critical-band rate patterns into specific loudness versus critical-band rate pattern. This relationship will be discussed in the next section.

Coming back to the beginning, we want to compare the loudness produced by the equally intense noises, i.e. a uniform exciting noise and a critical-band wide noise. This comparison can be made using the Fig. 8.16 right top diagram, where the loudness pattern produced by the narrow-band noise is indicated by a dotted curve. There is no question that the uniform exciting noise produces a much larger area than the narrow-band noise, even though the overall sound pressure levels of the two noises are equal (64 dB) in both cases. Judging the ratio of the two areas, we can see that the uniform exciting noise produces a loudness that is about 4 times larger than that produced by the narrow-band noise, in agreement with the measured data.

The fundamental assumption of our loudness model is that loudness is not produced from spectral lines or from the spectral distribution of the sound *directly*, but that total loudness is the sum of the specific loudnesses that are produced at different critical-band rates. It should be pointed out here that transforming the physical spectra (level versus frequency) into specific loudness versus critical-band rate yields the best hearing equivalent psychoacoustical values. Both the frequency related transformation into critical-band rate and the amplitude related transformation into specific loudness are of crucial importance for evaluating sound in terms of the eventual receiver, the human hearing system.

Specific loudness as a function of critical-band rate may be called a loudness distribution or loudness pattern. This distribution, although related to the spectral distribution of the sound, also takes into account the nonlinear relationship between excitation and specific loudness and the non-ideal frequency selectivity of the human ear, expressed in critical bands and in the slopes found in masking patterns.

8.7.1 Specific Loudness

Stevens' law says that a sensation belonging to the category of intensity sensations grows with physical intensity according to a power law. In consequence of that law, we have to assume that a relative change in loudness is proportional to a relative change in intensity. Specific loudness is developed using this power law. Instead of using the total loudness as the value in question, specific loudness is used. Additionally, instead of intensity, excitation has to be used. With a constant of proportionality k, relating the specific loudness N' and its corresponding increment $\Delta N'$ to the excitation E and its corresponding increment ΔE, we can formulate the power law as an equation using differences:

$$\frac{\Delta N'}{N'} = k \frac{\Delta E}{E} , \quad \text{or} \quad \frac{\Delta N'}{N' + N'_{\mathrm{gr}}} = k \frac{\Delta E}{E + E_{\mathrm{gr}}} , \tag{8.2}$$

if values N'_{gr} and E_{gr} corresponding to the internal noise floors at very low values of N' and E, respectively, are added. Threshold in quiet is assumed to be produced by such internal noise. Using the threshold factor, which was helpful in describing the relationship between masking sound and the masked threshold produced by that sound, we are able to calculate the excitation E_{gr} producing threshold in quiet by the test-tone excitation E_{TQ}, using the equation

$$E_{\mathrm{gr}} = E_{\mathrm{TQ}}/s , \tag{8.3}$$

where s is the ratio between the intensity of the just-audible test tone and the intensity of the internal noise appearing within the critical band at the test tone's frequency. This ratio was effectively used for calculating masked threshold produced by external sounds. Here it is used to estimate the internal noise using the knowledge of threshold in quiet.

Treating this equation of differences as a differential equation, it can be solved by using the reasonable boundary condition that excitation equal to zero leads to specific loudness equal to zero, i.e. $E = 0$ leads to $N' = 0$. We find the relation

$$N' = N'_{\mathrm{gr}} \left[\left(1 + \frac{sE}{E_{\mathrm{TQ}}} \right)^k - 1 \right] , \tag{8.4}$$

which can be transformed into the final expression using a reference specific loudness N'_0

$$N' = N'_0 \left(\frac{E_{\mathrm{TQ}}}{s\,E_0} \right)^k \left[\left(1 + \frac{sE}{E_{\mathrm{TQ}}} \right)^k - 1 \right] . \tag{8.5}$$

The most important value in this equation is the exponent k. It can be approximated quite simply by using the dependence of the loudness of uniform-exciting noise as a function of its level, as mentioned in the preceding section. Because (8.5) can be approximated for large values of E, for which the influence of threshold in quiet is negligible, as

$$N' \sim \left(\frac{E}{E_0} \right)^k , \tag{8.6}$$

the exponent of 0.23 is the value of k we are searching for. For frequencies in the neighbourhood of 1 kHz, for which the threshold factor is $s = 0.5$, and with the additional boundary condition that a 1-kHz tone with a level of 40 dB has to produce exactly 1 sone as total loudness, the equation

$$N' = 0.08 \left(\frac{E_{\mathrm{TQ}}}{E_0} \right)^{0.23} \left[\left(0.5 + 0.5 \frac{E}{E_{\mathrm{TQ}}} \right)^{0.23} - 1 \right] \frac{\mathrm{sone_G}}{\mathrm{Bark}} \qquad (8.7)$$

is the final equation from which specific loudness can be calculated quantitatively. In this equation, E_{TQ} is the excitation at threshold in quiet and E_0 is the excitation that corresponds to the reference intensity $I_0 = 10^{-12}\,\mathrm{W/m^2}$. The value "1" in the parenthesis of (8.5) has been exchanged by the value $(1-s)$, i.e. 0.5, with the consequence that the specific loudness reaches asymptotically the value $N' = 0$ for small values of E. The index G at the unit "sone" is added as a hint for the user that the loudness given in this value is produced using the critical-band levels.

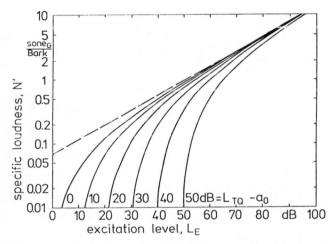

Fig. 8.17. Specific loudness as a function of excitation level with the difference between level at threshold in quiet, L_{TQ}, and transmission loss, a_0, as parameter. The broken line indicates a power law with the exponent of 0.23

The linear value of the excitation is rarely useful. We have seen that excitation level is much more effective and handier. Therefore, the relation between specific loudness, N', and excitation level, L_{E}, is outlined in Fig. 8.17. The parameter in the curves is the value $L_{\mathrm{TQ}} - a_0$, which was chosen in order to exclude dependencies on frequency or critical-band rate. This way, threshold in quiet on one hand, and the logarithmic transmission factor a_0 (representing the transmission between freefield and our hearing system) on the other, are incorporated. The gain factor, a_0, holds for the free-field condition. The other extreme condition is a diffuse field, for which we have to use a different transmission factor leading to the value $a_{0\mathrm{D}}$. Both values are given in Fig. 8.18 as a function of critical-band rate and frequency. Up to about 8 kHz, both values deviate less than $\pm 4\,\mathrm{dB}$. This seems to be not very much; however, for calculating loudness as precisely as possible, such values become important.

As outlined in Fig. 8.17, the specific loudness, N', rises rapidly for low values of L_E. For large values of L_E, all curves reach an asymptote, given by the dashed line, which corresponds to the power law with an exponent of 0.23. The curves with different values of parameter $L_{TQ} - a_0$ are produced one from the other by shifting the curve with the parameter 0 dB to the right and upwards along the broken asymptote in such a way that the vertical asymptote towards low levels reaches the value of L_E at a place equal to the value of the parameter.

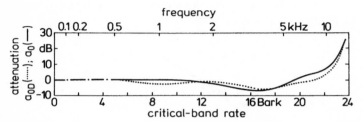

Fig. 8.18. Attenuation corresponding to a transmission factor needed in the free-field condition (a_0, *solid curve*) or for diffuse-field condition (a_{0D}, *dotted curve*) as a function of critical-band rate (*lower scale*) and frequency (*upper scale*)

In previous chapters, it was shown that excitation level is a useful and interesting value for describing hearing effects. Specific loudness is an even more hearing equivalent value of similar importance that is very useful for describing other effects in hearing such as sharpness.

8.7.2 Loudness Summation (Spectral and Temporal)

In the introduction to Sect. 8.7, we ignored the influence of threshold in quiet and of the transmission factor, a_0. To calculate loudness quantitatively as exactly as possible, it is necessary to take into account these two values as a function of critical-band rate. This is done in Fig. 8.19 for free field conditions and the two sounds and levels as used in Fig. 8.16 (uniform exciting noise and critical-band wide noise, each with a level of 64 dB). The distribution of excitation level does not change for the narrow-band noise since a_0 is 0 dB at 1 kHz. The excitation-level distribution of the uniform exciting noise, however, shows the influence of a_0. Corresponding influences are seen in the specific loudness versus critical-band rate pattern. The distribution for narrow-band noise is exactly the same as shown in Fig. 8.16. The distribution of the uniform exciting noise, however, shows not only the peak corresponding to the large sensitivity of our ear in the region around 4 kHz, but also the decrement of specific loudness towards low critical-band rates corresponding to the influence of threshold in quiet on specific loudness, as outlined in Fig. 8.17.

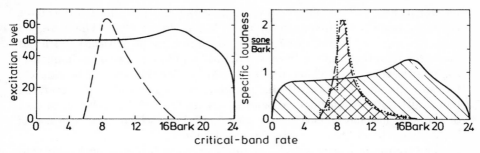

Fig. 8.19. Excitation level and specific loudness, each as a function of critical-band rate for uniform exciting noise (*solid lines*) and narrow-band noise centred at 1 kHz with the same overall level (*broken curves*). The differently hatched areas indicate the corresponding total loudness. The dotted line represents the approximation used for the loudness calculation procedure

This influence of threshold in quiet becomes clear when the curves outlined in Fig. 8.17 are used. For large excitation level, L_E, the same specific loudness is reached independent of the threshold in quiet. For lower levels, however, the influence of threshold in quiet becomes very strong as the excitation level approaches the level given as the parameter. This means that the influence of threshold in quiet is most prominent for relatively faint sounds. Another effect also becomes visible: the influence of threshold in quiet, which is often assumed to play an important role, is smaller than expected for medium and high levels. The reason for this is that the three octaves between 20 and 160 Hz, within which the threshold in quiet changes drastically when plotted as a function of frequency, are reduced to the relatively small value of only 1.5 Bark if plotted as a function of critical-band rate. Therefore, the influence of threshold in quiet is usually effective in everyday loudness calculation only at the left end of the loudness distribution.

One important factor may be mentioned again. When the integration of specific loudness leads to the total loudness, we have to be aware of the fact that integration (corresponding to measuring the area below a certain curve) can only be performed if linear scales are used on the abscissa as well as on the ordinate. Therefore, it is not possible to do any integration process on the left part of Fig. 8.19. This can only be done on the right part, where specific loudness is plotted on a linear scale as a function of critical-band rate, also on a linear scale.

Specific loudness as a function of critical-band rate is given in all of the figures as a continuous curve. This corresponds to the very fine frequency resolution of our hearing system. Critical bands have been used as if they can be shifted along the critical-band rate to any place. In many practical cases however, it may not be necessary to have as many as 640 critical-band filters (corresponding to the 640 audible frequency-modulation steps along the pitch scale). For practical purposes, it is sufficient to use 24 critical-band filters, for which the lower cut-off frequency of the adjacent upper filter corresponds to the upper cut-off frequency of the adjacent lower filter. Therefore, we can approximate the excitation-level distribution by 24 critical-band levels measured with these filters. Because critical-band filters are not as frequently available as third-octave band filters, we can also use third-octave band filters for centre frequencies above 315 Hz. For lower centre frequencies, however, the bandwidths of the third-octave band filters are too small in relation to that of the critical-band filters. Useful approximations, which allow the use of third-octave band filters, will be discussed later.

The comparison between the loudness of a narrow-band noise and that of uniform exciting noise represents the extreme types of spectral characteristics. In Fig. 8.7, the continuous rise of loudness is plotted as a function of bandwidth for constant sound pressure level. This increment of loudness as a function of bandwidth can be described as a spectral effect. For the same overall intensity and centre frequency, any stimulus narrower than a critical band, even a pure tone, produces a pattern of specific loudness corresponding to that shown for the noise one critical band in width. The SPL falling into one critical band is responsible for the specific loudness produced. Therefore, bands narrower than a critical band are represented by a pattern of specific loudness that is independent of bandwidth. However, the loudness pattern begins to change as soon as the width of the noise exceeds the width of the critical band. For a fixed overall level, the height of the pattern decreases but at the same time it extends over a greater portion of the critical-band rate scale. The extension along the critical-band rate produces an increment of the area that is larger than the decrement produced by the reduction of height. This way, the total loudness increases with increasing bandwidth even though the height of the pattern decreases. This effect is especially pronounced at medium levels. At very low levels, the slope of the specific-loudness function becomes steep, and therefore the decrement in height is stronger than the increment produced by the extension over the critical-band rate. Therefore, loudness remains constant or even decreases at very low levels as a function of bandwidth. When exceeding the critical bandwidth at very high levels, loudness does not increase strongly; only for bandwidths larger than about an octave, corresponding to three critical bands, does loudness increase. The reason for this effect, also shown in Fig. 8.7, is the flatter upper slope of the excitation pattern, and consequently also the flatter loudness pattern, towards higher critical-band rates. In this case, not only the peak value of the specific-loudness pattern decreases with increasing bandwidth but also the upper slope, with the result that the area is almost constant, i.e. constant loudness, for bandwidths between critical band and octave band at levels near 100 dB SPL. The loudnesses calculated using the model and those measured psychoacoustically are very close to one another; they both show the invariance of loudness for spectral extension within one critical band and the same dependence on bandwidth beyond the critical band.

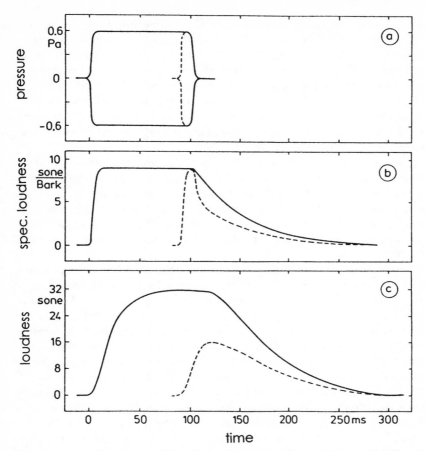

Fig. 8.20a–c. Processing of loudness of tone impulses of 100 ms (*solid*) and 10 ms (*dashed*) duration at 5 kHz. (**a**) temporal envelopes of the two tone impulses, (**b**) corresponding time functions of specific loudness, (**c**) corresponding time functions of loudness

Temporal summation is another factor in the development of loudness. Although spectral summation produces larger effects in everyday sounds, the temporal effect is also quite important. The model can be extended so that temporal effects are taken into account. In doing so, the excitation level and specific loudness have to be treated as time-dependent values. Excitation is correlated with masking, therefore the masked thresholds produced in premasking, simultaneous masking and postmasking can be used and transformed into time functions for specific loudness.

Along these lines, the development of loudness for short tone bursts is outlined in Fig. 8.20. Tone bursts at 5 kHz of 100 ms (solid) as well as 10 ms (dashed) duration are considered. The upper panel in Fig. 8.20 illustrates the temporal envelope of the long versus the short tone impulse. The middle panel shows the time function of specific loudness for both tone impulses. It is clearly seen that for longer tone impulses (solid) the decay is more gradual than for short impulses (dashed) showing a rather steep gradient at the beginning of the decay process. The lower panel in Fig. 8.20 shows the time function of loudness averaged across all channels and filtered by a 3rd order low-pass filter. In line with psychoacoustic data (see Fig. 8.12) the tone impulse of 100 ms duration produces a loudness which is about twice as large as the loudness of the 10 ms impulse.

In terms of models, this means that the total loudness, the sum of all the specific loudnesses over the critical-band rate, is transferred through a special low-pass system, the impulse response of which corresponds to the temporal response of the loudness-processing neural system. The effect of such a low-pass system, optimized using the data of loudness as a function of duration and data on temporal partial masking, is shown in the lower panel of Fig. 8.20.

The approximation of the temporal effects in the model represents a kind of incomplete temporal integration. Corresponding to the cut-off frequency of the special low-pass system, the integration time lasts roughly about 45 ms. The spectral distribution of the sound leads also to an integration, namely the summation along critical-band rate. Because of the two kinds of integration, spectral and temporal, we have to know which one takes place first. Searching for an answer to this question in anatomy, it seems reasonable that spectral summation comes first because this distribution is available along the basilar membrane, where the stimuli of the sensory cells are transferred quickly into the corresponding neural excitation. Temporal summation seems to take place in more central areas, a fact which has been also clarified for postmasking using otoacoustic emissions. The results of such experiments confirm that the decay of the cochlear excitation is much quicker than the decay of postmasking, which may be assumed to be a counterpart of loudness. Therefore it was assumed that temporal summation comes second and spectral summation first.

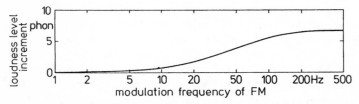

Fig. 8.21. Loudness-level increment as a function of modulation frequency for frequency-modulated tones with a total frequency deviation of ± 700 Hz. Centre frequency of the FM-tone is 1.5 kHz, sound pressure level is 60 dB

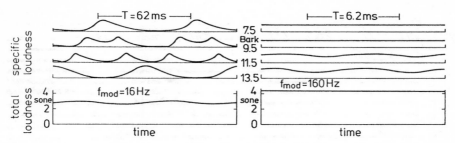

Fig. 8.22. Examples of the specific loudnesses produced by frequency modulated tones at a modulation frequency of 16 Hz (*left part*) and 160 Hz (*right part*). The two lower curves indicate total loudness as a function of time

This can be proven by using strongly frequency-modulated pure tones and judging their loudness as a function of modulation frequency. Using a frequency shift of ±700 Hz and a centre frequency of 1500 Hz, loudness comparisons show that the loudness level increases as a function of modulation frequency only for modulation frequencies above about 20 Hz. The results are outlined in Fig. 8.21. The model predicts that loudness level should increase for similar modulation frequencies near about 20 Hz. In this range of modulation frequency, postmasking becomes effective so that the specific loudnesses in neighbouring critical-band-rate regions remain almost independent of time (Fig. 8.22). This way, the frequency-modulated tone acts as a broadband sound. In the range of low modulation frequencies up to about 15 Hz, however, the specific loudness in each channel shows strong temporal fluctuations, so that the loudness increment remains relatively small. The reason is that the specific loudness is mostly concentrated in one or two neighbouring critical bands during a time at which the specific loudness is almost zero for far-away critical bands. For such low frequencies of modulation, we do not expect a loudness increment in the model, although the critical duration in the development of loudness as a function of duration is nearly 100 ms.

The effect is illustrated in Fig. 8.22 for modulation frequencies of 16 and 160 Hz. For 16 Hz, the specific loudness is given at four critical-band rates in the upper left part. The total loudness is plotted in the lower part and indicates a loudness almost independent of time, although the specific loudnesses at each critical-band rate vary strongly as a function of time. The total loudness remains at a value somewhat below 3 sone. For 160 Hz modulation frequency, however, specific loudnesses in each channel remain almost constant. Even though their peak values remain somewhat less than the peak values reached for low modulation frequency, the sum of all these time independent specific loudnesses (upper right parts in Fig. 8.22) leads to a total loudness near 4.5 sone, as indicated in the lower part of the figure. This loudness is larger than the peak value reached for a modulation frequency of 16 Hz and corresponds to the increment of loudness level shown in Fig. 8.21.

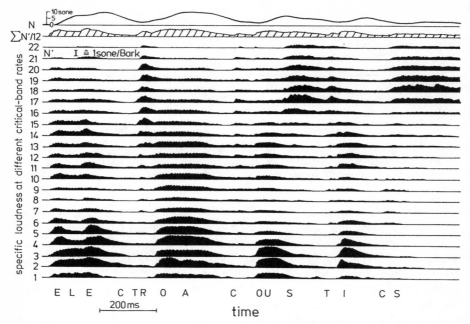

Fig. 8.23. Specific loudness versus critical-band rate versus time pattern of the spoken word "electroacoustics". Specific loudness is plotted for 22 discrete values of critical-band rate. The ordinate scale used is marked at the panel related to 21 Bark and indicates 1 sone/Bark. Time is the abscissa and 200 ms are marked. The sum of the specific loudnesses divided by 12 is indicated as the hatched area in the second curve from the top. Total loudness as a function of time is plotted in the uppermost drawing

This result indicates that the model simulates temporal and spectral effects very well.

The specific loudness versus critical-band rate versus time functions are those functions that best illustrate the information processing in the human hearing system. Such three-dimensional patterns contain all the information that is subsequently processed and which leads to the different hearing sensations. In order to illustrate the complex structure of such patterns the word "electroacoustics" is used. The specific loudness produced by this spoken sound is outlined in Fig. 8.23 for 22 places along the critical-band scale, each 1 Bark distant from the next. The sum of all these specific loudnesses without temporal weighting, is given in the curve second from the top. It indicates the spectral summation that produces a curve following the stimulus relatively quickly at the rising regions, but showing the decay corresponding to postmasking. The time function of perceived loudness is given on the top curve. It is strongly smoothed, but shows single syllables with relatively clear separation. It becomes clear from this curve that peak loudness, normally assumed to be perceived loudness, is produced by the vowels in speech. Consonants and plosives are very important for understanding, and are clearly visible in the specific loudness versus critical-band rate versus time patterns; their contribution to the total loudness, however, is almost negligible.

The model of loudness determination uses most of the characteristics of our hearing system. Such a model realized in electronic networks or in computer programs represents an excellent loudness meter. If the model simulates all, or at least most of the important characteristics responsible for our loudness sensations, then it measures loudness almost as well as our hearing system. Because loudness plays an important role in noise abatement, the realization of such a model also becomes important for technical use.

8.7.3 Loudness Calculation and Loudness Meters

Critical-band filters are rarely available, whereas third-octave band filters are often used. A loudness calculating procedure was therefore elaborated to make use of the third-octave band levels produced by a sound. The graphical procedure was also implemented in a computer program published in DIN 45 631 using third-octave band levels, by transferring them into values corresponding approximately to specific loudness, adding the slope towards higher critical-band rates and summing up the area below the distribution of specific loudness. This way, the total loudness can be calculated from the third-octave band levels produced by the sound in question. The approximation of the ear's frequency selectivity by third-octave filter bands led to two compromises in order to be as accurate as possible. Firstly, the differences in bandwidth were transferred into critical-band rate distances that were not 1 Bark. Secondly, the additional specific loudness produced by the cut-off slope in the abutting filter towards lower frequencies is taken into account by changing the exponent of specific loudness slightly from 0.23 to 0.25. Because the measurement of all the third-octave band levels needs a certain time, the graphical procedure and that using the computer program are meant only for relatively steady-state sounds. However, the loudness of sounds of any kind of spectral distribution are thus very precisely calculated.

An example is given in Fig. 8.24. Even though the abscissa gives cut-off or centre frequencies, it is scaled according to critical-band rate. In order to simplify the procedure for ordinate values, only the main specific loudness and the upper slope of the specific loudness are used in the procedure; the specific loudness towards lower critical-band rates is ignored. Therefore, the shape of the specific loudness, which is produced by a tone or by a narrow-band noise, is approximated by three parts. It starts with a vertical rise up to the third-octave band level measured, comes to a main value corresponding to the third-octave band level in question and rises again vertically for the next higher third-octave band, if its level is larger than that of its neighbouring lower third-octave band. If the next higher third-octave band level is lower, the decrease towards higher centre frequencies follows the broken lines, corresponding to the upper slope of specific loudness. This way, the final specific

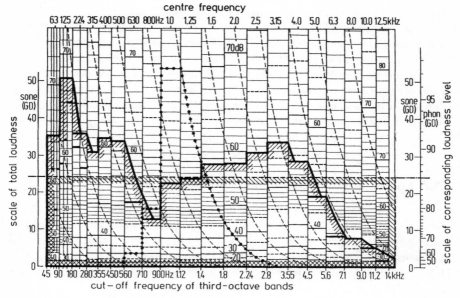

Fig. 8.24. Example of the loudness calculation procedure using the charts indicating the measured third-octave band levels of a factory noise. Specific loudness is on the ordinate while the critical-band rate expressed in cut-off frequencies of the third-octave bands is on the abscissa. The area surrounded by the thick solid line and hatched from lower left to upper right indicates the total loudness of the noise. This area is approximated by a rectangular area of the same basis but of a height indicated by the area hatched from upper left to lower right. The height of this rectangular area marks the total loudness on the left scale and the corresponding loudness level on the right scale

loudness versus critical-band rate pattern, shortened to "loudness pattern", is determined and indicated by the highest thick solid lines in Fig. 8.24.

The upper slope of specific loudness corresponds closely to the transformed upper excitation slopes, an example of which is shown in Figs. 8.16 and 8.19. For narrow-band sounds, this upper slope contributes strongly to the total loudness, i.e. to the total area below the curve. Therefore, it contributes especially to the total loudness of pure tones. An example is given in Fig. 8.24 by the dotted line for a 70-dB, 1-kHz tone. Generally used third-octave band filters show a leakage towards neighbouring filters of about −20 dB. This means that a 70 dB, 1-kHz tone produces the following levels at different centre frequencies: 10 dB at 500 Hz, 30 dB at 630 Hz, 50 dB at 800 Hz and 70 dB at 1 kHz. Therefore, in agreement with the model of loudness, the lower slope of the loudness pattern becomes less steep.

This loudness calculation procedure is part of an international standard (ISO 532 B). Its use may be illustrated in detail by a noise produced in a wood factory. The unweighted sound level of the noise produced by a machine is 73 dB, the A-weighted sound pressure level is measured to 68 dB(A). All the thin lines plotted in Fig. 8.24 belong to the normalized patterns of the loudness calculation procedure that are available for five different maximal levels of third-octave-bands (35, 50, 70, 90, and 110 dB) and for two kinds of sound field (free and diffuse). The optimal choice for using such a graph is to select that which just contains the largest third-octave-band level measured in a spectrum. In factory halls, a diffuse field may be a better approximation than a free field. Very close to the sound source, however, or in an area with very little reflection, a free-field condition may be assumed.

In order to specify the calculated loudnesses, the index "G" is used as a hint for the calculation procedure, while the indices "F" or "D" are hints for the free and diffuse sound field, respectively. The graphs show, at the lower abscissa, the cut-off frequencies of the third-octave band filters, while the upper abscissa gives their centre frequencies. The measured third-octave-band levels of the machinery noise are given in the diagram as thick horizontal lines, which can be drawn by using the steps indicating the third-octave band level in the third-octave band filter in question. At the left end of the horizontal bar, a vertical line downwards is drawn. At the right end of the horizontal thick bar, if the next band level is lower, a downward slope is added which runs in parallel to the curves, which are shown in the diagram as broken lines. This way, an area is formed extending from low to high frequencies. It is bordered by the straight line upwards at the left and right sides of the whole diagram, and also by the horizontal lower abscissa. The area within these boundaries is marked by hatching. To calculate the area quantitatively, a rectangular area of equal surface is drawn, which has the width of the diagram as a basis. The height of this rectangle is a measure of the total area, which is marked by shadowing from upper left to lower right. Using its height (the dashed-dotted line), the loudness or the loudness level can be read from the scales on the right or the left of the diagram. In the case outlined in Fig. 8.24, a calculated loudness N_{GD} of 24 sone (GD) and a corresponding loudness level L_{NGD} of 86 phon (GD) is found. The factory noise has a relatively broad spectrum, therefore quite a difference appears between the measured sound level of 73 dB or the A-weighted sound pressure level of 68 dB(A) on one hand and the calculated loudness level of 86 phon (GD) on the other.

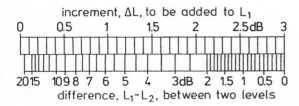

Fig. 8.25. Nomogram for a quick calculation of the increment in level to be added to the larger level as a function of the difference between the levels to be added

The approximation of critical bands by third-octave bands is only acceptable for frequencies above about 300 Hz. For lower frequencies, third-octave bands are too small in relation to the critical bands, so two or three third-octave bands have to be added in order to approximate critical bands. This is the case for the two third-octave bands between 180 and 280 Hz and for the three third-octave bands between 90 and 180 Hz and between 45 and 90 Hz. In these cases, the critical-band level has to be approximated by adding up the sound intensities falling in the given third-octave bands. Without recalculating sound intensities from the sound pressure levels, the addition of these intensities is possible by simply using a nomogram shown in Fig. 8.25. It indicates the sound level differences between the two levels, the intensities of which have to be added, and is given as $L_1 - L_2$, where L_1 is the larger level. The upper scale indicates the level increment which has to be added to the level L_1 in order to get the new total sound pressure level. This new SPL corresponds to the sum of the two former sound intensities. For equal sound pressure level $L_1 = L_2$, we reach an increment $\Delta L = 3$ dB; for a difference $L_1 - L_2 = 2$ dB, we get ΔL of about 2 dB; for $L_1 - L_2 = 6$ dB, the increment is only 1 dB. For differences $L_1 - L_2$ larger than 10 dB, the influence of L_2 can be ignored in most of the cases. In order to add a third or a fourth level, this method can be repeated as often as necessary. The thick horizontal lines shown in Fig. 8.24 with the large width of about a critical band at low centre frequencies of the third-octave bands, were produced in this way from the smaller horizontal thin bars which correspond to the measured third-octave band levels. For precise measurements, especially in cases where the components at very low frequencies are dominant, the procedure can gain accuracy by using a weighting of the lower frequency components corresponding to the equal loudness contours. Table 8.1 gives the data for such a procedure.

Table 8.1. Weighting of third-octave band levels, L_T, for centre frequencies, f_T, below 250 Hz

	f_T	25	32	40	50	63	80	100	125	160	200	250 Hz
range		deductions ΔL in dB										
I	$L_T + \Delta L \leq$ 45 dB	−32	−24	−16	−10	−5	0	−7	−3	0	−2	0
II	$\leq$ 55 dB	−29	−22	−15	−10	−4	0	−7	−2	0	−2	0
III	$\leq$ 65 dB	−27	−19	−14	−9	−4	0	−6	−2	0	−2	0
IV	$\leq$ 71 dB	−25	−17	−12	−9	−3	0	−5	−2	0	−2	0
V	$\leq$ 80 dB	−23	−16	−11	−7	−3	0	−4	−1	0	−1	0
VI	$\leq$ 90 dB	−20	−14	−10	−6	−3	0	−4	−1	0	−1	0
VII	$\leq$ 100 dB	−18	−12	−9	−6	−2	0	−3	−1	0	−1	0
VIII	$\leq$ 120 dB	−15	−10	−8	−4	−2	0	−3	−1	0	−1	0

The graphical procedure which finally leads to a loudness pattern has the advantage that partial loudnesses correspond to partial areas in the diagram. Therefore, in many cases the diagram clearly shows which partial area is

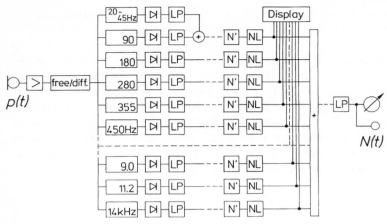

Fig. 8.26. Block diagram of a loudness meter

dominant or which part contributes strongly to the total loudness. In noise reduction, it is often very important to reduce firstly that part of the noise which produces the largest area in the loudness pattern. On the other hand, the diagram shows which parts of the spectrum are so small in relation to the neighbouring parts that they are partially or even totally masked. In Fig. 8.24, for example, the third-octave band level of 51 dB at the centre frequency of 630 Hz does not contribute to loudness because it is totally masked, as indicated by the fact that this third-octave band level lies below the shaded curve, limiting the total area and arising, at this frequency, from the third-octave band level at 500 Hz.

Using a filterbank of third-octave-band filters and electronic equipment which simulates the behaviour of the human hearing system, a loudness meter can be constructed. The advantage of such a loudness meter in comparison to the graphical procedure is that both spectral effects and temporal effects can be simulated. The block diagram of such a meter is shown in Fig. 8.26. It corresponds closely to the graphical procedure outlined in Fig. 8.24, with regard to the spectral effects. The sound pressure time function $p(t)$ is picked up by the microphone, fed to an amplifier and to a filter appropriate for free versus diffuse sound field. Then follows a filterbank, a rectifier and a low-pass with 2 ms time constant producing the temporal envelope of the filter outputs. In the next section N' specific loudness is produced according to data displayed in Fig. 8.17. With great simplification it can be said that in this section the fourth root of sound intensity or the square root of sound pressure is calculated. In the section NL, the nonlinear temporal decay of specific loudness as illustrated in the middle panel of Fig. 8.20 is realized. After longer signals, a more gradual decay of specific loudness occurs than after very short signals. In the following section the plus sign indicates the spectral integration taking into account effects of nonlinear spectral spread, i.e. flatter upper slopes of the loudness pattern for louder sounds. The following section LP comprises a third-order low-pass, the effects of which are illustrated by Fig. 8.20c.

Loudness meters as illustrated in Fig. 8.26 were realized as analog as well as digital instruments. The loudness patterns indicated are helpful guides for efficient noise reduction, since the area under the pattern is a *direct* measure of perceived loudness. In addition, the loudness-time functions represent valuable tools for the evaluation of the loudness of music, speech, or noise. In the meantime several companies manufacturing acoustic measurement instruments also provide loudness meters. Detailed examples for practical applications of loudness meters are given in Chap. 16.

9. Sharpness and Sensory Pleasantness

Previously, there has been a tendency to transfer everything in steady-state sounds not related to the sensations of loudness or pitch, to a residual basket of sensations called timbre. Using this definition of timbre, it is necessary to extract from the mixture of sensations those that may be important. The sensation of "sharpness", which may be related to what is called "density", seems to be one of these. Closely related to sharpness, however inversely, is a sensation called sensory pleasantness. This sensation, however, also depends on other sensations such as roughness, loudness, and tonalness.

9.1 Dependencies of Sharpness

Sharpness is a sensation which we can consider separately, and it is possible, for example, to compare the sharpness of one sound with the sharpness of another. Sharpness can also be doubled or halved if variables are available that really change sharpness. The variability of sharpness judgements is comparable to that of loudness judgements. One of the important variables influencing the sensation of sharpness is the spectral envelope of the sound. Many comparisons have indicated that the spectral fine structure is relatively unimportant in sharpness. A noise producing a continuous spectrum, for example, has the same sharpness as a sound composed of many lines if the spectral envelopes measured in critical-band levels are the same.

Sharpness increases for a level increment from 30 to 90 dB by a factor of two. This means that the dependence on level can be ignored as a first approximation, especially if the level differences are not very large. Another small effect is the dependence on bandwidth, as long as the bandwidth is smaller than a critical band. No difference in sharpness can be detected whether one tone or five tones fall within one critical band or even when a critical-band noise is used for comparison.

The most important parameters influencing sharpness are the spectral content and the centre frequency of narrow-band sounds. In order to give quantitative values, a reference point and a unit have to be defined. In Latin, the expression "acum" is used for sharp. The reference sound producing 1 acum is a narrow-band noise one critical-band wide at a centre frequency of 1 kHz having a level of 60 dB. This reference point is marked in Fig. 9.1

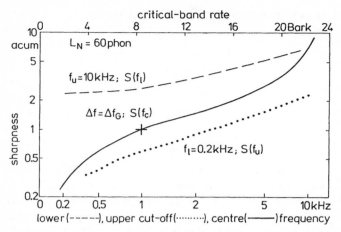

Fig. 9.1. Sharpness of critical-band wide narrow-band noise as a function of centre frequency (*solid*), of band-pass noise with an upper cut-off frequency of 10 kHz as a function of the lower cut-off frequency (*broken*) and of band-pass noise with a lower cut-off frequency of 0.2 kHz as a function of the upper cut-off frequency (*dotted*). The critical-band rate corresponding to the centre frequency or to the cut-off frequencies is given on the upper abscissa. The cross marks the standard sound producing a sharpness of 1 acum. Loudness level of all noises is 60 phon

by a cross. The solid line in this figure gives the sharpness of critical-band wide noises as a function of their centre frequency shown on the abscissa. It should be noted that the abscissa has a nonlinear scale that corresponds to the critical-band rate given as a linear scale on the upper abscissa. Sharpness, the ordinate, is given on a logarithmic scale. The solid curve may be considered as a stimulus versus sensation function for sharpness. For narrow-band noises, sharpness increases with increasing centre frequency. At low centre frequencies, sharpness increases almost in proportion to critical-band rate. This may be seen because the reference point, which belongs to a critical-band rate of 8.5 Bark, gives a sharpness of 1 acum, which is about four times larger than that produced by a narrow-band noise centred at 200 Hz, corresponding to 2 Bark and leading to about 0.25 acum. For a critical-band rate of 16 Bark corresponding to a centre frequency of about 3 kHz, a sharpness is reached that is about twice as large as that for 1-kHz centre frequency. This means that the proportionality of sharpness and critical-band rate lasts at least up to about 3 kHz. For higher frequencies, however, where critical-band rate does not increase much more because of its limit at 24 Bark, sharpness increases faster than the critical-band rate of the centre frequency of the narrow-band noise. This effect seems to be the reason that very high-frequency sounds produce a sensation that is dominated by their sharpness. From very low frequencies near 200 Hz to high frequencies near 10 kHz, sharpness increases by a factor of 50.

Bandwidth is the other variable that strongly influences sharpness. To limit the many ways this effect can be measured, two fixed values and two variables are used. One fixed value is the lower cut-off frequency of the noise which was kept constant at 200 Hz. The variable in this case is the upper cut-off frequency that is used as the abscissa. Loudness is kept constant, changing the upper cut-off frequency, and has a value that corresponds to that produced by a 60-dB critical-band wide noise at 1 kHz. Increasing the upper cut-off frequency from about 300 Hz increases sharpness continuously and, when plotted on the scales of Fig. 9.1, along an almost straight line between 0.3 and 2.5 acum. The dotted line shows that dependence. In the other manipulation, the fixed value is a high cut-off frequency of 10 kHz and the variable is the lower cut-off frequency. The dashed line shown in Fig. 9.1 corresponds to the data measured when lower cut-off frequency (or the corresponding critical-band rate) is used as the abscissa. To discuss the dependence on this variable, we start with a narrow-band noise near 10 kHz and shift the lower cut-off frequency so that, finally, a broad-band noise from 200 Hz to 10 kHz is produced. Decreasing the lower cut-off frequency of this noise band decreases sharpness. The decrement reaches a factor of about 2.5 when the lower cut-off frequency is decreased to 1 kHz. For further reduction of the cut-off frequency, sharpness remains almost constant, provided the overall loudness is kept constant.

The data shown as a function of the lower or upper cut-off frequency may not look meaningful but they contain important effects that are useful in developing models of sharpness. The sharpness of a narrow-band noise at 1 kHz to which more noise is added at higher frequencies to produce a spectral width from 1 to 10 kHz, increases from 1 acum to about 2.5 acum. This effect might be expected because adding more noise might increase the sharpness. However, if the upper cut-off frequency at 1 kHz is kept constant and the noise is widened downward to produce a band between 200 Hz and 1 kHz, sharpness decreases. This means that sharpness can be decreased by adding sound at lower frequencies. This is an effect that – at least at first glance – may be unexpected.

9.2 Model of Sharpness

It is helpful in developing a model of sharpness to treat sharpness as being independent of the fine structure of the spectrum; the overall spectral envelope is the main factor influencing sharpness. In Chaps. 6 and 8, it was shown that the spectral envelope is psychoacoustically represented in the excitation level versus critical-band rate pattern, or in the specific loudness versus critical-band rate pattern. For narrow-band noises, the solid curve in Fig. 9.1 showed that sharpness increases with critical-band rate for centre frequencies below about 3 kHz (16 Bark). The increment at higher frequencies can be taken into account by using a factor that is unity for all critical-band rates below 16 Bark but increases strongly for higher critical-band rates. Keeping in mind that sharpness decreases if lower cut-off frequency is reduced (see dotted curve in Fig. 9.1), we may assume that a kind of moment is responsible for the development of sharpness. Taking into account the factor, g, which increases above 16 Bark to values larger than unity, using the specific loudness versus critical-band rate pattern as the distribution in question, and using the boundary condition that a narrow-band noise centred at 1 kHz has to produce a sharpness of 1 acum, an equation of the following form can be given:

$$S = 0.11 \frac{\int_0^{24\,\mathrm{Bark}} N' g(z) z \, dz}{\int_0^{24\,\mathrm{Bark}} N' \, dz} \text{ acum .} \tag{9.1}$$

In this equation, S is the sharpness to be calculated and the denominator gives the total loudness N, which has already been discussed in Chap. 8. The upper integral is like the first moment of specific loudness over critical-band rate, but uses an additional factor, $g(z)$, that is critical-band-rate dependent. This factor is shown in Fig. 9.2 as a function of critical-band rate. Only for critical-band rates larger than 16 Bark does the factor increase from unity to a value of four at the end of the critical-band rate near 24 Bark. This takes into account that sharpness of narrow-band noises increases unexpectedly strongly at high centre frequencies. Equation (9.1) is only a weighted first moment of the critical-band rate distribution of specific loudness.

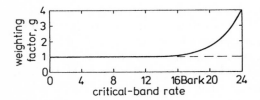

Fig. 9.2. Weighting factor for sharpness as a function of critical-band rate

Many psychoacoustically measured sharpnesses of different sounds have been compared with calculated sharpnesses, using the above formula. The agreement of the two is good in view of the relative simplicity of the equation. Three examples of the use of this model may be discussed. Figure 9.3 shows on the left the distribution of the critical-band level, L_G, as a function of critical-band rate for three different sounds: a 1-kHz tone (solid), uniform exciting noise (hatched from lower left to upper right) and a high-pass noise above 3 kHz (hatched from upper left to lower right). On the right of the figure, the corresponding weighted specific loudnesses as a function of critical-band rate are shown together with the location of their first moment (centre of gravity). The arrows show that a 1-kHz tone produces a much smaller sharpness in comparison with that produced by a high-pass noise with a cut-off frequency of 3 kHz. When the lower cut-off frequency is shifted towards lower values

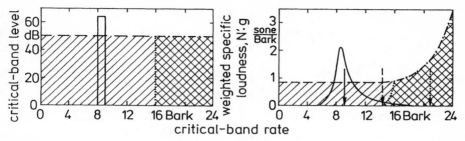

Fig. 9.3. Calculation of sharpness for narrow-band noise centred at 1 kHz (*solid*), uniform exciting noise (*broken* and *hatched* from lower left to upper right) and high-pass noise (*dotted* and *hatched* from upper left to lower right). The left-hand drawing indicates critical-band levels as a function of critical-band rate, while the right-hand drawing indicates the weighted specific loudness, again as a function of critical-band rate. The calculated sharpnesses are indicated by the three vertical arrows

and the noise is changed into a uniform exciting noise, sharpness decreases markedly in agreement with psychoacoustical results. It remains, however, clearly above that of the 1-kHz tone.

9.3 Dependencies of Sensory Pleasantness

Sensory pleasantness is a more complex sensation that is influenced by elementary auditory sensations such as roughness, sharpness, tonality, and loudness. Because of these influences, which make it almost impossible to extract sensory pleasantness as a single elementary sensation, it is necessary to measure the dependence of this sensation in relative values, using the techniques of magnitude estimation with an anchor.

The dependence of sensory pleasantness on sharpness was measured using sinusoidal tones, narrow-band noise of 30-Hz bandwidth and band-pass noise with a 1-kHz bandwidth as a function of centre frequency. Relative sharpness and relative sensory pleasantness were determined psychoacoustically in separate sessions. The data show some scatter. The relationships between sensory pleasantness and other sensations are given as curves in Fig. 9.4. Figure 9.4a shows pleasantness against relative roughness and Fig. 9.4b pleasantness against sharpness for the three sounds. It becomes clear that sensory pleasantness decreases with increasing sharpness. Pure tones already show the largest sensory pleasantness, while the band-pass noise seems to be a sound with low sensory pleasantness.

Similar to the dependence on sharpness, sensory pleasantness depends on roughness, a hearing sensation described in Chap. 11, although the dependence is not as strong. Because the data are again measured using the

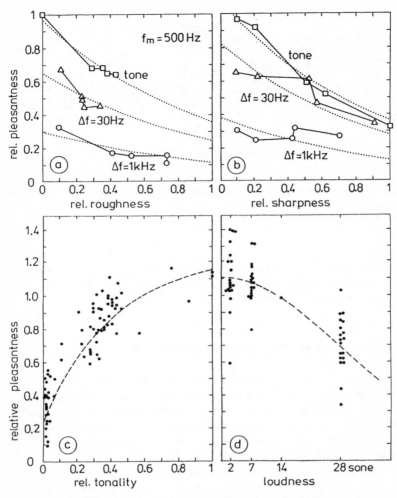

Fig. 9.4a–d. Relative pleasantness as a function of relative roughness with bandwidth as the parameter in (**a**); as a function of relative sharpness in (**b**); as a function of relative tonality in (**c**) and as a function of loudness in (**d**)

method of magnitude estimation with an anchor, only relative values can be given.

The relationship between sensory pleasantness and tonality, i.e. a feature distinguishing noise versus tone quality of sounds, is indicated in Fig. 9.4c. This dependence indicates that sensory pleasantness increases with tonality. Small tonality means small sensory pleasantness. For relative tonality larger than about 0.4, sensory pleasantness does not increase much.

The dependences of sensory pleasantness described so far have been determined using a constant loudness of 14 sone. Sensory pleasantness, however, depends also on loudness in such a way that up to about 20 sone, there is little influence. For values larger than 20 sone, sensory pleasantness decreases. This dependence of sensory pleasantness on loudness cannot be seen in isolation because roughness and sharpness also depend on loudness. If this influence is eliminated, then the relationship between sensory pleasantness and loudness, as given in Fig. 9.4d, remains.

In summary, it can be seen that sensory pleasantness depends mostly on sharpness. The influence of roughness is somewhat smaller and is similar to that of tonality. Loudness, however, influences sensory pleasantness only for values that are larger than the normal loudness of communication between two people in quiet.

9.4 Model of Sensory Pleasantness

Whether a sound is accepted as sounding pleasant or unpleasant depends not only on the physical parameters of the sound but also on the subjective relationship of the listener to the sound. These nonacoustic influences cannot be anticipated and have therefore to be ignored and eliminated if possible. This was done in the experiments, so that a model of sensory pleasantness can only be developed relating the characteristics of the human hearing system on one hand and the physical parameters of the sound on the other.

The relationship between relative values of sensory pleasantness and those of sensation sharpness, roughness, tonality, and loudness can be approximated using equations. A model for the calculation of sharpness has already been given. The sensation of roughness can also be calculated (see Chap. 11). A procedure for calculating the sensation of tonality does not yet exist so that it may be subjectively estimated as a first approximation, while loudness can be calculated relatively precisely. Therefore, it is possible to put the dependencies given in Fig. 9.4 into an equation that contains relative values of sharpness, roughness, tonality, and loudness. The result leads also to a relative value, P/P_0, of sensory pleasantness. The equation, based on relative values of sharpness S, roughness R, tonality T, and loudness N, reads:

$$\frac{P}{P_0} = e^{-0.7R/R_0} e^{-1.08S/S_0} \left(1.24 - e^{-2.43T/T_0}\right) e^{-(0.023N/N_0)^2} \tag{9.2}$$

Using this equation, it is possible to calculate the sensory pleasantness of any sound, if the sharpness, roughness, and loudness are calculated using the procedures given in Sects. 9.2, 11.2, and 8.7, respectively. Tonality has to be judged subjectively. It has been shown that tonality depends neither on the critical-band rate nor on the loudness. Relative tonality, however, depends on the bandwidth expressed in critical-band-rate spread, such that it decreases with increasing critical-band-rate spread starting from a sinusoidal tone producing relative tonality of unity to about 0.6, 0.3, 0.2, and 0.1 for critical-band-rate spreads of 0.1, 0.2, 0.57, and 1.5 Bark, respectively. Using these values it is possible to estimate the relative sensory pleasantness of different sounds. Results calculated according to (9.2) are indicated in Fig. 9.4 by dotted or dashed lines.

10. Fluctuation Strength

In this chapter, the fluctuation strength of amplitude-modulated broad-band noise, amplitude-modulated pure tones and frequency-modulated pure tones is addressed, and the dependence of fluctuation strength on modulation frequency, sound pressure level, modulation depth, centre frequency, and frequency deviation is assessed. In addition, the fluctuation strength of modulated sounds is compared to the fluctuation strength of narrow-band noises. Finally, a model of fluctuation strength based on the temporal variation of the masking pattern or loudness pattern is proposed.

10.1 Dependencies of Fluctuation Strength

Modulated sounds elicit two different kinds of hearing sensations: at low modulation frequencies up to a modulation frequency of about 20 Hz, the hearing sensation of fluctuation strength is produced. At higher modulation frequencies, the hearing sensation of roughness, discussed in detail in Chap. 11, occurs. For modulation frequencies around 20 Hz, there is a transition between the hearing sensation of fluctuation strength and that of roughness. It is a smooth transition rather than a strong border that exists between the two sensations.

Figure 10.1 shows the dependence of fluctuation strength on modulation frequency. The different panels represent the data for amplitude-modulated broad-band noise (AM BBN), amplitude-modulated pure tone (AM SIN) and frequency-modulated pure tone (FM SIN). In each panel, the fluctuation strength was normalized to the maximum value for that sound (left-hand ordinate). Because fluctuation strength is a sensation which one considers separately from other sensations, both absolute and relative values are useful. A fixed point is therefore defined for a 60-dB, 1-kHz tone 100% amplitude-modulated at 4 Hz , as producing 1 vacil (from vacilare in Latin, or vacillate in English). Using the data of fluctuation strength shown in Fig. 10.7, it is possible to give the absolute values as indicated in the right-hand ordinate scales.

All three panels of Fig. 10.1 clearly show that fluctuation strength shows a band-pass characteristic as a function of modulation frequency, with a maximum around 4 Hz. This means that sounds with a 4-Hz modulation frequency

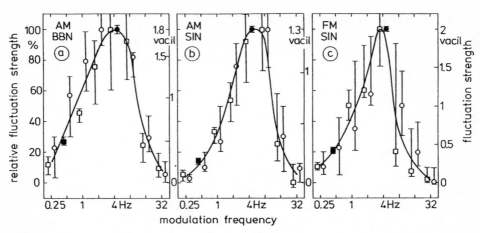

Fig. 10.1a–c. Fluctuation strength of three modulated sounds as a function of modulation frequency. (**a**) Amplitude-modulated broad-band noise of 60-dB SPL and 40-dB modulation depth; (**b**) amplitude-modulated 1-kHz tone of 70-dB SPL and 40-dB modulation depth; (**c**) frequency-modulated pure tone of 70-dB SPL, 1500-Hz centre frequency and ± 700-Hz frequency deviation

elicit large fluctuation strength, whether amplitude modulation or frequency modulation is used or whether broad-band or narrow-band sounds are modulated. The maximum fluctuation strength for a modulation frequency of about 4 Hz finds its counterpart in the variation of the temporal envelope of fluent speech: at normal speaking rate, 4 syllables/second are usually produced, leading to a variation of the temporal envelope at a frequency of 4 Hz. This may be seen as an indication of the excellent correlation between speech and hearing system.

The dependence of fluctuation strength on sound pressure level is displayed in Fig. 10.2. Again, the three panels show the results for amplitude-modulated broad-band noise, amplitude-modulated pure tone and frequency-modulated pure tone. In each panel, the fluctuation strength is normalized with respect to the corresponding maximum value on the left ordinate scales, and given in absolute values on the right. With increasing sound pressure level and for all sounds considered, the fluctuation strength increases. For amplitude-modulated sounds (Fig. 10.2a and b), the increase is somewhat more prominent than for the frequency-modulated pure tone (Fig. 10.2c). With an increase of 40 dB in sound pressure level, fluctuation strength of modulated sounds increases on average by a factor of about 2.5 (1.7 to 3).

Figure 10.3 shows the dependence of the fluctuation strength of amplitude-modulated broad-band noise and amplitude-modulated pure tones on both the modulation depth and the modulation factor. In each panel, the fluctuation strength is normalized with respect to the corresponding maximum value on the left ordinate scales and given in absolute values on the right ordinate

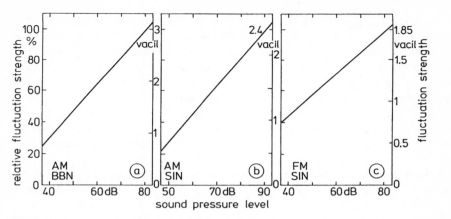

Fig. 10.2a–c. Fluctuation strength of modulated sounds as a function of sound pressure level. Stimulus parameters are the same as in Fig. 10.1, but the modulation frequency is 4 Hz

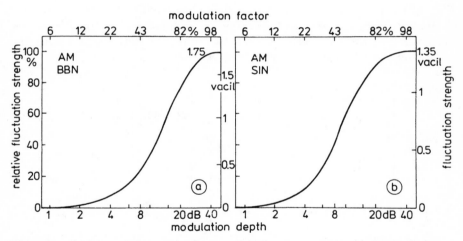

Fig. 10.3a,b. Fluctuation strength of two amplitude-modulated sounds as a function of modulation depth (or modulation factor). (**a**) Amplitude-modulated broadband noise of 60-dB SPL and 4-Hz modulation frequency; (**b**) amplitude-modulated 1-kHz tone of 70-dB SPL and 4-Hz modulation frequency

scales. According to the results displayed in Fig. 10.3, fluctuation strength is zero until a modulation depth of about 3 dB, after which it increases approximately linearly with the logarithm of modulation depth. To produce the maximum fluctuation strength of either sound, a modulation depth of at least 30 dB (modulation factor 94%) is necessary. Above that modulation depth, fluctuation strength remains constant at its maximal value.

Figure 10.4 shows the dependence of the fluctuation strength produced by modulated pure tones on centre frequency. In the left panel, results for amplitude-modulated pure tones are shown; the right panel indicates data for frequency-modulated pure tones. In each panel, the data are normalized relative to the respective maximal median fluctuation strength, but are also given in absolute values. The data displayed in Fig. 10.4a suggest that the fluctuation strength of amplitude-modulated pure tones depends very little on their centre frequency; despite the large interquartile ranges, the medians indicate the tendency for amplitude-modulated tones at very low (125 Hz) and very high (8 kHz) frequencies to produce less fluctuation strength than AM tones at medium frequencies. However, the results shown in Fig. 10.4b for FM tones indicate a clear dependence of fluctuation strength on centre frequency so that, although the fluctuation strength of FM tones is almost constant up to about 1 kHz, it decreases approximately linearly with the logarithm of the centre frequency towards higher frequencies.

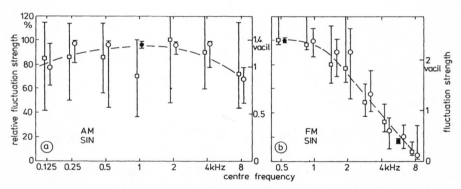

Fig. 10.4a,b. Fluctuation strength of modulated tones as a function of frequency. (a) Amplitude-modulated pure tone of 70-dB SPL, 4-Hz modulation frequency and 40-dB modulation depth; (b) frequency-modulated pure tone of 70-dB SPL, 4-Hz modulation frequency, and ± 200-Hz frequency deviation

This decrease can be understood in terms of the number of critical bands encompassed at different centre frequencies by FM tones with a constant frequency deviation of 200 Hz. As an example, the FM tone at 0.5 kHz sweeps between 300 and 700 Hz, i.e. between critical-band rates of 3 and 6.5 Bark, respectively. At the 8-kHz centre frequency, the modulation occurs between frequencies of 7.8 and 8.2 kHz, corresponding to 21.1 and 21.3 Bark, respec-

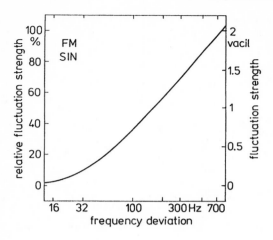

Fig. 10.5. Fluctuation strength of a frequency-modulated tone as a function of frequency deviation. The sound pressure level is 70 dB, the centre frequency 1500 Hz and the modulation frequency 4 Hz

tively. This means that at a centre frequency of 0.5 kHz, the FM tone varies over a critical-band interval of 3.5 Bark, whereas at 8 kHz it varies only over 0.2 Bark. Hence, the critical-band interval at 8 kHz is a factor of 17.5 smaller than the critical-band interval at 0.5 kHz. Regarding Fig. 10.4b, it can be seen that this factor of 17.5 is also found for the difference in the relative fluctuation strength at 0.5 and 8 kHz. This result can be taken as an indication that fluctuation strength of modulated sounds can be described on the basis of the corresponding excitation patterns.

Figure 10.5 shows the dependence of fluctuation strength on frequency deviation of an FM tone at a centre frequency of 1.5 kHz. Fluctuation strength is initially perceived at a frequency deviation of about 20 Hz and increases approximately linearly with the logarithm of frequency deviation. This result applies for an FM tone at 1500 Hz with 70-dB SPL and 4-Hz modulation frequency. For such a tone, the JND for frequency modulation corresponds to about $2\Delta f = 8$ Hz (see Fig. 7.8). Significant values of fluctuation strength (say 10% relative fluctuation strength) are achieved for frequency deviations larger than about 10 times the magnitude of the JNDFM at 4 Hz. This rule seems to apply also for AM sounds: the modulation depth at which about 10% relative fluctuation strength is reached (see Fig. 10.3), is about 10 times larger than the JNDAM at 4 Hz of about 0.4 dB for a 70-dB AM tone, and about 0.7 dB for the AM broad-band noise as displayed in Fig. 7.1.

Not only modulated sounds can elicit the hearing sensation fluctuation strength but also unmodulated narrow noise bands.

Figure 10.6 shows the dependence of relative fluctuation strength of narrow-band noise as a function of its bandwidth as well as effective modulation frequency. According to (1.6) this frequency can be calculated as follows: $f^*_{\mathrm{mod}} = 0.64 \cdot \Delta f$.

The comparison of the data displayed in Figs. 10.1 and 10.6 reveals that – irrespective of periodic or stochastic sound fluctuations – fluctuation strength

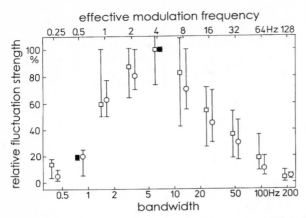

Fig. 10.6. Fluctuation strength of narrow-band noise as a function of its bandwidth. Center frequency 1 kHz, level 70 dB

shows a bandpass characteristic with a maximum around 4 Hz (effective) modulation frequency.

For unmodulated narrow-band noise, fluctuation strength increases with level and shows large values for center frequencies around 1 kHz similar to the data displayed for modulated sounds in Figs. 10.2 and 10.4a.

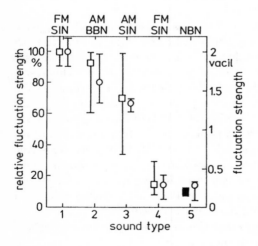

Fig. 10.7. Fluctuation strength of the sounds 1–5 as described in Table 10.1

Figure 10.7 enables a comparison of the fluctuation strength of five different sounds whose characteristics are listed in Table 10.1. The largest fluctuation strength is produced by the 70-dB tone with large frequency deviation. Some 10% less fluctuation strength is elicited by a 60-dB, amplitude-modulated broad-band noise. A 70-dB, amplitude-modulated 2-kHz tone pro-

Table 10.1. Physical data of sounds 1–5

Sound	1	2	3	4	5
Abbreviation	FM SIN	AM BBN	AM SIN	FM SIN	NBN
Frequency [Hz]	1500	–	2000	1500	1000
Level [dB]	70	60	70	70	70
Modulation frequency [Hz]	4	4	4	4	–
Modulation depth [dB]	–	40	40	–	–
Frequency deviation [Hz]	700	–	–	32	–
Bandwidth [Hz]	–	16000	–	–	10

duces a fluctuation strength about 30% down from that of the FM tone. Sound 4, a 70-dB FM tone with small frequency deviation, produces only about 1/10 of the fluctuation strength of sound 1. This result is expected on the basis of the data displayed in Fig. 10.5. Sound 5 represents a narrow-band noise with a bandwidth of 10 Hz. The fluctuation strength elicited by this narrow-band noise can be estimated as follows: as a first approximation, the narrow-band noise can be regarded as an AM tone at 1 kHz with 6.4 Hz modulation frequency (see Sect. 1.1). If an effective modulation factor of 40% for narrow-band noise is assumed, then according to the results displayed in Fig. 10.3, the fluctuation strength of narrow-band noise should be a factor of about 2.5 smaller than the fluctuation strength of AM tones with a 98% modulation factor. A comparison of the relative fluctuation strength of sound 3 and sound 5 in Fig. 10.7, however, reveals that the fluctuation strength of narrow-band noise is smaller by a factor of about 5 than the fluctuation strength of an AM tone. It is apparently the periodic fluctuation of AM tones, in contrast to the random amplitude fluctuations of the noise, that enhances the perceived fluctuation strength of the AM tone.

The large fluctuation strength of amplitude-modulated broad-band noise and frequency-modulated pure tones with large frequency deviation (sounds 2 and 1) can be related to the fact that excitation varies to a large extent along the critical-band rate scale. Therefore, it can be postulated that fluctuation strength is summed up across critical bands. This concept will be explained in more detail in the following section.

10.2 Model of Fluctuation Strength

A model of fluctuation strength based on the temporal variation of the masking pattern can be illustrated by Fig. 10.8, where the temporal masking

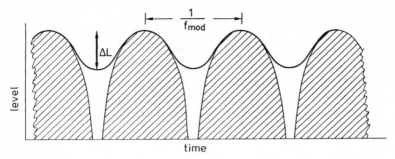

Fig. 10.8. Model of fluctuation strength: temporal masking pattern of sinusoidally amplitude-modulated masker leading to the temporal masking depth ΔL

pattern of a sinusoidally amplitude-modulated masker is schematically indicated by the thick solid line. The hatched areas indicate the envelope of a sinusoidally amplitude-modulated masker plotted in terms of sound pressure level. The interval between two successive maxima of the masker envelope corresponds to the reciprocal of the modulation frequency. The temporal variation of the temporal masking pattern can be described by the magnitude ΔL, which represents the level difference between the maxima and the minima in the temporal masking pattern. This so-called temporal masking depth, ΔL, should not be confused with the modulation depth, d, of the masker's envelope; the masking depth ΔL of the temporal masking pattern is smaller than the modulation depth d of the masker's envelope due to post-masking.

The equation

$$F \sim \frac{\Delta L}{(f_{\mathrm{mod}}/4\mathrm{Hz}) + (4\mathrm{Hz}/f_{\mathrm{mod}})} , \qquad (10.1)$$

shows the relationship between fluctuation strength, F, and the masking depth of the temporal masking pattern, ΔL, as well as the relationship between F and modulation frequency f_{mod}. The denominator clearly shows that a modulation frequency of 4 Hz plays an important part in the description of fluctuation strength: for faster modulation frequencies, the ear exhibits the integrative features evidenced in postmasking; for modulation frequencies lower than 4 Hz, effects of short-term memory become important.

For an amplitude-modulated broad-band noise, the magnitude of the masking depth of the temporal masking pattern, ΔL, is largely independent of frequency. For amplitude-modulated pure tones, some frequency dependence occurs because of the nonlinearity of the upper slope in the masking pattern. In addition to those factors, the magnitude of the masking depth shows a strong frequency dependence with frequency-modulated tones. This means that, for describing the fluctuation strength of AM tones and FM tones, the model can be modified as follows: instead of *one* masking depth of the temporal masking pattern ΔL, all of the magnitudes of ΔL occuring

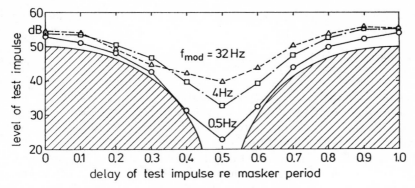

Fig. 10.9. Temporal masking pattern of 100% sinusoidally amplitude-modulated broad-band noise. *Hatched*: temporal envelope of masker; *parameter*: modulation frequency. All data plotted as a function of the temporal location within the period of the modulation of the masker

are integrated along the critical-band rate scale. A more detailed description of this procedure is given in Chap. 11, where the model for roughness is discussed.

A basic feature of the model of fluctuation strength, namely the masking depth of the temporal masking pattern, can be illustrated in Fig. 10.9. The temporal envelope of a sinusoidally amplitude-modulated broad-band noise is indicated by the hatched areas. The different curves represent the temporal masking patterns for the different modulation frequencies indicated. The maximum of the masking pattern shows up at a delay relative to the masker period of 0, the minimum at a delay of 0.5. The difference between maximum and minimum, called the temporal masking depth, clearly decreases with increasing modulation frequency. This means that as a function of modulation frequency, the temporal masking pattern shows a low-pass characteristic, whereas fluctuation strength shows a band-pass characteristic. Equation (10.1) explains how the low-pass characteristic of the temporal envelope is transformed into the band-pass characteristic of fluctuation strength. This bandpass characteristic describes the influence of modulation frequency on fluctuation strength. However, the value ΔL in the formula decreases approximately linearly with increasing frequency of modulation. Taking this into account, a relatively simple formula can be given for the fluctuation strength of sinusoidally amplitude-modulated broad-band noise:

$$F_{\mathrm{BBN}} = \frac{5.8(1.25m - 0.25)[0.05(L_{\mathrm{BBN}}/\mathrm{dB}) - 1]}{(f_{\mathrm{mod}}/5\mathrm{Hz})^2 + (4\mathrm{Hz}/f_{\mathrm{mod}}) + 1.5} \ \mathrm{vacil} \ , \tag{10.2}$$

where m is the modulation factor, L_{BBN} the level of the broad-band noise and f_{mod} the frequency of modulation. For amplitude- or freqency-modulated tones, fluctuation strength may be approximated by integrating the temporal masking depth, ΔL, along the critical-band rate. This leads to the following approximation:

$$F = \frac{0.008 \int_0^{24\text{Bark}}(\Delta L/\text{dB Bark})dz}{(f_{\text{mod}}/4\text{Hz}) + (4\text{Hz}/f_{\text{mod}})} \text{ vacil}, \tag{10.3}$$

where the masking depth, ΔL, may be picked up from the masking patterns described in Chap. 4 for the different critical-band rates of 1 Bark distance. The integral is then transformed into a sum of, at most, 24 terms along the whole range of the critical-band rate.

While for most sounds described in this chapter values of ΔL are available this is usually not the case for sounds typical for practical applications. Therefore, a computer program was developed which uses instead of the ΔL values the corresponding differences in specific loudness. Since the computer models of fluctuation strength and roughness are very similar some more detail is given in Chap. 11 and the related literature.

11. Roughness

Using a 100% amplitude-modulated 1-kHz tone and increasing the modulation frequency from low to high values, three different areas of sensation are traversed. At very low modulation frequencies the loudness changes slowly up and down. The sensation produced is that of fluctuation. This sensation reaches a maximum at modulation frequencies near 4 Hz and decreases for higher modulation frequencies. At about 15 Hz, another type of sensation, roughness, starts to increase. It reaches its maximum near modulation frequencies of 70 Hz and decreases at higher modulation frequencies. As roughness decreases, the sensation of hearing three separately audible tones increases. This sensation is small for modulation frequencies near 150 Hz; it increases strongly, however, for larger modulation frequencies. This behaviour indicates that roughness is created by the relatively quick changes produced by modulation frequencies in the region between about 15 to 300 Hz. There is no need for exact periodical modulation, but the spectrum of the modulating function has to be between 15 and 300 Hz in order to produce roughness. For this reason, most narrow-band noises sound rough even though there is no periodical change in envelope or frequency. Roughness is again a sensation which we can consider while ignoring other sensations.

11.1 Dependencies of Roughness

In order to describe roughness quantitatively, a reference value must be defined. In Latin, the word "asper" characterizes what we call "rough". To define the roughness of 1 asper, we have chosen the 60-dB, 1-kHz tone that is 100% modulated in amplitude at a modulation frequency of 70 Hz. Three parameters are important in determining roughness. For amplitude modulation, the important parameters are the degree of modulation and modulation frequency. For frequency modulation, it is the frequency modulation index and modulation frequency.

Figure 11.1 shows the roughness of a 1-kHz tone at a modulation frequency of 70 Hz, as a function of the degree of modulation. Values of the degree of modulation larger than 1 are not meaningful here. The data, indicated by the solid line, can be approximated by the dashed line. Because the ordinate and abscissa are given in logarithmic scales, the straight dashed line

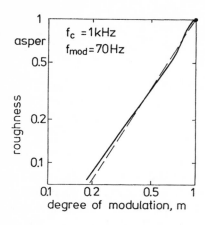

Fig. 11.1. Roughness as a function of the degree of modulation for a 1-kHz tone, amplitude-modulated at a frequency of 70 Hz. The dot in the right upper corner indicates the standard sound, which produces the roughness of 1 asper. The broken line indicates a useful linear approximation

represents a power law. The exponent is near 1.6 so that a roughness of only 0.1 asper is produced for a degree of modulation of 25%. This roughness is quite small and some subjects classify this as "no longer rough".

This dependence of roughness on the degree of modulation holds for tones of other centre frequencies also, although the modulation frequency at which the maximum roughness is reached depends on centre frequency. Figure 11.2 shows the dependence of roughness on modulation frequency at different centre frequencies for 100% modulation. This dependence has a band-pass characteristic. Roughness increases almost linearly from low modulation frequencies, in the double-logarithmic coordinates of Fig. 11.2, before it reaches a maximum. The maximum only depends on carrier frequency below 1 kHz where the maximum is shifted towards lower frequency of modulation with decreasing carrier frequency. The lower slope of this band-pass characteristic remains the same for frequencies of modulation below 1 kHz, even though the maximum decreases with decreasing centre frequency. For centre frequencies

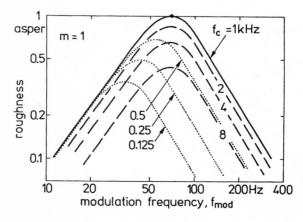

Fig. 11.2. Roughness of 100% amplitude-modulated tones of the given centre frequency as a function of the frequency of modulation

above 1 kHz, the height of the maximum is reduced although the frequency of modulation at which this maximum is reached remains unchanged. This means that above 1-kHz centre frequency there is a parallel shift downwards of the characteristic with increasing centre frequency.

The upper part of the band-pass characteristic can again be approximated by a straight line, with a relatively quick decay of roughness with increasing modulation frequency. It seems that the width of the critical band at lower centre frequencies plays an important role. At 250 Hz, the critical bandwidth is only 100 Hz. For a modulation frequency of 50 Hz, the two sidebands already have a separation of 100 Hz. For even higher modulation frequencies, the two sidebands fall into different critical bands. For centre frequencies above about 1 kHz, all the dependencies on frequency of modulation have the same shape. Here, the maximal roughness seems to be limited by the temporal resolution of our hearing system. Thus, two characteristics of the ear seem to influence the sensation of roughness: at low centre frequencies, it is frequency selectivity; at high centre frequencies, it is the limited temporal resolution.

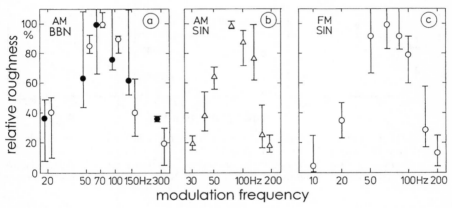

Fig. 11.3a–c. Roughness of three modulated sounds as a function of modulation frequency. (**a**) Amplitude modulated broadband noise of 60 dB SPL and 40 dB modulation depth; (**b**) amplitude modulated 1 kHz tone of 70 dB SPL and 40 dB modulation depth; (**c**) frequency modulated pure tone of 70 dB SPL, 1500 Hz center frequency, and ±700 Hz frequency deviation

Figure 11.3 shows the dependence of relative roughness on modulation frequency for amplitude modulated broadband noise, amplitude modulated pure tones, and frequency modulated pure tones. In line with the data presented in Fig. 11.2 the maximum of roughness occurs near a modulation frequency of 70 Hz irrespective of bandwidth or type of modulation. As with amplitude modulated pure tones also for amplitude modulated broadband noise or frequency modulated pure tones roughness vanishes for modulation frequencies above about 300 Hz.

Bands of noise often sound rough, although there is no additional amplitude modulation. This is because the envelope of the noise changes randomly. These changes become audible especially for bandwidths in the neighbourhood of 100 Hz, where the average rate of envelope change is 64 Hz (see Sect. 1.1). Therefore, roughness is particularly large at such bandwidths. For increasing bandwidth, the creation of roughness is limited by frequency selectivity. Nonetheless, at very high centre frequencies noises can be produced which are still within the critical band but have a bandwidth of 1 kHz. Such a noise, although it is randomly amplitude modulated, sounds relatively steady and produces only a very small sensation of roughness.

Figure 11.4 shows the dependence of roughness on sound pressure level. Data are given for amplitude-modulated broad-band noise, amplitude-modulated pure tone, and frequency modulated pure tone.

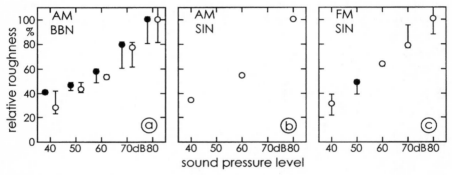

Fig. 11.4a–c. Roughness of modulated sounds as a function of sound pressure level. Modulation frequency is 70 Hz. (**a**) Amplitude-modulated broad-band noise with 40 dB modulation depth, (**b**) amplitude-modulated 1 kHz-tone with 40 dB modulation depth, (**c**) frequency-modulated pure tone at 1500 Hz with ±700 Hz frequency deviation

For an increase in sound pressure level by 40 dB roughness increases by a factor of about 3. This dependence of roughness on level is similar to the increase of fluctuation strength with level as displayed in Fig. 10.2.

An increment of roughness becomes audible for an increment in the degree of modulation of about 10%, which corresponds to an increment of about 17% in roughness. For amplitude-modulated 1-kHz tones and a modulation frequency of 70 Hz, a threshold of roughness is reached for values close to 0.07 asper. One asper is close to the maximum roughness for amplitude-modulated tones; there are thus only about 20 audible roughness steps throughout the total range of roughnesses.

Frequency modulation can produce much larger roughness than amplitude modulation. A strong frequency modulation over almost the whole frequency range of hearing produces a roughness close to 6 asper. Only amplitude modulation of broad-band noises is able to produce such a large roughness.

Figure 11.5 shows the dependence of roughness of an FM-tone on frequency deviation. An FM-tone centered at 1500 Hz with a level of 70 dB was modulated by a modulation frequency of 70 Hz.

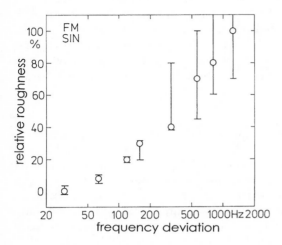

Fig. 11.5. Roughness of a sinusoidally frequency modulated pure tone as a function of frequency deviation. Center frequency 1500 Hz, sound pressure level 70 dB, modulation frequency 70 Hz

The results displayed in Fig. 11.5 indicate that for frequency deviations up to 50 Hz only negligible values of roughness show up. For larger values of the frequency deviation Δf roughness increases almost linearly with the logarithm of Δf.

11.2 Model of Roughness

As mentioned above, there are two main factors that influence roughness. These are frequency resolution and temporal resolution of our hearing system. Frequency resolution is modelled by the excitation pattern or by specific-loudness versus critical-band rate pattern.

It is assumed that our hearing system is not able to detect frequency as such, and is only able to process changes in excitation level or in specific loudness at all places along the critical-band rate scale; thus the model for roughness should be based on the differences in excitation level that are produced by the modulation. Starting with amplitude modulation, we can refer to data that describe the masking effect produced by strongly temporally varying maskers. Because masking level is an effective measure for determining excitation, it can be used to estimate the excitation-level differences produced by amplitude modulation. This procedure incorporates the two main effects already discussed, i.e. frequency and temporal resolution.

The temporal masking patterns outlined in Figs. 4.24, 4.25, and 4.27 show the temporal effects and indicate values of ΔL that can be used as a measure to estimate differences between the maximum and the minimum of the temporal masking pattern. This temporal masking depth, ΔL, becomes larger for lower modulation frequency. If roughnesses were determined only by this masking depth, then one would expect the largest roughness at the lowest modulation frequencies. This is not the case, indicating that roughness is a sensation produced by temporal changes. A very slow change does not produce roughness; a quick periodic change does, however. This means that roughness is proportional to the speed of change, i.e. it is proportional to the frequency of modulation. Together with the value of ΔL, this leads to the following approximation:

$$R \sim f_{\mathrm{mod}}\Delta L \ . \tag{11.1}$$

For very small frequencies of modulation, roughness remains small although ΔL is large. In this case f_{mod} is small, so that the size of the product remains small. For medium frequencies of modulation around 70 Hz, the value of ΔL is smaller than at low modulation frequencies. However, f_{mod} is much larger in this case, so that the product of the two values reaches a maximum. At high frequencies of modulation, f_{mod} is a large value but ΔL becomes small because of the restricted temporal resolution of our hearing system. Thus the product diminishes again. In this context, it should be realized that a modulation frequency of 250 Hz corresponds to a period of 4 ms, and the effective duration of the valley is only about 2 ms. In this case, the temporal masking depth, ΔL, becomes almost zero. Consequently, the product of ΔL and f_{mod} becomes very small and the roughness disappears.

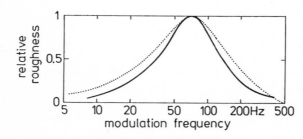

Fig. 11.6. Relative roughness as a function of the frequency of modulation as subjectively measured (*solid*) and calculated (*dotted*)

This way, roughness can be approximated and compared with the measured data. This is done in Fig. 11.6 by using the maximum roughness at 70 Hz as a reference. Roughness relative to the maximal value reached is plotted as a function of modulation frequency. The solid line corresponds to the data shown in Fig. 11.2 for centre frequencies above 1 kHz. The dotted line corresponds to the calculated values based on the assumption that roughness is proportional to the product of modulation frequency and masking depth. The calculated dependence agrees well with the subjectively measured one, although there are some differences especially at lower frequencies of modulation. There, subjectively measured roughness disappears more quickly than calculated roughness. In this region, subjects have difficulties differentiating between sensations of roughness and fluctuation strength, and concentrate mostly on one or the other.

For more precise calculations, it has to be realized that the value ΔL depends on the critical-band rate. The nonlinear rise of the upper masking slope outlined in Fig. 4.9 produces $\Delta L's$ in the region of the upper slope that are much larger than those corresponding to the main excitation. These effects can also be seen in Fig. 4.24. Comparing ΔL produced at a test-tone frequency identical to the frequency of the masker (1 kHz), with that produced on the upper slope near 1.6 kHz, it is evident that ΔL on the slope is much larger. To account for this effect, the given approximation is changed to:

$$R \sim f_{\mathrm{mod}} \int_0^{24\mathrm{Bark}} \Delta L_{\mathrm{E}}(z)\, dz \; . \tag{11.2}$$

Using the boundary condition that a 1-kHz tone at 60 dB and 100%, 70 Hz amplitude-modulated, produces the roughness of 1 asper, the roughness R of any sound can be calculated using the equation

$$R = 0.3 \frac{f_{\mathrm{mod}}}{\mathrm{kHz}} \int_0^{24\mathrm{Bark}} \frac{\Delta L_{\mathrm{E}}(z)\, dz}{\mathrm{dB/Bark}} \; \mathrm{asper} \; . \tag{11.3}$$

Unfortunately, we do not have data for ΔL as a function of critical-band rate that are as numerous as the data for excitation level or specific loudness. Therefore, the calculations are somewhat limited. Using the data available, however, we have been able to demonstrate that roughness can be calculated precisely as a function of the degree of modulation. In this case, the calculated value is influenced mainly by the dependence of ΔL on the degree of modulation, but not by the frequency of modulation. For modulated tones, the nonlinear rise of the upper slope of the masking versus critical-band rate pattern, creates larger contributions to roughness than those produced at the main excitation. This leads to the prediction – in agreement with psychoacoustical data – that the slope of the relationship between roughness and degree of modulation (both on logarithmic scales) is larger for sinusoidal tones (1.6) than for broad-band noises (1.3).

Some data concerning the value ΔL with frequency-modulated sounds are available. Approximations based on the equation given follow qualitatively and in many cases even quantitatively the psychoacoustically measured dependencies.

It is of advantage to transfer the ΔL values necessary for the calculation of roughness into the corresponding variations of specific loudness. As input to the model, the specific loudness-time function in each channel of a loudness-meter as illustrated in Fig. 8.26 is necessary. Moreover, the correlation between signals in neighboring channels has to be taken into account. On this basis, a computer program was developed which nicely accounts for the measured psychoacoustic data and can also describe quantitatively the roughness of noise emissions.

A variant of this program which essentially is based on the same features can also quantitatively assess the dependencies of fluctuation strength on relevant stimulus parameters (cf. Chap. 10).

12. Subjective Duration

When we talk about duration, we normally think of objective duration, i.e. physical duration measured in seconds, milliseconds or minutes. This is so, although we often check durations by listening to them in music, for example, or by giving a talk, where a short silence can add emphasis. If such durations can be measured by listening, they cannot be objective durations but must be subjective because they correspond to sensations. Subjective duration is not drastically different from objective duration if the durations of long-lasting sound bursts are compared. Therefore, it is often assumed that subjective duration and objective duration are almost equal. This is not so, however, when the duration of sound bursts is compared with the duration of sound pauses. In this case, drastic differences appear which indicate the need to consider subjective duration as a separate sensation.

12.1 Dependencies of Subjective Duration

The scale of subjective duration can be quantified by fixing a reference value and a unit. We have chosen as the unit the "dura" and have determined that a 1-kHz tone of 60 dB sound pressure level and 1 s physical duration, produces a subjective duration of 1 dura. By halving and doubling, we can produce the relationship between subjective duration and physical duration for 1-kHz tone bursts. The result is plotted in Fig. 12.1 in double logarithmic scales. The subjective duration, D, is the ordinate, and physical duration, T_i, is the abscissa. The reference point is marked by an open circle. Proportionality between the two values would be indicated by the broken 45° line.

This proportionality is effective over a wide range, starting at large durations of 3 s down to a duration of about 100 ms. Below that physical duration, subjective duration deviates from this proportionality with the tendency that subjective duration decreases less than physical duration. However, these results may be influenced by the fact that reducing the duration of the 1-kHz tone produces a different spectrum. Therefore, white noise was used to produce shorter sound bursts without changing the spectrum. For large durations, white-noise bursts show the same proportionality between physical and subjective duration as found for 1-kHz tones. The physical duration of

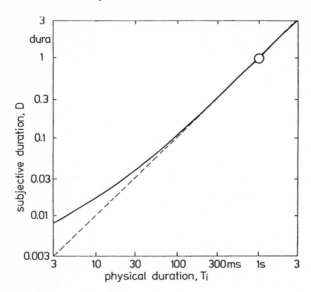

Fig. 12.1. Subjective duration as a function of the physical duration of 1-kHz tones at 60 dB SPL (*solid line*). The broken line indicates equality of physical and subjective duration. The open circle marks the standard sound producing a subjective duration of 1 dura

white noise can be reduced to 0.3 ms without much influence on the spectral shape. The results in this short duration area confirmed the tendency of 1-kHz tones to show little difference in subjective durations when physical duration is changed from 1 ms to 0.5 ms. From these measurements, it can be concluded that the effect shown in Fig. 12.1 is not a side effect based on the spectral broadening of the shorter 1-kHz tone bursts, but an effect that is based on the behaviour of human hearing. This means that one cannot expect subjective duration and objective duration to be equal for durations shorter than 100 ms.

This finding is somewhat astonishing; however, the results produced by comparing subjective durations produced by pauses with those produced by tone bursts are even more surprising. In this case, the method of comparison is used and pauses are changed to sound as long as tone bursts, and vice versa. The inset in Fig. 12.2 shows the temporal sequence of the test sounds to be compared. A sound burst of duration T_i is followed by a pause that lasts at least 1 s. After this, the tone is switched on again for 0.8 s, followed by a pause of duration T_p, switched on again for another 0.8 s and followed by a pause lasting at least 1 s. The subject compares the perceived duration of the tone burst with the perceived duration of the pause. In one experiment, the physical duration of the tone burst is changed, in the next the physical duration of the pause is changed by the subject to produce equal subjective durations. The use of the method of adjustment produces the same results as the use of the method of constant stimuli. The results plotted in Fig. 12.2 show curves of equal subjective duration plotted in the plane with physical duration of the pause as the ordinate and physical duration of the burst as

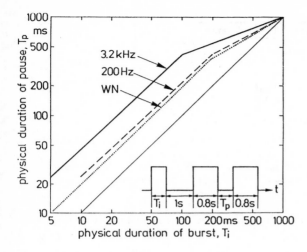

Fig. 12.2. Comparison of subjective durations produced by pauses and those produced by sound bursts. The physical duration of pauses (*ordinate*) producing the same subjective durations as a sound burst of given physical duration (*abscissa*) are shown for white noise (WN), 200-Hz and 3.2-kHz tones. The inset indicates the sequence of the sounds

the abscissa, both on logarithmic scales. The parameter is the type of sound used.

Tones of 3.2 kHz produce the largest effects while low-frequency tones of 200 Hz, or white noise, produce less pronounced effects. Both show, however, strong deviations from the 45° line that indicates equality of the durations of the burst and the pause. The expected result, that subjective duration of a pause and subjective duration of a burst are equal, is true for physical durations larger than 1 s but is certainly not true for smaller physical durations. Using 3.2-kHz tones as the sound from which the bursts and the pauses are extracted, it can be seen from Fig. 12.2 that a burst duration of 100 ms produces the same subjective duration as a pause which actually lasts as long as 400 ms. The difference in this case is a factor of four! For 200 Hz or white noise, the effect is smaller but still about a factor of two. The difference between physical burst duration and physical pause duration necessary to produce equal subjective durations, is true for burst durations below 100 ms down to durations as short as 5 ms.

The effect, which must play a strong role in music and speech perception, is not too dependent on loudness level if loudness levels larger than 30 phon are used.

12.2 Model of Subjective Duration

There are few psychoacoustical data available with regard to subjective duration. Thus, it is not possible to give an extensive model. The results so far, however, allow us to start with a model that is relatively simple, and makes use of the effect of premasking and postmasking. In most cases, the effect of

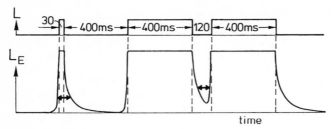

Fig. 12.3. Sound pressure level and excitation level as a function of time. The physical duration of the tone burst is chosen in such a way that it produces the same subjective duration as the pause. The double arrows mark the intervals in excitation level value that correspond to the respective subjective duration

premasking may be neglected, while the effect of postmasking becomes dominant with short pauses. Using the temporal masking effects as an indication of the excitation level versus time course produced by sound bursts and pauses, the comparison between the time course of sound pressure level and that of excitation level may lead to meaningful interpretations. Fig. 12.3 shows, in the upper panel, sound pressure level L as a function of time, and in the lower part excitation level L_E as a function of time. Premasking and postmasking, the latter being dependent on the duration of the preceeding masker burst, produce the time course shown in the lower panel. The functions outlined in Fig. 12.2 indicate for a 3.2-kHz tone, that subjective duration of a burst of 30 ms is equal to the subjective duration of a pause of 120 ms. The subjective equality of the two durations can be extracted using the assumption that subjective duration derives from the excitation level versus time course, by searching for those durations that are determined by levels 10 dB above the minima surrounding the time region in question. This means that the subjective duration of the 30-ms tone burst can be assumed as being picked up 10 dB above the excitation level corresponding to threshold in quiet, as indicated by a double arrow. The duration of the pause of 120 ms is picked up 10 dB above the minimum that is produced during the pause in the excitation level versus time course, again indicated by a double arrow. Comparison of the two double arrows indicates that these are almost equal in size. This means that equal subjective duration is produced by such a presentation of burst and pause.

Subjective duration plays an important role in music. The model assumes that subjective duration derives from the excitation level versus time course by picking up the duration 10 dB above the surrounding minima. In music, the notation of tone length and pause length is given by certain symbols, which are read by the musicians. An example is given in Fig. 12.4 in row (a). The black and white bars in row (b) show the correlated physical durations, which might be played by a non-musician. What is actually played physically by musicians is indicated by the black and white bars in (d). This is completely

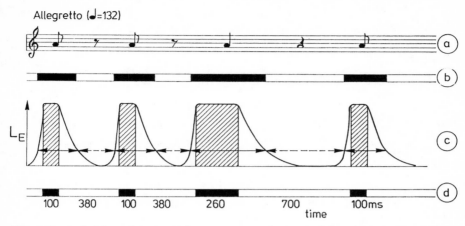

Fig. 12.4a–d. Musical notation for a sequence of tones in (**a**) and the corresponding expected sequence of durations in (**b**). The actually played sequences of tone bursts and pauses are given in (**d**), the corresponding excitation level versus time pattern is given in (**c**). The double arrows with solid lines indicate the subjective durations of the bursts, the double arrows connected with broken lines the subjective durations of the pauses

different from the notation transformed mathematically into the durations outlined in (b). Using the model, we can plot the excitation level versus time course, look for the length of the double arrows plotted 10 dB above the minima and see that the length of the first four double arrows is equal, and that the lengths of the fifth and sixth double arrows are equal to one another but about twice as long as the previous four. This means that bursts and pauses are perceived as being equally long, as required by the musical notation (a). This illustrates clearly that subjective duration plays an important role for rhythm and that musicians produce what is intended musically, although the notations for durations shorter than 1 s are physically wrong.

Another similar effect occurs with speech and is known to most of those working in speech analysis. The pronounciation of plosives is connected with the production of pauses that last between 60 and 150 ms. These pauses, physically needed before a plosive can be pronounced, are not obvious in speech perception – if one does not pay much attention to them, they are not perceived. This means that these pauses, although physically existent, are very short in subjective duration and play only a secondary role in speech perception. The model for subjective duration again shows that excitation level versus time pattern can be of great help in describing hearing sensations.

13. Rhythm

In this chapter, the physical data of the temporal envelope of sounds eliciting the perception of a subjectively uniform rhythm are shown. In addition, the rhythm of speech and music is discussed. For music, the hearing sensation of rhythm is compared with the hearing sensations of fluctuation strength and subjective duration. Finally, a model based on the temporal variation of loudness is proposed for the hearing sensation rhythm.

13.1 Dependencies of Rhythm

It is frequently assumed that sound bursts of equal temporal spacing elicit the sensation of a subjectively uniform rhythm. However, this simple rule holds only for very short sound bursts with steep temporal envelopes. Sound bursts with a more gradual rise in the temporal envelope often require systematic deviations from physically uniform temporal spacing, in order to produce a subjectively uniform rhythm. In Fig. 13.1, several examples are illustrated. The temporal envelope of the sound bursts is indicated schematically in the left column panels. The durations T_A and T_B indicate the physical durations of the bursts in the next two columns. The temporal shift, Δt, (relative to equal spacing of the intervals) that is required to produce subjectively uniform rhythm is given in the right column. A comparison of the data shown in Fig. 13.1f and 13.1h reveals the main influence of the temporal envelope on the perception of subjectively uniform rhythm: in Fig. 13.1h, both burst A and burst B show a steep increase of the temporal envelope. In this case, the time shift, Δt, is zero. This means that for a steeply rising temporal envelope, the equality of the temporal distance between the rising parts of the envelope leads to the perception of subjectively uniform rhythm. However, as illustrated in Fig. 13.1f, a temporal shift as large as 60 ms occurs for a gradual increase of the temporal envelope of burst B. In this case, it is not the physical starting point of burst B, but a temporal position 60 ms later which produces subjectively uniform rhythm. A comparison of the steepness of the temporal envelopes and the magnitude of the time differences Δt for all cases displayed in Fig. 13.1 confirms this trend: for the steep rise of the temporal envelope of burst B, small values of Δt occur, whereas gradual increases in the temporal envelope lead to large values of Δt.

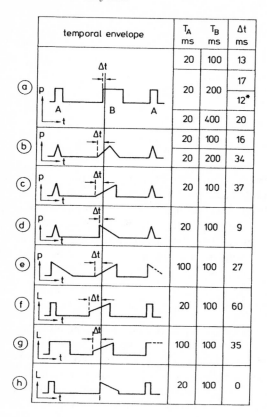

temporal envelope	T_A ms	T_B ms	Δt ms
(a)	20	100	13
	20	200	17
			12*
	20	400	20
(b)	20	100	16
	20	200	34
(c)	20	100	37
(d)	20	100	9
(e)	100	100	27
(f)	20	100	60
(g)	100	100	35
(h)	20	100	0

Fig. 13.1a–h. The temporal characteristics of sound bursts eliciting a perception of subjectively uniform rhythm. Schematic representation of the temporal envelopes are given in the left-hand column; durations T_A and T_B of the sound bursts A and B are indicated in the centre columns, and in the right column the temporal shift Δt with respect to a sequence of physically equidistant onset is found. The bursts are extracted from 3-kHz tones and in case (**a**) from white noise (*asterisk*)

In Fig. 13.2, an example of the rhythm of fluent speech is given. The task of the subject was to tap on a Morse key the rhythm of the sentence "he calculated all his results". In the lower part of Fig. 13.2, the distribution of the perceived rhythm, i.e. the histogram of the temporal distance between the strokes on the Morse key is given. From the lower part of Fig. 13.2, it is clear that the perceived rhythm of fluent speech is based on the syllables: each syllable leads to a rhythmic event, and hence to a stroke on the Morse key. In the upper part of Fig. 13.2, the loudness-time function of the test sentence is given, as measured by a loudness meter (see Chap. 8). A comparison of the upper and lower parts of Fig. 13.2 reveals that each syllable is correlated with a maximum in the loudness-time function. This result will be discussed in more detail in Sect. 13.2.

Figure 13.3 illustrates the relationship between the hearing sensations rhythm and fluctuation strength. For 60 pieces of music, each with a duration of 20 s, the perceived rhythm was measured by tapping the rhythm on a Morse key. The intervals between successive strokes were sampled, and the corresponding histograms calculated. The three thick arrows at the bottom

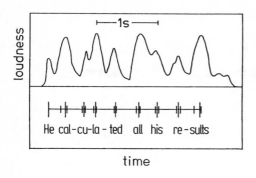

time

Fig. 13.2. The rhythm of fluent speech. *Lower part*: histogram of the rhythmic events produced by subjects listening to the sentence, "he calculated all his results." *Upper part*: corresponding loudness-time function

of Fig. 13.3 represent the results: the most frequent interval between successive strokes was 250 ms; this leads to a modulation frequency of 4 Hz (large arrow in Fig. 13.3). Other maxima occur at 8 and 2 Hz, corresponding to temporal distances between successive strokes of 125 and 500 ms. The broken line in Fig. 13.3 represents the dependence of the hearing sensation fluctuation strength on modulation frequency (compare Chap. 10). The solid curve gives the variation of the temporal envelope averaged across all 60 pieces of music. The results diplayed illustrate the remarkable correlation between the sensation of fluctuation strength (broken line) and the sensation of rhythm (arrows). In addition, the solid line in Fig. 13.3 illustrates that both fluctuation strength and rhythm depend primarily on variations in the temporal envelope of the sound.

The correlation between the hearing sensation of rhythm and the hearing sensation of subjective duration can be illustrated by means of Fig. 13.4. In Fig. 13.4a, the musical notation of a rhythmic sequence is given. Figure

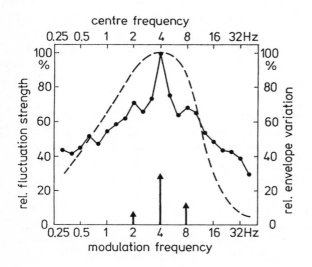

Fig. 13.3. The correlation between rhythm and fluctuation strength. *Arrows*: histogram of the temporal distance of the rhythmic events of 60 pieces of music. *Broken line*: dependence of fluctuation strength on modulation frequency. *Solid curve*: envelope variation of the 60 pieces of music as a function of the centre frequency of the third-octave filters used to analyze the envelope variation of the music

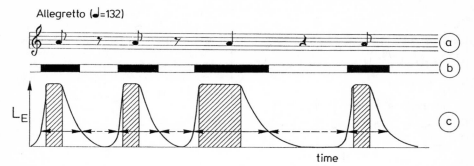

Fig. 13.4a–c. The relationship of rhythm and subjective duration. (**a**) Musical notation of a rhythmic pattern; (**b**) illustration of the subjective duration of notes (*black bars*) and pauses (*white bars*); (**c**) physical durations of the sound bursts (*hatched*) and pauses necessary to produce the rhythm indicated by musical notation. Subjective duration of notes and pauses is illustrated by double arrows according to the model of subjective duration of bursts (*solid lines*) and pauses (*broken lines*)

13.4b illustrates the corresponding expected subjective durations by black bars for notes and white bars for pauses. However, as shown in Fig. 13.4c, the physical duration of bursts and pauses performed by musicians differ quite substantially from the pattern shown in Fig. 13.4b, in order to subjectively produce the rhythm expressed in the notation. As discussed in more detail in Chap. 12, the physical durations of bursts and pauses have to differ by a factor of about 4 in order to produce the *same* subjective duration. In addition, the subjective doubling of musical notes has to be produced by increasing the physical burst duration by a factor of 2.6. Figure 13.4 therefore gives an excellent example of the fact that musically meaningful rhythmic patterns can only be produced when the relationship between physical duration and subjective duration is taken into account.

13.2 Model of Rhythm

The main features of a model for rhythm are explained in Fig. 13.5. The rhythm of sounds such as speech and music can be calculated on the basis of the temporal pattern of their loudness, measured by a loudness meter (see Chap. 8). Basically, each maximum of the loudness-time function indicates a rhythmic event. More specifically, the model postulates that only maxima above a value $0.43N_M$ are considered, where N_M represents the loudness of the highest maximum within a relevant time, for example that of a phrase. This means that the maxima around 1 s in Fig. 13.5, for example, are ignored. A second condition is that only relative maxima of sufficient height are considered. As indicated in Fig. 13.5, the maximum at the right of the

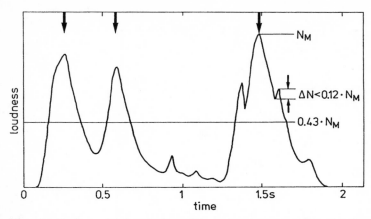

Fig. 13.5. A model for rhythm. The rules for extracting the three maxima that are taken as indications of rhythmic events (arrows at the upper abscissa) from the loudness-time function are given in the text

main maximum N_M produces an increase in the loudness-time function of only $\Delta N/N_M < 0.12$, and is therefore ignored. Only relative increments of loudness greater than 12% are significant. As a third condition, only maxima which show a temporal separation greater than 120 ms are considered to be separate rhythmic events. Therefore, the maximum at the left of the main maximum N_M is ignored as being too small relative to the largest inside the 120-ms window. This means that when calculating the rhythm produced by the loudness-time function given in Fig. 13.5, only three maxima are taken into account; they are more than 120 ms apart, lie above $0.43 N_M$ and produce an increase in loudness with respect to the preceding maximum of at least $0.12 N_M$. Often, it may be better not to use only the total loudness versus time function, but also parts of the loudness or even the specific loudness versus time functions as the basis for calculating rhythm.

Whereas most rhythmic events can be detected by means of variations in the (specific) loudness-time functions, variations in pitch can sometimes lead to rhythmic events despite constant loudness. In such cases, the variations in pitch have to be calculated according to the models of spectral pitch or virtual pitch described in Chap. 5.

14. The Ear's Own Nonlinear Distortion

A transmission system containing a nonlinearity produces harmonics when transmitting a pure tone. Such distortion products are often neither annoying nor audible, because they match almost totally with the fundamental. Musical tones are complex tones and consist of several harmonics. In this case, the nonlinearity changes the amplitude of these harmonics somewhat. This change is also rarely detected. Further, masking, which is stronger above the frequency of the tone, leads to the fact that strong fundamentals mask the higher harmonics produced by the nonlinear transmission system. Difference tones, produced by the nonlinearity when two tones (the primaries) are presented, are easier to detect. The difference tones are produced at frequencies below those of the primaries. The masking effect at medium and high levels is much smaller, and the difference tones are more easily detected. The frequencies of the difference tones may appear unharmonically related to the primaries and may therefore be annoying, if audible. Even when using high-quality electroacoustic equipment that does not produce audible nonlinear distortion products, we still hear difference tones that are produced by the nonlinearity of our hearing system.

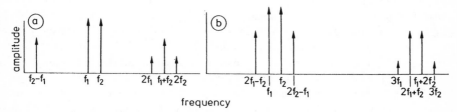

Fig. 14.1a,b. Distortion products produced by two primaries with frequencies f_1 and f_2 as a consequence of pure quadratic nonlinear distortion (**a**) and of pure cubic nonlinear distortion (**b**)

The spectra of primaries f_1 and f_2, and of the distortion products appearing with quadratic nonlinear distortion are shown in Fig. 14.1a, and with cubic distortions in Fig. 14.1b. The spectra indicate that the quadratic combination tones at $(f_2 - f_1)$ and $(f_2 + f_1)$ are a factor of two larger in amplitude than the quadratic harmonics at $2f_1$ and $2f_2$, and that the cubic difference tones at $(2f_1 - f_2)$ and $(2f_2 - f_1)$ are a factor of three larger than the cubic harmonics. This contributes to the fact that difference tones are the most easily detectable distortion products. For regular quadratic distortion, i.e. following a square law, the level of the quadratic difference tone follows the equation

$$L_{(f_2-f_1)} = L_1 + L_2 - C_2 \,, \tag{14.1}$$

where C_2 depends on the relative amplitude of the quadratic distortion. The level of the lower cubic difference tone, which is the most prominent distortion product produced by regular cubic (i.e. following a cubic law) distortion, follows the equation

$$L_{(2f_1-f_2)} = 2L_1 + L_2 - C_3 \,, \tag{14.2}$$

while the level of the upper cubic difference tone follows the equation

$$L_{(2f_2-f_1)} = 2L_2 + L_1 - C_3 \,. \tag{14.3}$$

Here, C_3 depends on the relative amplitude of the cubic distortion.

These equations hold for regular distortions, which are the usual kind produced in electroacoustic networks. In such equipment, the deviation from exact linearity is not very strong, and the equations given above are mostly fulfilled. Knowledge of these kinds of dependencies is helpful when the non-linear distortion products produced in our hearing system are discussed.

For quantitative psychoacoustical measurement of the nonlinear distortion products (mainly difference tones) produced in our hearing system, two methods are available – the method of cancellation and the pulsation-threshold method. The cancellation method uses an additional tone produced by the electronic equipment to have the same frequency as the difference tone, but being variable in amplitude and phase. This separately produced difference tone is added to the two primaries delivered to our hearing system. The subject adjusts the level and phase of this additional tone until the audible difference tone totally disappears. This way, the audible difference tone produced by our hearing system is cancelled. The level of the additional tone necessary for cancellation is a measure of the amplitude of the difference tone produced internally by our hearing system.

The pulsation-threshold method (see. Sect. 4.4.6) uses a non-simultaneous presentation, in which the pulsation threshold of the additionally presented electronically produced tone at the difference frequency is measured. The cancellation method has been partially opposed in the literature, using the argument that the additional cancellation tone may change the nonlinearity and may therefore influence the result. In principle such an effect appears in any nonlinear device. However, this influence has been studied carefully, and it was shown that, for the worst case (equal amplitude of the primaries), it results in only small 1- or 2-dB effects, which are about the accuracy of measurement. The method of pulsation threshold has other strong disadvantages. It compares two completely different conditions – that of simultaneous presentation (two primaries and the internally produced difference tone) with that of nonsimultaneous presentation of the additional tone. The difficulty is increased by the rather doubtful assumption that the decay of postmasking and the rise of premasking behave linearly. Therefore, the data presented below are produced using the method of cancellation. All tones are produced electronically and separately. The measurements were made monaurally using a DT48 earphone with a free-field equalizer. The distortion products of the equipment show such low values that they can be ignored.

14.1 Even Order Distortions

Our hearing system shows the best frequency selectivity in a frequency range around 2 kHz. There, the critical bandwidth is relatively small. Good frequency selectivity is necessary to measure the cancellation levels not only over a large range of frequency differences $(f_2 - f_1)$, but also over a large range in the level of the two primaries L_1 and L_2. In order to avoid harmonic conditions, frequencies with unusual values are chosen.

The upper part of Fig. 14.2 shows the level of the $(f_2 - f_1)$ tone needed to cancel the audible quadratic difference tone produced by the primary levels as indicated. The lower part of the figure shows the phase needed for cancellation. The frequency of the lower primary, f_1, is 1620 Hz, that of the upper primary, f_2, is 1944 Hz. The difference frequency, which is also the frequency $f_2 - f_1$ of the quadratic difference tone and of the cancellation tone, is 324 Hz. Figure 14.2b shows the effect for a similar condition, with the frequency $f_2 = 2592$ Hz and $(f_2 - f_1) = 972$ Hz. Cancellation level and cancellation phase are given in both cases as a function of the level of the primary, L_2. Cancellation level and cancellation phase show an almost regular behaviour, i.e. with increasing L_1 or L_2, the cancellation level, $L_{(f_2-f_1)}$, indicated by solid lines increases by almost the same amount (broken lines). The phase remains almost constant, independent of both L_1 and L_2. These relationships appear both for the small frequency separation and the large frequency separation of the primaries. This means that the effect of quadratic distortion produced in the hearing system of this subject (and many other subjects) follows approximately the relatively simple rule that is indicated by the broken lines. This rule corresponds to what is expected for a transmission system, that acts with an ideal quadratic distortion. The characteristic, expressed by the broken lines, follows the equation

$$L_{(f_2-f_1)} = L_1 + L_2 - 130 \, \text{dB} . \tag{14.4}$$

The fact that the two parts of Fig. 14.2 can be described well by the same set of dashed lines indicates that, in this case, the nonlinearity is independent of frequency distance. This means that the transmission characteristic of the hearing system of such a subject differs relatively little from regularity.

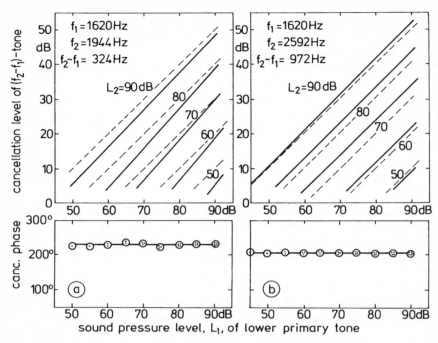

Fig. 14.2a,b. Sound pressure level *(upper part)* and phase *(lower part)* of the tone at the frequency of the quadratic difference tone needed for cancellation, as a function of the level L_1 of the lower primary. The level of the upper primary, L_2, is the parameter, the frequencies of the primaries and of the difference tone are indicated. The broken lines correspond to a behaviour that would be expected for regular quadratic distortion

Difference tones can be cancelled only for levels of one of the two primaries above about 70 dB. For lower values of the primaries, the difference tone remains inaudible. For $L_1 \approx L_2 = 70$ dB, the cancellation level remains at about 10 dB. This means that it is 60 dB below the level of the primaries, corresponding to an amplitude of the sound pressure of the cancellation tone of about one thousandth that of the primaries. This difference becomes smaller for increasing level of the primaries and reaches about 40 dB (corresponding to 1 %) for $L_1 \approx L_2 = 90$ dB.

Most of the subjects show the behaviour described above. However, a smaller group shows a dependence of the cancellation level on the level of the primaries that cannot be approximated by the dashed lines. Additionally, for these subjects, the phase is no longer independent of level. An example is indicated in Fig. 14.3. The coordinates and parameters are the same as for Fig. 14.2. A comparison between the two sets of data shows that there is some tendency for the broken lines indicated in Fig. 14.2 to appear in

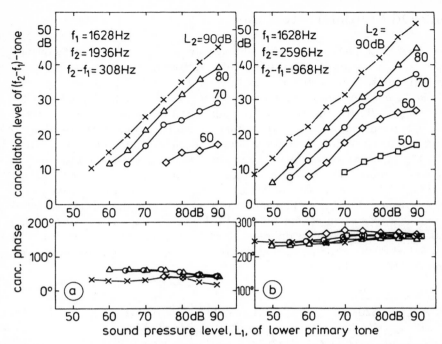

Fig. 14.3a,b. Cancellation data of quadratic difference tone as outlined in Fig. 14.2 but for a subject who shows strong deviations from regular quadratic distortion

Fig. 14.3. However, there is also strong deviation from the broken lines, visible especially in Fig. 14.3a.

Measurements of the cancellation level and cancellation phase for the quadratic difference tone as a function of the frequency region of the lower primary, have shown frequency independence at least for the first group showing almost regular quadratic distortion behaviour. This indicates that the quadratic difference tone measured by the cancellation method seems to be produced before the frequency-selective mechanism becomes effective, i.e. in the middle ear. The development of the quadratic difference tone for the second group seems to be rather complicated. A mixture of the nonlinear distortion produced in the middle ear and of that produced in the inner ear may be the reason.

Distortions of the fourth order are almost inaudible. They can be produced and cancelled by adding the additional tone only at levels above 85 to 90 dB. In this region, the stapedius muscle is already active and may strongly influence the results.

14.2 Odd Order Distortions

The level and phase of the simultaneously presented tone of frequency $(2f_1 - f_2)$ necessary to cancel the cubic difference tone produced by the two primaries, is indicated in Fig. 14.4 for the same parameters as given in Fig. 14.2. Using the data point for the cancellation level produced for $L_1 = L_2 = 90\,dB$, the broken line indicates the characteristics expected from a transmission system with regular cubic distortions (corresponding to (14.2)). It is clear that the data produced by this subject are very different from the expected values. In addition, phase depends strongly on the level of the lower primary. Furthermore, the values obtained with a higher frequency f_2 (Fig. 14.4b), show much smaller levels and different phases relative to the results given in Fig. 14.4a. This indicates that the cubic distortions produced in our hearing system cannot be described by a regular cubic transfer function. There seems to be a different effect that, even for very low levels of only 40 or 30 dB, produces cubic difference tones that are audible and can therefore be cancelled out.

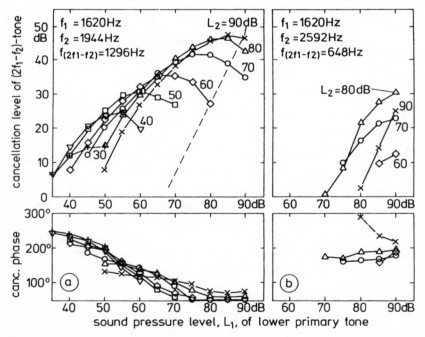

Fig. 14.4a,b. Sound pressure level (*upper part*) and phase (*lower part*) of a tone at the frequency of the cubic $(2f_1 - f_2)$-difference tone needed for cancellation as a function of the level of the lower primary. The level of the upper primary is the parameter. The frequencies of the primaries and of the cubic difference tones for the two frequency separations are indicated. The dashed line follows (14.2)

The largest difference between the data expected from the regular cubic distortion and the measured data, is that the cancellation level eventually decreases with increasing levels of the lower primary. This decrement, for example in Fig. 14.4a with $L_2 = 60\,$dB and L_1 larger than 65 dB, indicates that the hearing system acts less nonlinearly for higher input levels, i.e. indicating the existence of a distortion source that cannot be described by simple regular nonlinear characteristics. The unusual amplitude and phase behaviour of the cancellation tone was also measured for other subjects. Although the effect is irregular and seems not to be describable by a simple system, most of the subjects show similar results so that, at least in certain parameter regions, averaging was possible.

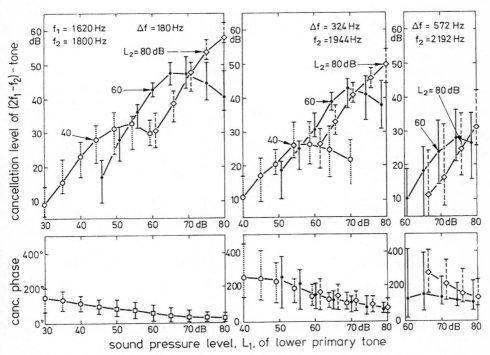

Fig. 14.5. $(2f_1 - f_2)$-cancellation values as in Fig. 14.4, but showing the medians and interquartile ranges from six subjects; parameters for the three frequency separations as indicated. The phase ranges given in the lower left drawing indicate all data at a particular value of the lower primary level, regardless of the level of the upper primary

Figure 14.5 shows the medians and interquartile ranges of cancellation level and cancellation phase of the $(2f_1 - f_2)$-difference tone from six subjects. Frequency of the lower primary is kept constant $(f_1 = 1620\,$Hz) while frequency of the upper primary is changed from 1800 Hz to 1944 Hz and

2192 Hz in parts a–c, respectively. The abscissa is again the level of the lower primary, while the level of the upper primary is the parameter. A comparison of the three parts clearly shows that level and phase of the cancellation tone depend strongly on the frequency separation $\Delta f = f_2 - f_1$. The six subjects show behaviour that is similar to that of the subject whose results are outlined in Fig. 14.4. The interquartile ranges of cancellation level and cancellation phase seem to increase with increasing frequency separation of the two primaries. If the individual data sets are compared with the median values, it is clear that the individual differences are caused by a parallel shift upwards or sidewards; the general shape of the curves remains similar for all subjects. Only for the largest frequency separation of the primaries ($\Delta f = 572$ Hz) does individually different behaviour appear. Some subjects produce data that are similar to the medians; other subjects show level minima and phase jumps that make averaging a doubtful procedure.

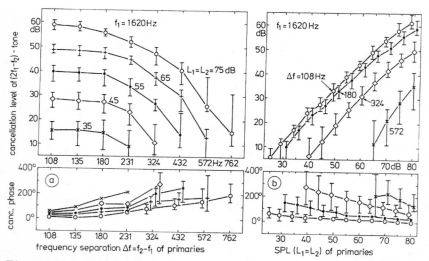

Fig. 14.6a,b. Level and phase of a tone cancelling the cubic difference tone, as a function of the frequency separation of the primaries. Medians and interquartile ranges of six subjects are indicated. The left part shows the data as a function of the frequency separation of the primaries; the right part gives the data as a function of the level of the primaries. The symbols for parameters are given in the upper panels

The frequency difference, $\Delta f = f_2 - f_1$, seems to play an important role. Therefore, data have been measured for the parameter configuration where $L_1 = L_2$, while the frequency separation, Δf, is the abscissa. Figure 14.6 again shows medians and interquartile ranges from the same six subjects, in part (a) as a function of the frequency separation with $L_1 = L_2$ as the parameter and in part (b) with $L_1 = L_2$ as the abscissa and Δf as the parameter. These data show clearly that cancellation level decreases markedly with increasing Δf, while cancellation phase increases. As a function of the level of the two primaries ($L_1 = L_2$), cancellation level rises continuously while cancellation phase decreases. It is interesting to note that cancellation level increases almost in the same way as $L_1 = L_2$. For a regular cubic distortion it would be expected, when $L_1 = L_2$ (see (14.2)), that cancellation level would increase three times faster than the level of the primaries. The interquartile ranges of the six subjects seem to again increase with increasing frequency separation, because in this range minima in the cancellation level may depend on the level of the lower primary.

Such effects are more clear at higher frequencies of the primaries. Figure 14.7 shows, for one subject, the dependence of cancellation level and cancellation phase on the level of the lower primary, for a lower primary at 4800 Hz. The frequency of the upper primary, and consequently the frequency separation, Δf, is changed from 436 Hz to 750 and 1000 Hz for parts a, b, and c, respectively. Part (a) shows a behaviour of cancellation level and cancellation phase that is comparable to that outlined in Fig. 14.4a. Cancellation phase changes over the whole range by about 200°. Increasing Δf to 750 Hz results in quite different behaviour from that shown for smaller frequency differences. Cancellation level no longer increases monotonically with L_1 before decreasing again, while phase dependence becomes an inverted S-shape. It seems that phase decreases most quickly in the ranges of L_1 for which the cancellation level shows either a minimum, or a tendency towards a minimum. These tendencies become striking for $\Delta f = 1000$ Hz. Cancellation level as a function of L_1 shows not just one notch but two or perhaps three notches; the phase decreases in a step-like manner at almost exactly those values of L_1 at which the cancellation level shows a minimum.

Because there are large individual differences in such conditions, and because the minima are at different places, averaging is meaningless. All subjects show an increasing phase range for increasing frequency separation of the primaries. The subjects that show minima in the dependence of the cancellation level on primary levels also show phase steps at the same points. Other subjects show relatively smooth increments or decrements of the cancellation level, and also relatively smooth decrements of phase with increasing level of the lower primary. Further investigations have indicated that the correlation between the steps of phase and the minima in the cancellation level holds in all cases. Detailed variations of levels and frequencies of the primaries have shown that in some cases a dependence on L_1 can be found that indicates a step upwards in cancellation phase, where the cancellation level itself shows a minimum. In these cases, the dependence of the cancellation phase on the lower primary level follows a curve that is almost 360° lower that expected. This means that the phase pattern depends on the individual or on the parameters, and can produce phase jumps downwards at the levels of L_1 for

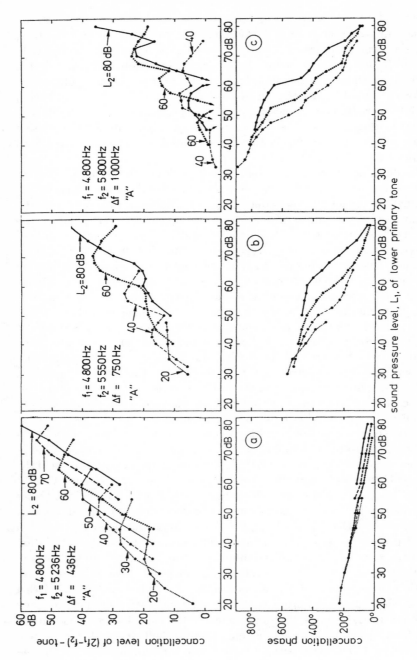

Fig. 14.7a–c. Level and phase for cancellation of the cubic difference tone as a function of the level of the lower primary, with the level of the upper primary as the parameter. The frequency of the lower primary is 4.8 kHz. The frequency of the upper primary changes from (a) to (b) to (c) as indicated

which the cancellation level reaches a minimum, or phase jumps upwards in order to reach a similar value but 360° higher.

Data measured with the method of cancellation show the following typical effects. For small frequency separations Δf between the primaries, the dependence of cancellation level on L_1 and on L_2 shows the same characteristic regardless of the frequency range of the primaries: cancellation level rises with increasing L_1 if L_2 is kept constant, and rises with L_2 if L_1 is kept constant; it reaches a maximum and then decreases slowly. For frequencies of f_1 larger than about 1 kHz and for frequency separations larger than a characteristic value, minima in the cancellation level pattern are accompanied by steps in cancellation phase. The characteristic boundaries are about 300 Hz for frequency $f_1 < 3$ kHz and about $0.1f_1$ for $f_1 \geq 3$ kHz. Cancellation phase varies somewhat more when L_1 is varied (L_2 kept constant) than with varying L_2 (L_1 kept constant).

14.3 Models of Nonlinear Distortions

Quadratic distortions indicated by $(f_2 - f_1)$-difference tones can be described relatively simply for the majority of subjects, by assuming a slightly nonlinear transfer characteristic of the quadratic behaviour in the middle ear. The irregular behaviour of quadratic difference tones shown by a smaller group may be a result of two sources, one in the middle ear and the other in the inner ear. The contributions of both together form the unexpected behaviour found in some subjects. Although the sum of the two contributions may not be so easy to describe quantitatively, the contribution of the inner ear seems to be created in a way similar to that discussed below for the production of cubic difference tones.

The nonlinear preprocessing model with active feedback in the inner ear was described in Sect. 3.1.5. Not only two-tone suppression can be treated in detail in a hardware model consisting of 150 sections each with a nonlinear feedback loop, but also the creation of the cubic $(2f_1 - f_2)$-difference tone. The basic data are the level and phase distributions along the sections produced in this system by single tones. An enhancement is created at medium and low input levels through the activity of the nonlinear feedback loops, which results in a more sharply peaked level-place pattern. These patterns become much less peaked ("de-enhancement") for high input levels. At high levels, the level and phase distributions along the sections are very similar to those produced by the linear passive system (feedback loop switched off). The creation of $(2f_1 - f_2)$-difference tones is based on the same nonlinear effect. In this case, two tones are used as input. The de-enhancement produced by the additional tone is described as a gain reduction in the saturating nonlinearity of the feedback loop, as a consequence of increasing input levels. In each section, difference-tone wavelets are created that travel – thereby changing level and phase – to their characteristic place, where they add up to the vector sum corresponding to the audible difference tone. In the case of cancellation, the vector sum has to be compensated by an additional tone of the same frequency and magnitude but of opposite phase. Based on this strategy for treating $(2f_1 - f_2)$-difference tones, the relevant levels of the difference-tone wavelets and the corresponding phases can be picked up at each section. This is done for two examples in the following paragraphs. Additionally, the vector sum of the wavelets that clearly illustrates the development of minima in the $(2f_1 - f_2)$-cancellation tone versus level L_1 of the lower primary is also shown.

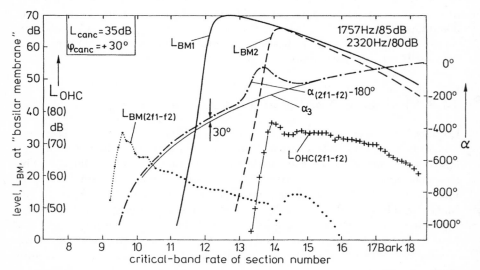

Fig. 14.8. Level-place patterns of L_{BM1}, L_{BM2}, $L_{BM(2f1-f2)}$, and $L_{OHC(2f1-f2)}$ (*left ordinates*) produced in the hardware model by the two primaries indicated in the right upper corner. Phase-place patterns (*right ordinate*) are given for $\alpha_{(2f1-f2)}$, the phase of the $(2f_1 - f_2)$-difference tone produced in the nonlinear feedback loops by the two primaries and for α_3, the phase of a $(2f_1 - f_2)$-tone presented alone. Note the difference of the two phase patterns by $30°$ for the region below 13 Bark, which corresponds to the cancellation phase, φ_{canc}, as indicated together with the cancellation level, L_{canc}, at the left upper corner

Primaries of 1757 Hz at 85 dB and 2320 Hz at 80 dB show characteristic places of $cz_1 = 12.3$ Bark and $cz_2 = 14$ Bark, while the characteristic place of the cubic difference tone is $cz_{(2f_1-f_2)} = 9.6$ Bark. This means that the level-place patterns are separated, as can be seen in Fig. 14.8. The level L_{BM1} (solid) reaches its maximum near $z = 12.5$ Bark, while the level L_{BM2} (dashed) reaches its maximum near $z = 14.2$ Bark because of the de-enhancement that takes place at the high input levels used.

These two levels produce similar values for the upper slopes of the two level-place patterns, a configuration in which large difference-tone wavelets

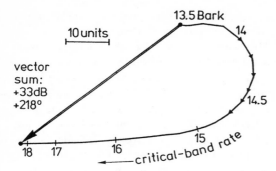

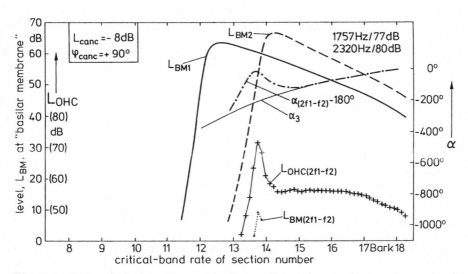

Fig. 14.9. Vector diagram calculated from the many wavelets created at the sections (with indicated critical-band rate), by the two primaries given in Fig. 14.8. All the wavelets travel to the characteristic place (9.6 Bark in this case) to produce a vector sum as indicated. It should be compared with the cancellation (opposite phase) data of Fig. 14.8

of similar values are produced. The levels of these wavelets are indicated as L_{OHC} (plus signs) and do not change much over a large range of the critical-band rate, z. The phase of the wavelets decreases smoothly between the critical-band rates of 18.5 and 15 Bark (dashed-dotted). Near 15 Bark, however, phase starts to increase again, reaches a peak near $z = 13.6$ Bark, decreases quickly between 13.5 Bark and 13 Bark, and finally follows the expected characteristic. All of these wavelets sum to create a $(2f_1 - f_2)$-difference tone at its characteristic place. The corresponding curve (dotted) shows a pattern that is typical for lower levels containing a strong peak at the characteristic place, which corresponds to 9.6 Bark in this case. A cancellation level at the input of the model of 35 dB and a phase of $+30°$ is

Fig. 14.10. Level-place pattern as in Fig. 14.8, but for the lower-primary level reduced by 8 dB as indicated. Note the smaller values of $L_{OHC(2f1-f2)}$ and $L_{BM(2f1-f2)}$ in comparison to Fig. 14.8

necesssary to cancel out the $(2f_1 - f_2)$-difference tone produced in the model by the two primaries at the characteristic place of 9.6 Bark.

This value can be constructed by using the vector length and vector phase angles of all wavelets to be summed. Using the corresponding data, the vector diagram outlined in Fig. 14.9 is constructed. It shows that the vector sum, with a length of 46 units corresponding to a level of 33 dB and with the phase of $+218°$, results from wavelet vectors of different length and – even more important – of different phases (directions). The vector sum is smaller than the sum of all the magnitudes of the vectors of the wavelets, because of the different vector directions. This reduces the total length. The phase of the vector sum depends mostly on the phase of the wavelets, which differs up to $180°$, while their amplitude (length) does not vary strongly. The constructed vector sum leads to values that are slightly different from the cancellation values. However, the differences are small and the phase is almost exactly $180°$ larger, i.e. opposite to cancellation phase as it should be.

Reducing the lower primary by 8 dB, the two primaries become 1757 Hz at 77 dB and 2320 Hz at 80 dB. Therefore, the place patterns indicated in Fig. 14.10 by L_{BM1}, L_{BM2}, and by the phase angles do not differ much from those shown in Fig. 14.8. The level of the wavelets, however, is reduced drastically for all places above 14 Bark; a pronounced peak at 13.7 Bark now appears. Surprisingly enough, almost nothing can be measured for a distribution of the summed-up $(2f_1 - f_2)$-difference tone indicated by $L_{BM(2f_1 - f_2)}$.

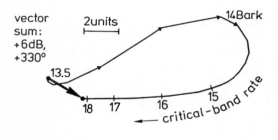

vector sum: +6 dB, +330°

Fig. 14.11. Vector diagram as in Fig. 14.9, but for the data outlined in Fig. 14.10. Note the large unit scale, i.e. the small vector sum

Fig. 14.12a,b. Comparison between human data (*left*), and model data (*right*). Psychoacoustically measured $(2f_1 - f_2)$-cancellation level and phase data, with L_1 as the abscissa and L_2 as the parameter in the left part (**a**). The four patterns correspond to the frequency separations between the two primaries of 135, 231, 324, and 572 Hz, as indicated. The frequency of the lower primary is 1620 Hz. Cancellation data produced in the model by adding a $(2f_1 - f_2)$-tone with level L_{canc} and phase φ_{canc} at the input, in such a way that L_{BM} at the characteristic place of frequency $(2f_1 - f_2)$ is reduced below noise level. The abscissa and parameter are the same as above. The frequency of the lower primary is 1757 Hz. Frequency separations between the two primaries in (**b**) are chosen to be comparable to those in (**a**) and amount to 105, 211, 352, and 563 Hz for the four right-hand patterns. Note the close agreement between the two sets of data, not only in general but also in detail

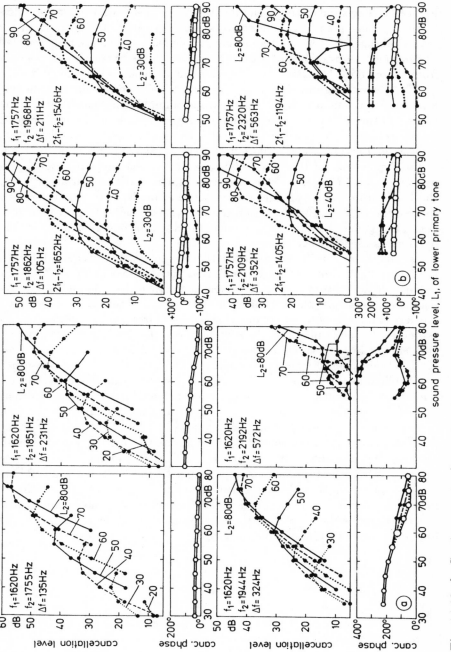

Fig. 14.12a,b. Caption see opposite page

The vector diagram of the wavelets and the sum vector are shown in Fig. 14.11. The sum vector of the many wavelets leads to a point very close to the starting point, i.e. the wavelets produced along the sections of the model cancel each other out almost completely; the large-amplitude wavelets originating from sections 13.6 to 13.8 show a phase angle almost opposite from those above 14.5 Bark, each of which has a much smaller magnitude.

The vector diagram shows the importance of a vector summation that leads to a very small residual vector, while summing the magnitudes of the vectors of the wavelets and neglecting their phase would lead, incorrectly, to a relatively large magnitude. The vector summation of the wavelets leads, as seen in Fig. 14.1, to a situation that would correspond to a minimum in the dependence of the $(2f_1 - f_2)$-cancellation level as a function of the primary level. In order to show the advantages and the limitations using the hardware cochlear nonlinear preprocessing model with active feedback to describe $(2f_1 - f_2)$-cancellation data, psychoacoustically measured data (Fig. 14.12a) and data from the model (Fig. 14.12b) may be compared. The parameters used in each case are comparable.

A superficial comparison between the two sets of data readily indicates that the general trends of corresponding patterns are very similar. More impressive, however, may be the agreement in characteristic details. The crossing points, the slopes, the level differences $L_1 - L_2$, and the peaks of the curves are very similar in both sets of data for the small frequency separation of the primaries. Absolute phase is arbitrary; however, the phase characteristics are very similar in both cases. For large frequency differences of the primaries, the level dependence shows the minima in both cases while the phase patterns split into two phase branches with almost 360° difference. Besides the remarkable agreement between many typical characteristics of the two data sets, there are also a few differences. These are caused mainly by the form of the nonlinear saturating transfer functions in each of the feedback loops. These functions are only rough approximations to physiology and they do not show much nonlinearity for input levels that correspond to values below about 40 dB. The appropriate correspondence should be at about 20 dB. The general agreement and the agreement in detail between the two data sets indicates, however, that the model's description of the cochlear nonlinear preprocessing as the reason for the unusual level dependence of $(2f_1 - f_2)$-difference tone is a useful approximation.

15. Binaural Hearing

Using both ears, we are able to acquire a sensation of the direction from which a sound comes. This can be done even when we close our eyes, or do not see the sound source. Using only one ear, such a sensation cannot be readily produced. This means that we are able to process and correlate the sounds coming to each ear. As always in psychophysics, we have to discriminate between the stimulus and the sensation. Stimulus would be the direction in which the sound source is actually located. The correlated sensation is that of the perceived direction of the sound. The two directions are not necessarily identical.

There are many psychoacoustical effects that can be traced back to the two different stimuli received by our ears. We will discuss just-noticeable interaural delay, binaural masking-level differences, lateralization, localization and binaural loudness. There are many more effects that are described in books on spatial hearing. Especially interesting for room acoustics are sensations such as clarity, separability, diffuseness, spaciousness, echo-content, and so on. These applications are discussed in books on room acoustics, whereas we are concentrating more on some fundamental effects.

15.1 Just-Noticeable Interaural Delay

If a sound source is located not directly in front of our nose, it produces sound signals at our ears that are delayed, one relative to the other. The maximal possible interaural delay is produced when the sound source is located 90° from frontal incidence. Thus a deviation of the incoming waves for the two sounds of about 21 cm is the maximum possible, corresponding to a delay time of 0.6 ms. The just-noticeable difference in localization, which means the just-noticeable change of the direction of the sound source that is necessary to produce a just-noticeable difference in the sensation of the direction, is about 5° for frontal incidence corresponding to a time delay of about 50 µs. This value can be assumed as being the just-noticeable interaural delay at frequencies below about 1.5 kHz. It should not be confused with temporal resolution, which belongs to the stimulation of only one ear and reaches a much larger value of about 2 ms. The just-noticeable interaural delay is

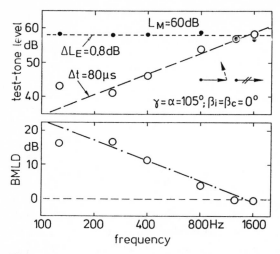

Fig. 15.1. Median test-tone thresholds (*upper panel*) as a function of the test-tone frequency. The masker level is 60-dB SPL and the signal leads the ipsilateral masker by 105° (see vectors in inset). The dots show the monaural results and the open symbols the binaural results. The maskers in the binaural case are in-phase as the vector diagrams in the upper panel indicate. The dotted line indicates the expected values based on a just-noticeable difference in level of 0.8 dB, the dashed line shows the expected values based on a just-noticeable interaural delay of 80 μs. The lower panel shows the corresponding binaural masking level differences (*open circles*) as a function of test frequency, with the predicted values (*dashed-dotted line*) derived from the expected values of the upper panel

somewhat different from subject to subject. Values between 30 μs on the low side and almost 200 μs on the high side have been measured.

Time delay in this context is often thought of as being the delay between the onset of a click, of a tone burst, or of some other short impulsive sound. Another way to produce a delay using steady-state sinusoidal tones is to produce a phase shift between the two ears. If the phase of a tone presented to one ear is different from that of a tone of the same frequency presented to the other ear, the time delay produces a sensation from which threshold can be measured. Assuming that the phase shift of the tone presented to one ear is produced by adding a tone of the same frequency, but out-of-phase by 105°, the phase shift of the sum of the two tones is changed in relation to the situation without the additional tone. If this complex is presented to one ear and a pure tone without an additional tone is presented to the other ear, it is possible to present two sounds to the ears that differ little in amplitude but differ in phase. The broken line in the upper part of Fig. 15.1 shows the results of such experiments. It indicates the level of the added 105° out-of-phase tone needed to reach threshold if the same tone with zero phase, which may be called the masker, is presented to the other ear. The two maskers at both ears are identical and have a level of 60 dB. The data also hold for other frequencies of the tones.

This experiment can be understood in two ways. One is to discuss the results as a measure of the just-noticeable interaural delay as a function of frequency, the other is to discuss the result as that of a binaural masking-level difference measurement. In the second case, only one additional experiment is needed, namely to measure the just-noticeable test-tone level as a function of frequency with a monaural presentation. In this case, the pure tone in the other ear is switched off. The results of this monaural condition are shown as solid symbols in the upper part of Fig. 15.1. The thin dashed horizontal line corresponds to a just-noticeable level increment $\Delta L_E = 0.8$ dB. This result is a median of four subjects and is in close agreement with the 1 dB discussed earlier (see Chap. 7). In the binaural case, the masker in the ear contralateral to the signal was in phase with the ipsilateral masker, as indicated in the vector diagrams of the inset in the upper part of Fig. 15.1. Again, the level L_T of the test tone added in order to be just audible – now in frequency of the three tones presented binaurally; the test tone has a phase of $105°$ relative to the masker in the ear to which it is presented. The results closely follow the broken line, which indicates a constant delay of $\Delta t = 80\,\mu s$. The difference between the two sets of measurements is a difference in the test-tone level needed to be just audible in the case of monaural presentation, and in case of the binaural presentation as indicated in the inset. This is an extreme condition of producing a binaural masking-level difference (often abbreviated as "BMLD"). This value is used as the ordinate in the lower part of Fig. 15.1, showing a decrease in the BMLD as a function of the frequency. The dashed-dotted line shows the results expected from the two lines drawn in the upper part of the figure.

These data show two effects: first, the just-noticeable interaural time delay – in this case $80\,\mu s$ – is a limit that is also effective for BMLDs. The second effect is that phase shifts of sinusoids producing a time delay of $80\,\mu s$ are limited to a frequency range that does not exceed 1500 Hz. Because the phase shift of $105°$ of the test tone, in relation to the phase of the masker, was chosen to produce the largest possible effect, we cannot expect binaural masking-level differences which are based on steady-state conditions or quasi steady-state conditions at frequencies larger that 1500 Hz. According to the lower part of Fig. 15.1, one finds large BMLDs up to frequencies of 250 Hz and then BMLDs decreasing to almost zero at about 1500 Hz.

15.2 Binaural Masking-Level Differences

Although an extreme condition for BMLDs was discussed in the former section, the basic effect of BMLDs was clear. There are, however, many other dependencies of BMLDs on the parameters of the stimuli, a few of which should be discussed before a model describing these effects is introduced. The

BMLD is that difference in just-audible test-tone level that appears when the presentation of signal or masker to one of the ears is changed. There are many stimulus conditions possible because of the two signals, one at each ear, and also the two masker conditions. Most often, the phases of the maskers are equal at the two ears while the phase of the signal is zero at one ear and inverted at the other. The opposite condition, i.e. maskers different and signals identical at the ears, is also used.

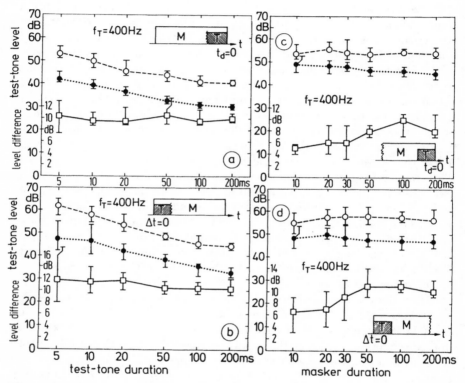

Fig. 15.2a–d. Threshold of test-tone bursts masked by uniform masking noise impulses of 300 ms duration, as a function of the test-tone duration, is indicated in (**a**). Open and closed circles belong to $N_0 S_0$- and $N_0 S_\pi$-conditions, respectively. The squares and the small ordinate scale indicate the medians of individually calculated binaural masking level differences. The bars show the interquartile ranges of eight and four subjects. Masker level is 60 dB, the frequency of the test signal is 400 Hz and the test signal ends together with the masker as indicated in the inset. The data in (**b**) are similar, however, for the condition that the test signal starts together with the masker. The parts (**c**)and (**d**) show corresponding data for a test signal duration of 10 ms as function of the masker duration (abscissa) for the conditions indicated by the insets

15.2.1 Dependencies of BMLDs

An interesting and fundamental dependence of BMLD is that on the duration of masker and test tone. The left side of Fig. 15.2 shows the dependence of the just-noticeable level of the test tone, masked by uniform masking noise, as a function of duration for a test-tone frequency of 400 Hz. One measurement was done with the test signal in both ears in phase (S_0, open circles) or 180° out of phase S_π, filled circles). Phase of the noise masker was kept constant at N_0 in both ears.

The results show very clearly that masked thresholds in both conditions depend on test-tone duration. The BMLD, however, which is plotted at the bottom of each panel, does not show any dependence. This occurs whether the test tone is presented at the end of the masker (Fig. 15.2a) or at the beginning of the masker (Fig. 15.2b). The duration of the masker was kept constant at 300 ms. In another experiment, similar conditions were produced, however, the masker duration was varied while the test-tone signal duration was kept constant at 10 ms. The results are indicated on the right side of Fig. 15.2. It shows that the BMLD increases from about 5 to 10 dB as masker duration increases from 10 to 200 ms, regardless of the temporal position of the test-tone burst (shown in the insets of Fig. 15.2c and d). These effects also hold for other frequencies of the test tone although the BMLD becomes smaller for higher frequencies.

All BMLDs described so far have been produced under simultaneous presentation, i.e. the masker and test signal are presented at the same time. It is very interesting to see whether BMLDs can also be produced under conditions of non-simultaneous masking. The results of such a condition are outlined in Fig. 15.3. The temporal conditions of masker and test sound are indicated in the insets. Postmasking thresholds L_T of the 10-ms, 800-Hz test tones masked by uniform masking noise bursts of 300 ms duration ($L_M = 70$ dB), are shown as a function of delay time t_d between the end of the masker and the end of the test signal for N_0S_0-configuration (open circles) and N_0S_π-configuration (dots) in part (a) of the figure. Part (b) shows premasking thresholds for which the time difference Δt between the onset of the test signal and the onset of the masker is the abscissa. The BMLDs calculated from these data are plotted as the lower curve on the separate ordinate scale. The BMLD shows the same dependence on the temporal position of the test signal as the masking. Additional measurements have shown that similar effects are produced at lower test-tone frequencies. Sequences of masker bursts produce masking patterns within which premasking and postmasking superimpose. This superposition takes place also in BMLDs. The dependence of postmasking decay on masker-burst duration is also found in BMLDs. BMLDs measured in both simultaneous and non-simultaneous masking are directly related to the amount of masking, in such a way that more masking produces larger BMLDs.

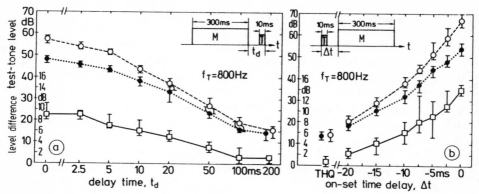

Fig. 15.3a,b. Post-masking thresholds of 10-ms-long test tones masked by uniform masking noise bursts of 300 ms duration, as a function of the delay time between the end of the masker and the end of the signal (see inset in the left part). The open circles correspond to the $N_0 S_0$-condition of test signal and masker, the closed circles to the $N_0 S_\pi$-condition and the squares indicate the medians of individual binaural masking level differences using the ordinate with the small scale. Test-tone frequency is 800 Hz. The right part shows the same set of data for pre-masking thresholds for which the time difference between the onset of the test tone and the onset of the masker is the abscissa. Threshold in quiet is marked by the left-hand data

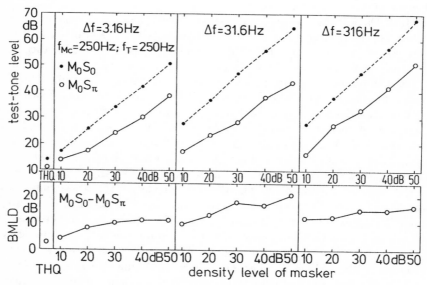

Fig. 15.4. The top three panels show median thresholds for 250-Hz tones, as a function of the spectral density level of the masker with a centre frequency of 250 Hz and bandwidths Δf of 3.16, 31.6 and 316 Hz. The solid symbols connected by dashed lines indicate the results from the $M_0 S_0$-condition and the open symbols the results from the $M_0 S_\pi$-condition. The bottom three panels show medians of the individual observers' BMLDs as a function of the spectral density level of the masker for the three masker bandwidths

The dependence of BMLDs on level was studied for different bandwidths of the masker at a frequency of 250 Hz. Figure 15.4 shows this dependence with three different bandwidths. The top three panels indicate thresholds for 250-Hz tones as a function of the spectral density level of the masker, with centre frequency, f_{Mc}, equal to 250 Hz and bandwidths of 3.16, 31.6, and 316 Hz. The solid symbols connected with dashed lines indicate the results from the M_0S_0-condition and the open circles those from the M_0S_π-condition. The bottom three panels illustrate the corresponding BMLDs. It seems that the BMLD increases only slightly as a function of the density level of the masker. BMLDs seem to reach larger values for a bandwidth of 31.6 Hz. Similar effects, although smaller in value, have been found for centre frequencies and test-tone frequencies of 800 Hz.

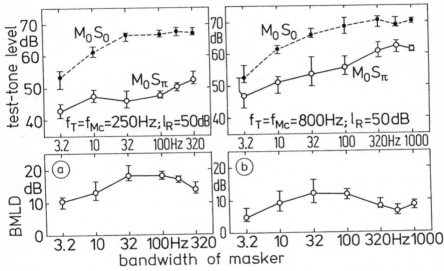

Fig. 15.5a,b. Median thresholds of 250-Hz test tones (**a**) and 800-Hz test tones (**b**) as a function of the masker bandwidth. The closed symbols connected by dashed lines indicate the results from the M_0S_0-condition, the open symbols connected by continuous lines those from the M_0S_π-condition. The spectral density level of the masker is 50 dB. The bottom panel shows the median of the individual observers' BMLDs as a function of masker bandwidth. Vertical bars indicate the interquartile ranges (six observers)

The data showing the dependence of the BMLD on bandwidth are outlined in Fig. 15.5. Again, the thresholds in the two conditions M_0S_0 and M_0S_π are plotted in the upper part. The frequency of the test tone and the centre frequency of the noise is 250 Hz for part (a) and 800 Hz for part (b); the density level was kept constant at 50 dB. BMLDs plotted in the lower part show a flat maximum between 32 and 100 Hz where values near 18 dB or

12 dB are reached. Towards lower and higher bandwidths, the BMLD seems to decrease. Data from other centre frequencies show similar effects.

For a density level of 60 dB, the BMLD was measured not only where the frequencies of the test tone were equal to the masker's centre frequency, but also at test-tone frequencies in the neighbourhood of the narrow-band masker. In Fig. 15.6, the upper panel illustrates masked threshold as a function of test-tone frequency, f_T. The bandwidth of the masking noise decreases from 100 Hz to 31.6 Hz to 10 Hz from parts (a) to (b) to (c). The BMLDs plotted on the bottom of each part show a maximum at the centre frequency of the noise and decrease as the test-tone frequency deviates from the centre frequency. This is similar to, but more rapid than, the changes in masked threshold and means that the BMLD drops rapidly when masker and signal have no frequency components in common.

In order to find out whether it is the masking effect itself which is responsible for the decreasing BMLD, a special condition was created so that the masking noise was shifted and centred on the frequency of the test tone. The spectral density level of the masking noise, however, was reduced from 60 dB to a lower value to produce identical masking in the $M_0 S_0$-case, where the masking noise at 60 dB spectral density level was centred at 250 Hz. The results of this experiment are shown in Fig. 15.7. In the upper panel, the medians of the level L_T at threshold are plotted as a function of the test-tone frequency f_T. The curves labelled $M_0 S_0$ and $M_0 S_\pi$ are replications of the results plotted in Fig. 15.6c. A comparison indicates the reproducibility of the BMLD's are shown with open symbols in the bottom panel. The filled symbols connected with dotted lines in the top panel of Fig. 15.7 show the results for the $M_0 S_\pi$-condition obtained when the noise was centred on the frequency of the test tone and adjusted in level as mentioned above. The difference between the results in this condition and the filled symbols of the $M_0 S_0$-condition in the top panel are the BMLDs that are shown as filled symbols in the bottom panel. These BMLDs depend only slightly on test-tone frequency, and much less than in the condition indicated in Fig. 15.6c. These additional data show very clearly that the reduction of BMLD is neither a result of changing signal frequency nor a result of changing the effective level of the masker.

Tonal maskers in relation to narrow-band maskers have the advantage that they do not show any random temporal change of envelope. Therefore, many BMLD data have been produced using 250-Hz masker tones with a 250-Hz test tone of different phase relations. Figure 15.1 is an example of such a condition. These results suggest that the just-noticeable difference in level (monaurally about 1 dB) and the just-noticeable interaural delay (about 100 μs) are primarily responsible for BMLDs.

The main difference between tonal maskers and narrow-band maskers is the temporal variation in envelope and phase. In order to extract the influence of these two variables, special frozen-noise maskers have been produced

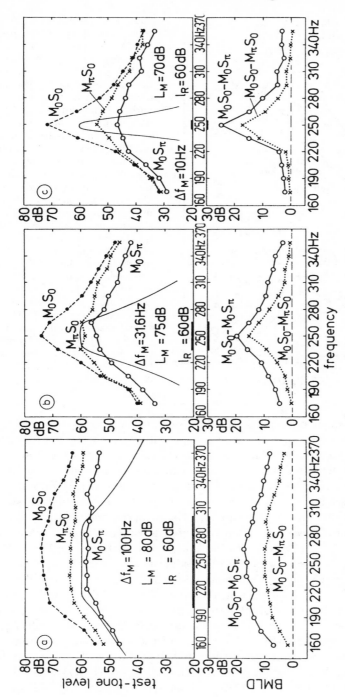

Fig. 15.6a–c. The upper panels show the median test-tone level at masked threshold for four observers as a function of test-tone frequency. The parameter denotes the interaural phase condition. The masking noise had a nominal band width of 100 Hz on the left, 31.6 Hz in the middle and 10 Hz on the right. The spectral density level within the bands was 60 dB. (The thin solid curve shows the actual spectral density level). The bottom panel shows the medians of the BMLDs of the individual observers, again as a function of test-tone frequency

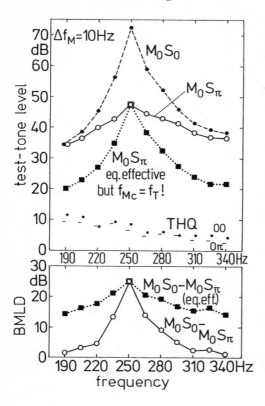

Fig. 15.7. Medians of the test-tone level at threshold are plotted as a function of test-tone frequency in the top panel. For M_0S_0- and M_0S_π-conditions, the masking noise of 60-dB spectral density level is centred at 250 Hz with a bandwidth of 10 Hz. The filled squares show the results in an M_0S_π-condition where a masking noise of 10 Hz bandwidth is centred on the test-tone frequency. It's spectral density level was adjusted separately for each observer, to produce the same masked threshold as produced at that frequency in M_0S_0-condition by the 60-dB masker centred on 250 Hz. In the lower part of the upper panel, the medians of the threshold in quiet (THQ) are indicated for M_0S_0- and M_0S_π-conditions. The bottom panel shows the medians of the individual observers' normal BMLDs ($M_0S_0 - M_0S_\pi$) as open symbols and the BMLDs for matched effective masking ($M_0S_0 - M_0S_\pi$) as closed symbols

from continuously repeated 500 ms-long sections. The top portion of Fig. 15.8 shows about 350 ms of such a noise. It is centred at 250 Hz and has a bandwidth of 31 Hz. In the BMLD experiment, the noise was identical in both ears, i.e. the M_0-condition. The trace below this noise shows a 300-ms burst of the 250-Hz sinusoid used as the signal. It is shown in the phase during which it was added in the S_0-condition. Instead of the 300-ms burst, short tone bursts of 10-ms duration and 5-ms Gaussian rise and fall times were used as test signals. These have been presented at different times, t, within the noise burst. The signal presentation time, t, is the abscissa in Fig. 15.8. One of these signals is shown in the romboidal aperture centred at $t = 100$ ms. The middle panel of Fig. 15.8 shows test-signal level, L_T, at threshold. The signal is either in phase (S_0) or 180° out of phase (S_π) as indicated. The bottom panel shows BMLDs for the 12-ms test signals. (The data points on the right side of the figure belong to test-signal durations of 300 ms). The masker level was 70 dB SPL. Thresholds of the 12-ms test-tone burst show large variations as a function of time, and the variations for M_0S_0 are larger than those for M_0S_π. The variations of the two threshold curves as a function of time seem to show no correlation. Therefore, the resulting BMLDs range

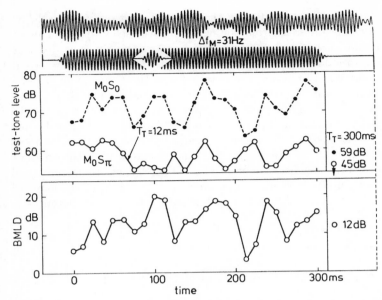

Fig. 15.8. A segment of a repetitive burst of noise which was centred at 250 Hz with a bandwidth of 31 Hz is shown at the top. The masker noise is identical for both ears. The trace below the noise is a 300-ms burst of the 250-Hz sinusoid used as the signal. It is shown in the phase in which it was added in the S_0-condition. One of the signals actually used is shown in the rhomboidal aperture; it is 12 ms long and has a Gaussian rise/fall-time of 5 ms. In the case illustrated, the 12-ms signal occured at $t = 100$ ms The middle panel shows median threshold levels as a function of the time at which the peak of the 12 ms signal occurred within the noise. The signal at the ears was either in phase, S_0, or $180°$ out-of-phase, S_π. The bottom panel shows the median of individual BMLDs also as a function of the time of occurrence of the signal (similar data for a 300-ms signal are shown at the right side of the figure)

from 3 dB to 20 dB. This means that the BMLD varies strongly as a function of time. When using long tone bursts, some form of averaging seems to take place. For a test-tone duration of 300 ms, this leads to a threshold of 59 dB for M_0S_0 and of 45 dB for M_0S_π. The corresponding BMLD would be 14 dB comparable to that of the individual medians, which result in an average of 12 dB.

15.2.2 Model of BMLDs

The model assumes devices or channels that are tuned to particular interaural delays. These devices, which exist only for frequencies below 1500 Hz, are assumed to be arrays of channels each tuned to a different characteristic interaural delay. Each channel then acts as an interaural delay filter. It transmits whatever signal arrives at its characteristic delay without atten-

uation, but attenuates more and more those signals that arrive with delays
that are increasingly different from the characteristic delay of the channel.
It is further assumed that each element of the array of channels is identical.
Using this assumption it is possible to measure the attenuation characteristic
directly. The derived attenuation characteristics of the assumed interaural-
delay-tuned channels are plotted in Fig. 15.9. Attenuation in dB is the ordinate and interaural delay in μs the abscissa. The straight line shows the
median derived from data at 250 Hz and 60-dB masker level. The broken
lines show the data derived from the individual observers who were most
and least sensitive to level differences, i.e. the observers with the smallest
and largest just-noticeable level differences, JNDLs. Because their individual
JNDLs have been used to produce these lines, the variation in the slope of
the mechanism occurs in addition to variability attributable to differences in
their JNDLs.

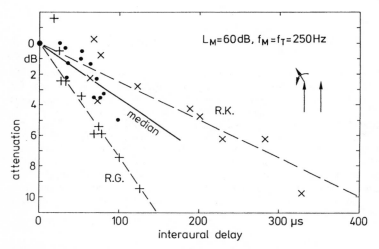

Fig. 15.9. Derived attenuation characteristics of interaural-delay-tuned channels,
with attenuation on the ordinate as a function of interaural delay on the abscissa.
The filled symbols (fitted by the solid line) are median data for 250-Hz, 60-dB
maskers. The other symbols show data for two extreme individual observers

The results of Fig. 15.9 suggest that the just-noticeable interaural delay,
$JN\Delta T$, for the median observer, for which a JNDL of 1 dB is assumed,
should be about 30 μs. This value is close to the $JN\Delta T$ measured for subjects
who have practice in binaural listening at low frequencies. In this context,
it should be noted that the model has only one single decision axis, which
is measured as ΔL. Therefore, all factors that influence JNDL should also
influence $JN\Delta T$. The model, therefore, accounts for the similarity in the
effects of level, duration, simultaneous masking or postmasking on JNDL and
$JN\Delta T$. It also accounts for the fact that JNDs are correlated for individual

subjects and thereby suggests the source of some of the residual individual variability in $JN\Delta T$.

The BMLD therefore results from the fact that the subject is able to use channels in which the masker's effect is reduced. This occurs because the differences in interaural delay between the masker and the resultant signal-plus-masker direct the two wave forms to different channels. All the information contained in the place of the signal (its frequency, phase or amplitude) is available at the improved signal-to-noise ratio in the channels whenever a BMLD is produced. The model also suggests that any change in the stimulus which increases the size of the interaural delay produced by adding a signal, should increase the size of the BMLD. Temporal effects, which limit the duration of a given interaural phase condition, should reduce the size of the BMLD.

The use of the model that relates $JN\Delta T$, JNDL, and JNDs to the BMLD, may be illustrated by two examples. Although the model reduces the observers' judgements in BMLD to a judgement of level, it may be helpful to keep JNDL and $JN\Delta T$ separate in discussions of its applications.

The first example is that of the frozen-noise masker with test-tone frequency equal to the centre frequency of the masking noise. Figure 15.10 shows the BMLDs predicted for the most interesting intervals of signal presentation in the frozen-noise experiment presented in Fig. 15.8. The top line shows a segment of the repetitive noise (centre frequency 250 Hz, bandwidth 32 Hz), and the second line shows the signal (12-ms, 250-Hz tone burst), both on an expanded scale in relation to Fig. 15.8. The relative phase of the signal and masker at the peak of the signal is shown in the next line. The fourth and fifth lines show the ratio of signal-to-masker level required to produce a JNDL at each phase of 3.5 dB for a $JN\Delta T$ of 300 µs. These values are consistent with the JNDs of the short-duration signals. From these two values, the corresponding ratio of signal-to-masker intensity for the phase angle given is calculated. These are plotted as $20 \log{(S_0/M)}$ and $20 \log{(S_\pi/M)}$. The difference between the two levels, the BMLD, is plotted in the bottom panel as a function of the time of occurence of a signal maximum together with the medians and interquartile ranges from Fig. 15.8. With one exception at 200 ms, the predictions agree closely in detail using the model. With longer duration signals, however, it is necessary to consider how information is combined from different temporal ranges of the signal. If one uses 200 ms as the integration time, an improvement of about 12 dB between an average threshold for the 12-ms signal and that for the 300-ms signal in the M_0S_0-case should be expected. The median difference between the thresholds obtained with 12-ms signals averaged across all temporal ranges for each observer separately, and the result with the 300-ms signal is 13.7 dB in the M_0S_0-condition. For M_0S_π, the improvement is 13.3 dB. Thus, for simultaneous masking with signals having frequencies within the band of the masking noise, a straight-

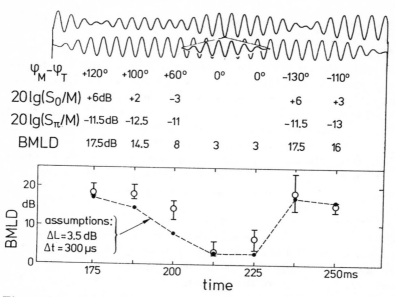

$\varphi_M - \varphi_T$	+120°	+100°	+60°	0°	0°	-130°	-110°
$20\lg(S_0/M)$	+6dB	+2	-3			+6	+3
$20\lg(S_\pi/M)$	-11.5dB	-12.5	-11			-11.5	-13
BMLD	17.5dB	14.5	8	3	3	17.5	16

Fig. 15.10. The top line shows a segment of the repetitive noise ($f_{MC} = 250$ Hz, $\Delta f_M = 31$ Hz) and the second line shows the signal (12-ms burst of a 250-Hz sinusoid), i.e. an expanded segment of Fig. 15.8. The relative phase of signal and masker at the peak of the signal is shown on the next line. The fourth and fifth lines show the ratio of signal-to-masker level required to produce, at each phase, a JNDL of 3.5 dB (the row $20\log(S_0/M)$) and/or a $JN\Delta T$ of 300 μs, whichever is smaller (the row $20\log(S_\pi/M)$). The difference between the two levels, the BMLD, is plotted in the bottom panel as a function of the time of occurrence of the signal maximum, together with the median results and interquartile ranges from Fig. 15.8.

forward temporal integration up to a duration of 200 ms seems to be a good approximation.

The second example deals with the effect of signal frequency. It describes, in other coordinates, results that are found in threshold measurements, in which a monaural sinusoidal signal is added either in phase or 90° out-of-phase to a continuous masker of the same frequency. The masker is presented binaurally and is always identical in each ear. The frequency is varied from 63 to 4000 Hz. Figure 15.11 shows the transformed data at $L_M = 50$ dB (left) and 70 dB (right). Results for the condition with in-phase addition of the signal are shown by the line labelled "ΔL", which gives the value of $\Delta L = 20 \log(1 + p_T/p_M)$ shown on the outer ordinates in dB and on a logarithmic scale. As a first approximation, the value of ΔL is independent of frequency between 250 and 4000 Hz but rises slightly towards lower frequencies. With the binaural masker and 90° out-of-phase addition of the signal, the subject may reach either a JNDL or a $JN\Delta T$. The results for this condition are shown in two curves, one labelled Δt and the other $\Delta L'$. The two curves are derived by taking the threshold signal and calculating first the corresponding

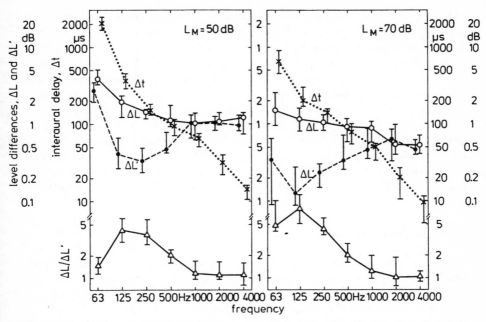

Fig. 15.11. In-phase addition of the monaural signal and masker produces a level difference, ΔL, related to the outer ordinates. Quadrature addition produces both level changes, $\Delta L'$ (outer ordinate) and interaural delays, Δt (inner ordinate). All three are shown as a function of frequency. The bottom curve ($\triangle$) shows the ratio of ΔL to $\Delta L'$. Results derived from two masker levels are shown separately in the two panels

interaural delay ΔT and then the corresponding level change $\Delta L'$. When $\Delta L'$ falls below ΔL, as it does for frequencies below about 1 kHz, the level cue from quadrature addition is inaudible at threshold. The observer uses the cue related to interaural delay. Above 1 kHz, this cue becomes unusable because the sensitivity to interaural delay decreases rapidly above 1 kHz or, in terms of the model, because the channels tuned to interaural delay are either not available or are too broadly tuned at these frequencies. Above about 1 kHz, the level change, $\Delta L'$, reaches the level of audibility before a detection based on interaural delay is possible. This may be seen more clearly by noting the ratio of ΔL to $\Delta L'$ (indicated in the lower curves). In the regions in which the ratio is much larger than unity, a large binaural masking-level difference is expected. The dependence of BMLD on frequency has often been noted; the diagram, however, shows clearly why the BMLD disappears for frequencies larger than about 1200 Hz.

From the applications of the model, three factors may be added. (1) The well known variability of $JN\Delta T$ is based on two factors: the variation in JNDL from subject to subject, and the differences in the slope of the individual observers' delay-tuned channels. It may be that these two factors are correlated. (2) BMLDs are conventionally measured in terms of signal level. There are, however, important nonlinearities between the conventional measure and the determining JNDs, i.e. JNDL and $JN\Delta T$. However, the effects of varying stimulus parameters on JNDs are reflected sometimes through nonlinearities such that either small differences and/or individual differences can produce what may be often called unusual dependencies. Many of these strange dependencies can be described by the model. (3) The model explains quite clearly that the BMLDs are in fact an improvement in signal-to-masker ratio which consequently leads to a release from masking in frequency and amplitude discrimination. The model also explains the reduction in magnitude of BMLDs when signal and masker frequencies differ from each other.

15.3 Lateralization

Effects of lateralization are mostly studied by presenting sounds through headphones. In this case, the sound image is usually located *within* the head. In such a situation, the task of the subject is to describe the perceived lateral displacement of the sound. This displacement can be projected onto a straight line, which connects the entrances to both ear canals. In contrast to localization, a three-dimensional phenomenon, lateralization is described along only one dimension.

Lateralization is governed by differences in level and time of the sounds presented to the two ears via headphones. Figure 15.12 shows the dependence of lateralization on interaural level differences. Pure tones of 1 kHz were used for the experiment. The levels of the tones presented to each ear were chosen such that their overall intensity corresponded to a sound pressure level

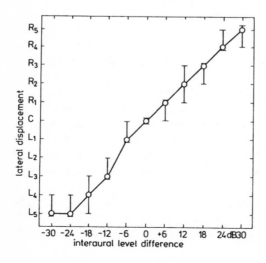

Fig. 15.12. Lateralization as a function of interaural level difference. 1-kHz tones, overall intensity corresponding to 80 dB SPL

of 80 dB. Positive values of the interaural level difference indicate that the earphone on the right ear of the subject was fed by a higher voltage. The ordinate in Fig. 15.12 shows the lateral displacement perceived by the subjects: the image is either shifted to the centre (C) or displaced to a variable degree to the right ($R_1 \ldots R_5$) or to the left ($L_1 \ldots L_5$). A lateralization that is "completely right" is denoted R_5, a lateralization that is "completely left" is indicated by L_5. As can be seen in Fig. 15.12, a centre image with no lateral displacement shows up for zero interaural level difference; this judgement ist seen to have very little variability. With increasing interaural level difference, the sound image shifts more and more to the right. For negative values of the interaural level difference, the sound image shifts more and more to the left. At interaural level differences of ±30 dB, the extreme lateralizations of the sound image are obtained. For a restricted range of interaural level differencies, ±10 dB, a just-audible displacement is achieved if the level difference is varied by 1 to 2 dB.

A similar dependence of perceived lateral displacement can be obtained if, instead of level differences, differences in the phase of the signals presented to the earphones are used. If, at frequencies below 1500 Hz, the left earphone delivers sound impulses about 1 ms later than the right earphone, the image shows a lateral displacement totally to the right (R_5). If the left earphone leads by 1 ms, the lateral displacement is completely to the left (L_5). Between these extreme positions there is a smooth transition.

The time shifts between the signals presented to each earphone can be produced in two different ways: on one hand, time shifts can be applied to the carrier signals; on the other, interaural time shifts can be applied to the envelopes of the signals presented to the left and right earphone, respectively. In the first case, with interaural time shifts of the carrier, lateral displacements occur up to a carrier frequency of about 1.6 kHz. This frequency corresponds to a critical-band rate of 11.5 Bark. Hence, lateral displacements can be obtained by interaural time shifts of the carrier only, if the carrier shows a critical-band rate within the lower half of the critical-band-rate scale. However, interaural time shifts of the envelope lead to a perception of lateral displacements for carriers within the middle and upper range of the critical-band-rate scale. At lower frequencies, time shifts of the envelope and time shifts of the carrier are not easy to distinguish.

There is a trade-off between interaural level differences and interaural time shifts. This means that a shift of the sound image towards the right, obtained by presenting the sound to the right earphone earlier, can be shifted back to a centre image by presenting the sound to the left earphone at a higher level.

15.4 Localization

In contrast to lateralization, which represents a one-dimensional phenomenon, localization encompasses three dimensions. With localization, two effects are important: localization as such, i.e. the perception of a sound with respect to its direction and distance, and localization blur, i.e. the precision by which the location of a sound image can be given.

With respect to localization blur, deviations of about 2° from a forward direction can be detected with sinusoidal signals. For sound sources facing one of the ears, i.e. at angles of ±90°, localization blur shows somewhat larger values around ±10°. For sounds presented behind the subject, localization again improves and localization blur amounts to about ±5°. With narrow-band signals, a special type of localization frequently occurs. In such a case, called inversion, the sound source is situated in front of the subject but the subject perceives the sound as coming from behind. This means that the sound source and the perception of the location of the sound source are symmetric about the axis of the two ears. In such situations, the following effects may occur: if the subject faces the direction 0° and the sound source is presented at an angle of 30°, the sound may be perceived at an angle of 150°. This means that the sound source is situated in front of the subject somewhat to the right, but that the sound is perceived as coming from a position *behind* the subject.

8kHz

300Hz, 3kHz 1kHz, 10kHz

Fig. 15.13. Schematic illustration of the localization of narrow-band sounds in the median plane, irrespective of the position of the sound source

Another interesting feature is the localization of narrow-band sounds in the median plane. This effect can be illustrated by means of Fig. 15.13. With some simplification, it can be stated that sounds are perceived as coming from a specific direction, irrespective of the location of the sound source. If narrow-band sounds with a centre frequency of 300 Hz or 3 kHz are presented, the sound image is always perceived in front of the subject. Narrow-band sounds centred at 8 kHz are perceived as coming from a location above the head of the subject, even if the sound source is located in front of the subject. Narrow-band sounds centred at 1 or 10 kHz are perceived to originate behind the head of the subject, again irrespective of the actual location of the sound source. This is an astonishing effect and was attributed by Blauert to the frequency characteristics of the hearing system, and was called "determining frequency bands".

So far, the localization of only one sound source has been considered. However, in practical applications such as the reproduction of music via a stereosystem, localization of sounds reproduced by multiple sources is of interest. In a traditional stereo arrangement, the subject faces two loudspeakers positioned in front, each at an angle of about 30° from straight ahead. If both loudspeakers radiate the same sound pressure level at the same time, a summing localization occurs and the sound is perceived as coming from a location midway between the two loudspeakers. If the sound pressure level from the right loudspeaker is 30 dB higher than that from the left speaker, the sound is perceived as originating in the direction of the right loudspeaker. This means that a dependence of the location of the perceived sound is obtained in a manner similar to that shown in Fig. 15.12 for lateralization. Also, as discussed with lateralization, localization of the perceived sound is governed by time differences between the sounds from the loudspeakers. If a low-frequency sound from one loudspeaker leads the sound from the other by 1 ms, the perceived localization of the sound corresponds to the first loudspeaker.

In this context, an effect termed the "law of the first wave front" deserves attention. If a sound is switched on with a click in the left loudspeaker, then it will be perceived as originating from that location, even if the left loudspeaker is later faded out and the sound is radiated only from the right loudspeaker! This effect of the "first wave front" works even if the level of the second loudspeaker is increased by some decibels above the level of the first loudspeaker. The "law of the first wave front" has great importance in room acoustics. The localization of the sound of musical instruments is in the direction of the stage, even if the level of sound reflected e.g. from a balcony is *higher* than the level of the direct sound wave. However, if the time difference between the direct sound and the reflected sound exceeds a critical value (for speech and music near 50 ms), echos may be perceived which can be highly annoying for values larger than 100 ms.

15.5 Binaural Loudness

If a subject listens over headphones to a sound using only one earphone, and suddenly the second earphone is also connected to the sound source, a distinct increase in loudness is perceived. The increase of binaural loudness in comparison with monaural loudness depends somewhat on actual loudness, and hence on the sound pressure level. For soft sounds (20 dB SL), switching from monaural listening to binaural listening produces an increase in perceived loudness by a factor of about 2. At high levels (80 dB SL), the increase from monaural to binaural loudness amounts only to a factor of about 1.4. This means that at low levels (20 dB SL), the increase from monaural to binaural loudness corresponds to an increase in (monaural) sound pressure level of 8 dB. At high levels (80 dB SL), the corresponding level increase amounts to 6 dB. An interesting case of binaural loudness is that of loudness

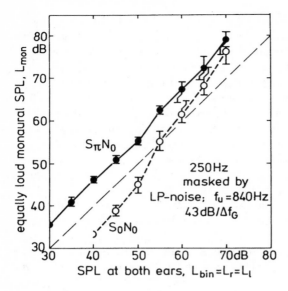

Fig. 15.14. The loudness of binaurally presented 250-Hz tones partially masked by binaurally presented low-pass noise with 840-Hz cut-off frequency and 43 dB SPL per critical band, expressed as the equally loud monaural 250-Hz tone, the level of which is the ordinate. The SPL of the equal-amplitude binaural tones is the abscissa. Two-phase conditions of the signal and masking noise are indicated. The half symbols indicated approximate the location of the masked thresholds in the two conditions

partially affected by noise. When the loudness of tones is measured, the interaural phase conditions in which the tones and the noises are presented to the ears becomes important, as already discussed in Sect. 15.2 in connection with the binaural masking-level difference for very small loudness, i.e. the masked threshold. The number of variables influencing the perception of loudness becomes large in the binaural situation. A few data from a preliminary study may give an idea of the expected effects.

250-Hz tones with the same level in each ear have been used as binaural signals, the loudness of which was measured by adjusting the level of a monaurally presented 250-Hz tone to sound equally loud. The background noise was low-pass filtered with a steep cut-off at 840 Hz and an SPL of 43 dB in each critical band, corresponding to a density level of 23 dB. The data in Fig. 15.14 show the level of a monaurally presented 250-Hz tone, necessary to sound as loud as the binaurally presented 250-Hz tones (abscissa). The masking noises were in phase at both ears (N_0), while the signals, the 250-Hz tones are in phase (S_0) for one set of measurements (open circles, lower curve) and 180° out of phase (S_π) for the second set (filled circles, upper curve). The data are medians and interquartile ranges of four subjects. The results at low signal-to-noise ratios indicate – as expected from BMLD data – that presenting the 250-Hz tones binaurally in opposite phase produces greater loudness than in-phase presentation. It is astonishing, however, that this increment is effective not only near masked threshold but up to 40 dB above it. Masked threshold for the two phase conditions of the signals is close to the leftmost symbols, which are drawn as half symbols because only two of the subjects were able to hear the 250-Hz tones at these levels.

The dependence of binaural loudness on the phase difference between the signals at the ears (abscissa) is outlined in Fig. 15.15 using the same method of monaural adjustment for two conditions of the masking noise: in phase at both ears (N_0, open circles) and 180° out of phase (N_π, filled circles). The masking noise was the same as in the former experiment; the level of 250 Hz-tones in the binaural presentation was set to 50 dB. The data indicated for $\Delta\varphi = 0°$ and $\Delta\varphi = 360°$ are replications, and may be compared with the data given in Fig. 15.14 at $L_{bin} = 50$ dB. The consistency of the data is good and the individual differences, as marked with the interquartile ranges in the two experiments, are remarkably small.

The data of the two figures indicate that phase differences in binaural hearing can influence loudness, so that out-of-phase presentation of tones increases loudness up to a factor of two (equivalent in effect to a 10-dB level difference for monaural presentation). Phase changes in both signals and maskers (noise, in this case) clearly influence binaural loudness.

Binaural loudness summation can also be studied with loudspeakers in a standard stereo arrangement. The loudness summation depends both on the frequency separation and time separation of the sounds from the two loudspeakers. For zero frequency separation and zero temporal separation, two loudspeakers produce a larger loudness than one loudspeaker, corresponding to a level difference of about 4 dB. For frequency separations as large as 4 kHz, the loudness increase for presentation over two loudspeakers corresponds to 12 dB at low loudness levels (45 phon) and to about 8 dB at high loudness levels (85 phon).

In summary, under favorable conditions binaural loudness may achieve almost twice the value of monaural loudness for the same physical level presented at about the same time.

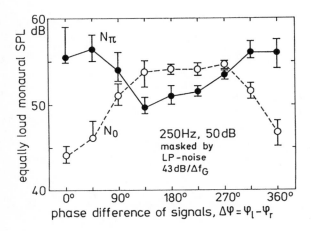

Fig. 15.15. Loudness of binaurally presented 250-Hz tones partially masked as in Fig. 15.14, but with an SPL of 50 dB and shown as a function of the phase difference of the signals (abscissa). Two phase conditions of the low-pass masker noise are indicated

16. Examples of Application

In our human society, acoustical communication plays a very important role, because besides our receiving system (our ears), we also have a transmitting system (our speech organ). Therefore, the applications of psychoacoustics are spread across many different fields. Most often, psychoacoustical data provide the fundamental basis from which solutions to problems (such as finding the limiting characteristics of transmitting systems, or the limits of noise production) are elaborated. Even in the region of art, to which music belongs, the characteristics of our hearing system as the receiver of music play *the* dominant role. Consequently, psychoacoustics is very important also in musical acoustics. In view of the diversity of the field, it is impossible to discuss a large number of different applications in detail. However, this chapter gives some impression of how much psychoacoustics is involved in the different fields that have been mentioned, and may give those who want to solve similar problems some hints on how they might proceed. The list of papers for each section of the chapter may also be of help.

16.1 Noise Abatement

Noise, or unwanted sound, has to be limited. In the 1960ties, there were many methods used to measure noise. Pushed by industry, the International Standardization Organization (ISO) was forced to standardize a method that was not only practical but that would also indicate correct and adequate values. Standardization is a time-consuming process, and it was finally decided to solve the problem in two steps. The first step produced a simple method that could easily be implemented using relatively cheap equipment. The ISO was aware of the fact that this initial method, measurement of A-weighted SPL, might produce inaccurate or even misleading results in noise control. Therefore, the ISO produced a second standard that was not as simple as the former but produced much more appropriate values based on the human sensation of loudness. Two corresponding methods have been described as loudness calculation procedures in ISO 532 and were published only a few years after the dB(A)-proposal. The method described in ISO 532 B is based on what is described in Chap. 8 on loudness and the examples will be elaborated using that method.

Two kinds of loudness distributions are often used. The first one is specific loudness versus critical-band rate pattern, which will be abbreviated here to "loudness pattern". The second distribution is based on the time function of total loudness, which can normally be measured only using loudness meters. Because loudness of noises often depends strongly on time, a cumulative loudness distribution is used. This distribution gives the percentage of time during which a given loudness is reached or exceeded, as a function of loudness.

16.1.1 Loudness Measurement

The graphical procedures for calculating loudness described in ISO 532B or in DIN 45 631 were implemented in a computer program published in the latest version of DIN 45 631.

The program shows a very good approximation of the loudness function for the 1 kHz standard tone; it is within 2% from 0.02 to 150 sone. It is shown in Fig. 16.1 together with the corresponding functions for uniform exciting noise (cf Fig. 8.4).

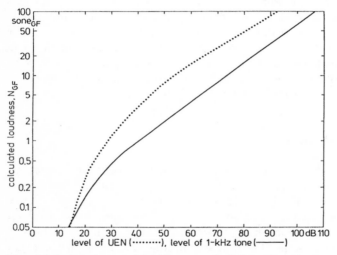

Fig. 16.1. Loudness of 1-kHz tones (*solid*) and uniform exciting noise (*dotted*) calculated by a computer program or measured by a loudness meter, as a function of sound pressure level

The use of the loudness-calculating procedure is most effective in noise reduction. An example of a woodworking factory may give insight into its advantages. Initially, the noise was mostly that of a circular saw, which produced a spectrally dominant narrow-band sound together with more broadly tuned noise. Figure 16.2a shows the original loudness pattern indicating the

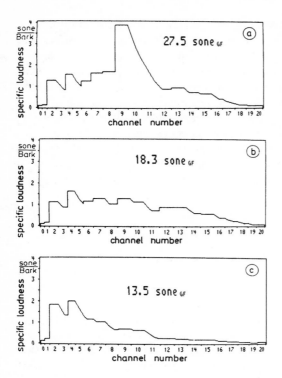

Fig. 16.2a–c. Loudness patterns, i.e. specific loudness as a function of critical-band rate expressed in channel numbers, calculated by a computer program for three stages of a noise reduction activity in a woodworking factory. Original situation in (**a**); using special circular saw blades in (**b**) and with additional absorption in (**c**). The total loudness is indicated in each panel

large partial area near 1.25 kHz (channel 9), which also produces a large contribution to the total loudness. First, this dominant partial loudness had to be reduced using special noise-reduced circular saw blades. The remaining noise, which is shown in Fig. 16.2b, was broad-band noise that had a relatively low A-weighted SPL but a relatively large loudness. Therefore, additional noise reduction was introduced in the higher frequency region, where it can be accomplished more easily than at lower frequencies. This way, the loudness could be reduced further, as indicated in the loudness pattern of Fig. 16.2c. The relative loudness reduction from 27.5 to 18.3 and finally to 13.5 sone (GF) corresponds to factors of 1.50 and 1.36. These ratios indicate that the second step was also quite effective, although the reduction of A-weighted SPL was 13 dB(A) in the first step and only 4 dB(A) in the second. The workers in the factory, however, were very pleased with the second step.

The loudness of steady-state sounds can be assessed by the computer program published in DIN 45 631. For sounds with strong temporal variations, however, in addition the nonlinear temporal processing of loudness in the human hearing system has to be simulated in loudness meters as illustrated in Fig. 8.26. Loudness meters calculate the loudness corresponding to ISO 532 B almost instantaneously, i.e. using the same temporal characteristics as our hearing system. In order to simulate the characteristics of our hearing system, critical-band analysis has to be done every 2 ms. This means that a filter bank has to be installed that simulates the frequency selectivity of our inner ear. In addition to loudness pattern and loudness-time function, loudness meters can also produce statistical loudness distributions, i.e. the percentage of time for which a given loudness is reached or exceeded as a function of loudness. Such a statistical distribution is of importance in cases where loudness varies strongly as a function of time and for the assessment of noise immissions.

Figure 16.3 shows an example for the loudness of speech. In psychoacoustic experiments the loudness of a continuous speech noise (CCITT Rec. G 227) was matched to the loudness of a test sentence. The corresponding loudness-time functions are shown in Fig. 16.3 for the sentence (left) and for the speech noise (right).

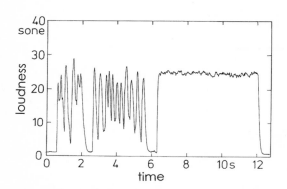

Fig. 16.3. Loudness-time function of a test-sentence and of speech noise for perceived equal loudness

The data displayed in Fig. 16.3 reveal that the loudness of a sentence does *not* correspond to an average value of the fluctuating loudness-time function but rather to a loudness value near the maximum.

A statistical treatment displayed in Fig. 16.4 as cumulative loudness distribution reveals that the continuous speech noise produces almost the same loudness during the whole presentation (dotted) whereas the loudness of the sentence varies considerably with time (solid). The crossing point of the two curves in Fig. 16.4 indicates that the loudness of speech can be described by the percentile loudness N_7, i.e. the loudness value reached or exceeded in 7% of the measurement time.

16.1.2 Evaluation of Noise Emissions

Noise emitted by different sound sources can be assessed subjectively in psychoacoustic experiments and physically by noise measurement devices like loudness meters.

For the subjective evaluations usually the method of magnitude estimation (with anchor) is used. For physical measurement of noise emissions, the

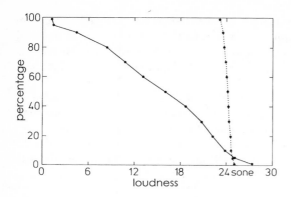

Fig. 16.4. Cumulative loudness distributions of test sentence (*solid*) and speech noise (*dotted*) for perceived equal loudness

percentile loudness N_5 is recommended; however, for a large number of emissions, the maximum loudness N_{max} gives nearly the same values.

Figure 16.5 shows an example for the evaluation of noise emissions by printers. Different needle printers (A, B, E, F, G), daisy wheel printers (C, D), and an inkjet printer (H) were included. Subjective loudness evaluations are illustrated by circles, loudness measured by a loudness meter by stars. The plus signs indicate A-weighted sound power which is used in many countries to describe the noise emission of products. All data are normalized relative to the values for printer F.

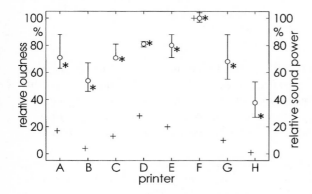

Fig. 16.5. Evaluation of the noise emission by different printers. Relative loudness evaluated subjectively (*circles*) or by a loudness meter (*stars*) in comparison to the relative A-weighted sound power (*plus signs*)

The data displayed in Fig. 16.5 indicate that the loudness of different needle printers may differ by a factor of 2 (F versus B). An inkjet printer can produce a loudness about only 1/3 of the loudness of a loud needle printer (H versus F).

A comparison of circles and stars in Fig. 16.5 reveals that the subjective evaluation can be predicted by physical measurements with a loudness meter. On the other hand, A-weighted sound power overestimates the differences drastically.

For example it is stated frequently in advertisings that the noise emission of printer G is by 90% less than that of printer F. This is correct in physical terms of A-weighted sound power. However, the loudness difference perceived by the customer is only about 35%, in line with the indications by a loudness meter.

Therefore, reductions in noise emissions of products should be given in loudness ratios since with the exaggerating sound power ratios, in the end the public may no longer trust predictions of acoustic improvements.

Results displayed in Fig. 16.6 illustrate the noise emission of different types of aircraft during take-off, normalized relative to the values for aircraft A.

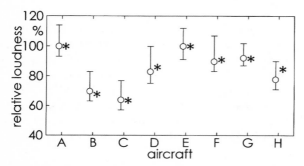

Fig. 16.6. Evaluation of noise emissions by different aircraft during take-off. Data from subjective evaluations (*circles*) and physical measurements by a loudness meter (*stars*)

A comparison of the data illustrated by circles versus stars again shows the predictive potential of physical measurements by a loudness meter with respect to the subjective evaluation of the loudness of noise emissions. Old aircraft, equipped with engines with small by-pass ratio, (e.g. A, E) are by a factor of about 1.5 louder than modern aircraft with engines with large by-pass ratio (e.g. B, C).

It should be mentioned that the concept of noisiness put forward by Kryter and used for the certification of aircraft is also in line with the subjective evaluation. On the other hand, A-weighted or D-weighted sound power again substantially overestimates the differences between noises produced by different aircraft.

As concerns railway noise, noises produced usually depend on the type of train, its length, and speed. Table 16.1 gives an overview of some noises studied.

The corresponding results of subjective evaluations in psychoacoustic experiments and physical ratings by a loudness meter are given in Fig. 16.7. All data are normalized relative to the values for train C.

Table 16.1. Overview of some train noises studied

	Type of train	Length [m]	Speed [km/h]
A	Freight train	520	86
B	Passenger train	95	102
C	Express train	228	122
D	ICE	331	250
E	ICE	331	250
F	Freight train	403	100
G	Freight train	175	90

The data displayed in Fig. 16.7 suggest that also for train noise, physical measurements of loudness by a loudness meter can predict the subjective loudness evaluation. The lowest loudness values are produced by relatively short trains (B, G) at medium speed. Large loudness values are produced by a long freight train (A) at low speed, or ICE-super-express-trains at high speed (D, E).

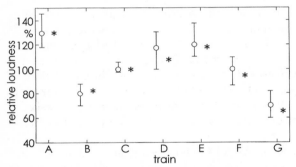

Fig. 16.7. Evaluation of noises produced by different trains as described in Table 16.1. Data from subjective evaluations (*circles*) and physical measurements by a loudness meter (*stars*)

The last example discussed concerns the noise emission of grass trimmers. Products of several manufacturers at different rpm were studied. Results are given in Fig. 16.8. All values were normalized relative to the data for noise number 9.

A comparison of the data illustrated in Fig. 16.8 by circles versus stars indicates that also for sounds in the category "leisure noise", perceived loudness can be predicted by loudness measurements with a loudness meter.

In summary, the loudness of noise emissions can be physically measured by a loudness meter in line with subjective evaluations for rather different kinds of noises. Ratios of A-weighted sound power – although frequently used in advertisings – usually overestimate differences in the loudness of noise emissions substantially.

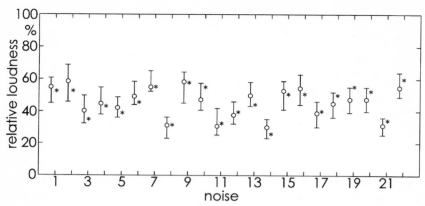

Fig. 16.8. Evaluation of noise emissions by grass trimmers. Data from subjective evaluations (*circles*) and physical measurements by a loudness meter (*stars*)

16.1.3 Evaluation of Noise Immissions

In psychoacoustic experiments, noise immissions are frequently assessed as follows: the subjects are presented stimuli of e.g. 15 minutes duration which include a soft background noise (e.g. road-traffic noise of 40 dB(A)) plus several louder events (e.g. fly-overs of aircraft). During the presentations, the subject has to indicate the instantaneous loudness by varying the length of a bar presented on the monitor of a PC. After the experiment is over (e.g. after 15 minutes) the subject has to fill in a questionnaire and is asked – among others – to indicate the overall loudness of the preceding 15 minutes by marking the length of a line.

This subjective evaluation of overall loudness of a noise immission (expressed in mm line-length) can be compared with physical magnitudes of the noise immission. It turns out that in many cases the perceived overall loudness of noise immissions corresponds to the percentile loudness N_5 measured by a loudness-meter. The value N_5 is the loudness which is reached or exceeded in 5% of the measurement time. This means that N_5 represents a loudness value close to the maxima of the loudness-time function of the noise immission.

Figure 16.9 shows as an example loudness-time functions of 15 minute noise immissions from aircraft noise when using old, louder aircraft (B) versus modern, softer aircraft (C).

The data displayed in Fig 16.10 as circles represent the subjective evaluation of the noise immissions illustrated in Fig. 16.9 and can be compared to data of physical measurements represented by stars. The line-length indicating the subjectively perceived overall loudness is given on the left ordinate, the correlated percentile loudness N_5 on the right ordinate.

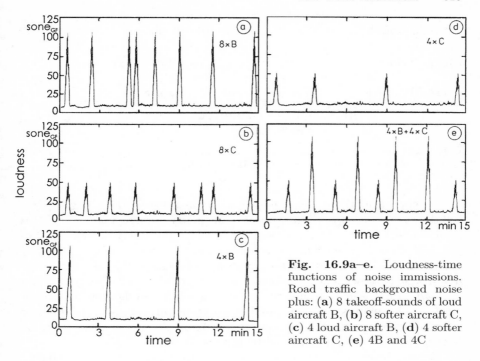

Fig. 16.9a–e. Loudness-time functions of noise immissions. Road traffic background noise plus: (**a**) 8 takeoff-sounds of loud aircraft B, (**b**) 8 softer aircraft C, (**c**) 4 loud aircraft B, (**d**) 4 softer aircraft C, (**e**) 4B and 4C

A comparison of subjective and physical data displayed in Fig. 16.10 suggests that the percentile loudness N_5 is a good predictor of the subjectively perceived overall loudness of noise immissions from aircraft noise.

As described in the references to this section percentile loudness values can also predict the overall loudness of noise immissions from road traffic noise, railway noise, industrial noise, and leisure noise like noise from tennis courts.

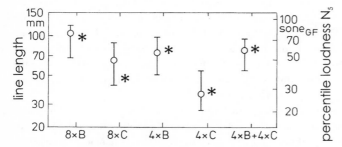

Fig. 16.10. Overall loudness of the noise immissions illustrated in Fig. 16.9. *Circles*: Line-length matched to subjectively perceived overall loudness. *Stars*: Percentile loudness N_5 physically measured by a loudness-meter

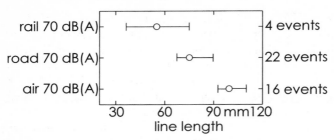

Fig. 16.11. Overall loudness of noise immissions from rail-traffic, road-traffic or air-traffic each with the same L_{eq} of 70 dB(A) within 15 minutes

In many regulations, noise immissions are assessed by the energy-equivalent A-weighted-level L_{eq}. However, even at same L_{eq} value, noise immissions from different means of transportation can be judged rather differently. Corresponding results from field studies were named "railway bonus" or "aircraft malus" indicating that at *same* L_{eq} railway noise can be *less* annoying than road-traffic noise while aircraft noise can be *more* annoying than road noise.

Figure 16.11 shows results from corresponding psychoacoustic experiments. Noise immissions of 15 minutes duration with noises from different means of transportation at the same L_{eq} of 70 dB(A) were judged according to their overall loudness indicated by the line length.

The psychoacoustic data displayed in Fig. 16.11 are in line with the concepts of "railway bonus" and "aircraft malus" put forward on the basis of results from field studies. For noise immissions with the same L_{eq} of 70 dB(A), overall loudness indicated in psychoacoustic experiments by line length increased from rail- over road- to air-traffic.

16.1.4 Evaluation of Sound Quality

With respect to sound quality evaluation, in addition to acoustic features of sounds in particular aesthetic and/or cognitive effects may play an essential part. Therefore, in laboratory situations not always all factors contributing to the annoyance or the pleasantness of sounds can be assessed. However, the psychoacoustic elements of annoying sounds can be described by a combination of hearing sensations called psychoacoustic annoyance. More specifically, psychoacoustic annoyance (PA) can quantitatively describe annoyance ratings obtained in psychoacoustic experiments.

Basically, psychoacoustic annoyance depends on the loudness, the tone colour, and the temporal structure of sounds. The following relation between psychoacoustic annoyance, PA, and the hearing sensations loudness, N, sharpness, S, fluctuation strength, F, and roughness, R can be given:

$$PA \sim N \left(1 + \sqrt{[g_1(S)]^2 + [g_2(F, R)]^2} \right) \tag{16.1}$$

In essence, (16.1) illustrates that for psychoacoustic annoyance in addition to loudness itself, loudness-dependent contributions of sharpness as well as fluctuation strength and roughness have to be taken into account using some type of RMS averaging.

A quantitative description of psychoacoustic annoyance is based on results of psychoacoustic experiments with modulated versus unmodulated narrow-band and broadband sounds of different spectral distribution. It reads as follows:

$$PA = N_5 \left(1 + \sqrt{w_S^2 + w_{FR}^2}\right) \tag{16.2}$$

with

$-$ N_5 percentile loudness in sone

$-$ $w_S = \left(\dfrac{S}{\text{acum}} - 1.75\right) \cdot 0.25 \lg \left(\dfrac{N_5}{\text{sone}} + 10\right)$ for $S > 1.75\,\text{acum}$

$$\tag{16.3}$$

describing the effects of sharpness S and

$-$ $w_{FR} = \dfrac{2.18}{(N_5/\text{sone})^{0.4}} \left(0.4 \cdot \dfrac{F}{\text{vacil}} + 0.6 \cdot \dfrac{R}{\text{asper}}\right)$ $\tag{16.4}$

describing the influence of fluctuation strength F and roughness R.

By means of (16.2) psychoacoustic annoyance can be described in line with data from psychoacoustic experiments not only for synthetic sounds but also for technical sounds like car noise, air conditioner noise, or noise from tools like circular saws, drills etc.

As an example, in Fig. 16.12 psychoacoustic annoyance of car sounds as measured in psychoacoustic experiments is compared to data calculated according to (16.2). The car sounds used in the experiments are listed in

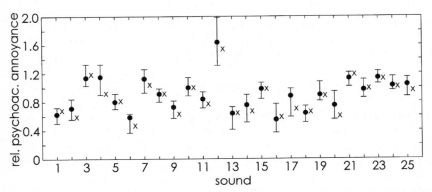

Fig. 16.12. Psychoacoustic annoyance of car sounds listed in Table 16.2. *Dots*: Data from psychoacoustic experiments. *Crosses*: Values calculated according to (16.2)

Table 16.2. Car sounds used for experiments on psychoacoustic annoyance

Sound	motor	speed in km/h	gear	distance in m
1	Diesel	60	3	7.5
2	Otto	30	1	7.5
3	Diesel	70	2	7.5
4	Otto	0	idle	0.9
5	Otto	80	3	7.5
6	Otto	80	3	15.0
7	Diesel	35	1	7.5
8	Diesel	70	4	7.5
9	Otto	50	2	7.5
10	Diesel	110	4	7.5
11	Diesel	acceleration	1	7.5
12	Diesel	racing start	1	7.5
13	Otto	60	3	7.5
14	Diesel	30	1	7.5
15	Otto	50	2	3.5
16	/	60	coast	7.5
17	Diesel	0	idle	7.5
18	Diesel	60	4	7.5
19	Otto	ISO 362	2	7.5
20	Diesel	0	idle	7.5
21	Diesel	80	3	3.75
22	Diesel	50	2	3.75
23	Diesel	80	3	3.75
24	Diesel	90	4	7.5
25	Otto	80	3	3.75

Table 16.2. The sounds stem from cars with Diesel- or Otto-engines driven at different speeds in different gears and recorded at different distances. Pass-by at constant speed, idling, coasting, acceleration, accelerated pass-by according to ISO 362 as well as a racing start are included.

Data for subjective and physical evaluation of psychoacoustic annoyance are given in Fig. 16.12. All data were normalized relative to the value for sound 10, a pass-by of a car with Diesel-engine driven in 4th gear at 110 km/h and recorded at a distance of 7.5 meter. The results of the psychoacoustic experiments are given by dots (medians and interquartiles), the calculated values by crosses.

A comparison of data illustrated by dots and crosses in Fig. 16.12 reveals the predictive potential of (16.2) for psychoacoustic annoyance. Large values of psychoacoustic annoyance are obtained for a racing start (sound 12), a Diesel car at high speed and low gear (sound 3) or Diesel cars at short distance (sounds 21 and 23).

In summary, psychoacoustic annoyance can assess the psychoacoustic elements of sound quality. Taking into account other elements like aesthetic and/or cognitive factors, fashionable acoustic trends etc., psychoacoustic annoyance offers great potential for sound engineering.

16.2 Applications in Audiology

The hearing system is a very important sensory organ for us humans. It is a receiver of sound information that, together with our speech producing system, enables us to participate in acoustical communication. Therefore the ability of hearing-impaired patients to discriminate speech is one of the important aspects to be measured quantitatively in clinical audiology. There are five fundamental features that are necessary in order to discriminate speech: sensitivity, frequency selectivity, amplitude discrimination, temporal resolution and information processing. To avoid serious social handicaps, the five features have to be good enough to discriminate and understand speech. The measure of sensitivity is most often produced by means of pure-tone audiometry where threshold in quiet is measured. Information processing is mostly measured by speech discrimination tests. The tests actually include measurements of the other three features because without these, speech discrimination is not possible. For diagnostic reasons, however, it is often of great interest to differentiate between a loss of these three features and a loss of information processing, especially for such cases in which speech discrimination fails. Audiology is interested in all of the single effects that may contribute in part to what is measured in total, namely speech discrimination. Most of the single factors that contribute to speech discrimination have been well-studied for normal-hearing subjects. The corresponding set of data for hard-of-hearing listeners is by no means as comprehensive. An important reason is that the methods that are used for normal-hearing listeners are very rarely useful in cases of hearing impairment. Simplified methods and simplified apparatus have to be applied, and the following paragraphs try to help solve the problems.

16.2.1 Otoacoustic Emissions

What is now known about otoacoustic emissions indicates that they are produced in the inner ear, i.e. in the peripheral part of our processing system. Peripheral in this context means that the information still deals with AC values that can either be acoustical, mechanical (in the middle ear), hydromechanical (in the inner ear) or electrical (within the haircells). In all cases, oscillations with positive and negative values carry the information. The change from AC values into action potentials takes place in the synapses of the inner haircells, an area defined as the border between peripheral information processing and retro-cochlear information processing. This definition may be different from that used in audiology. From the information processing point of view, however, this definition is very useful. We assume that the production of otoacoustic emissions and frequency resolution are typical features of our hearing system that are located in the inner ear, i.e. peripheral. We also assume that what is called recruitment may be produced primarily within the inner ear.

Neurophysiological data concerning signal processing in hearing stem almost exclusively from animals. Only from very rare operations is it possible to collect data from human beings. Most of the psychoacoustical data, however, stem from subjectively measured human data. Therefore, most decisions regarding the type of disease in hearing-impaired subjects are based on psychoacoustical data, although objective results might be more useful. Therefore, the purely physically measured otoacoustic emissions seem to be an attractive tool for obtaining information about signal processing in the human inner ear. The problem, however, is that otoacoustic emissions seem to be produced only in subjects who have a hearing loss no greater than about 20 dB. Out of more than 200 students in a school for the hearing-impaired, all having a hearing loss larger than 60 dB in the whole frequency range, not one showed delayed evoked otoacoustic emissions of the type most easily measured. For this reason, otoacoustic emissions rarely serve as a tool for obtaining information about the kind of hearing impairment that leads to a hearing loss of more than 20 dB.

Objective measurements, however, can be used in cases where it is not possible to communicate with the human subject. Babies only a few months of age are typical cases. It is very important to determine as early as possible whether the hearing system acts normally or abnormally. To make such a decision, the measurement of delayed evoked otoacoustic emissions in babies can produce helpful information. Almost 98% of normally hearing subjects, including babies, produce these emissions. Therefore, their lack can be used as an indication that the development of the inner ear is at least not normal, or that the middle ear strongly attenuates the sound transmission. The existence of emissions indicates that middle and inner ear act normally, at least in that frequency range in which the emissions can be measured.

Figure 16.13a shows a typical delayed evoked otoacoustic emission measured from an 18-month old child. The two traces are repeated measurements to give an idea of the reproducibility. The time function shows not only delayed emissions (at 2.3 kHz with 7 ms delay; at 1.25 kHz with 13 ms delay; at 0.9 kHz with 16 ms delay) but also a 2.95-kHz spontaneous emission that is triggered by the stimulating series of pulses. Figure 16.13b shows the corresponding time function for measurements from a subject with hearing problems. Because babies can not be forced to remain as quiet as necessary, the apparatus used for measuring evoked otoacoustic emissions needs a device for rejecting artefacts.

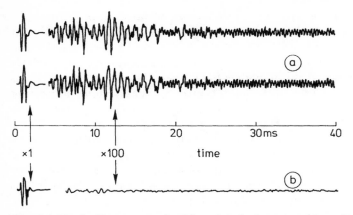

Fig. 16.13a,b. Two traces of a delayed evoked otoacoustic emission from a healthy inner ear, picked up in the closed ear canal of a baby (**a**). The traces show excellent reproducibility even during long delays where a spontaneous emission triggered by the evoker appears. The lower trace (**b**) shows a corresponding time function recorded from a subject with some hearing loss. Note that the amplitude is magnified for a factor of 100 after a delay of about 6 ms

16.2.2 Tuning Curves

Classical masking describes an effect that is more precisely defined as simultaneous spectral tone-on-tone masking. This effect was discussed in detail in Chap. 4. Figure 4.13 shows two possiblitites for measuring this effect using the method of tracking: the classical masking curve and the psychoacoustical tuning curve. The correlation of these two curves was outlined in the sequence of Figs. 4.14 – 16. The latter shows a set of psychoacoustical tuning curves, i.e. the level L_M of the masking tone necessary to just mask a test tone of level L_T (parameter), as a function of the frequency of the masker. This masker frequency can be expressed in Δz, the separation between masker- and test-tone frequencies measured as critical-band rate.

The method of tracking is relatively easy even for an untrained subject. However it is more convenient for clinical use to measure step-wise, not continuously. To achieve this, an electronic device was developed with which seven points of the tuning curve could be measured, three points below and three points above the test-tone frequency. The seventh point is obtained from the level of the test tone normally set 5 to 10 dB above threshold in quiet. In most clinical cases, data from both the low-frequency and high-frequency regions are needed. Therefore, the apparatus provides test frequencies of 500 Hz and 4 kHz, and 6 masker frequencies in the neighbourhood of each of the two test frequencies. The frequency spacings of the six masker frequencies in relation to the test frequency are chosen so that the simplified tuning curve consisting of the seven measured points, becomes a useful approximation of the continuously measured tuning curve.

The procedure begins with the determination of threshold in quiet for the test-tone frequency. In order to make the test tone more clearly audible, it is switched on and off smoothly every 600 ms. After determination of threshold in quiet, the test-tone level is set to a fixed value about 10 dB above threshold in quiet. For determining tuning curves, the subject again listens to the interrupted test tone, although a continuous masker tone of different frequencies is now simultaneously presented. Masker frequencies of 215, 390, 460, 540, 615, and 740 Hz are used for a test-tone frequency of 500 Hz. These frequencies are spaced unevenly to determine tuning curves with only seven measurements as well as possible. The level of the masker is determined at each masker frequency so that the continuous masker tone just masks the interrupted test tone. Therefore, the hearing-impaired listener always listens to the same interrupted tone and signals when the tone is heard. This way, seven masker-level data indicating the level where the test tone is just masked are produced. The same procedure is used for a test-tone frequency of 4 kHz, where masker frequencies of 1.72, 3.12, 3.68, 4.32, 4.92, and 5.92 kHz are produced in order to also obtain information about the "tail" of the tuning curve towards low frequencies.

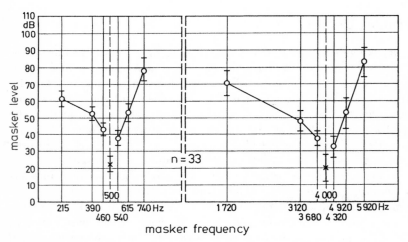

Fig. 16.14. Simplified tuning curves in the frequency regions of 0.5 (*left*) and 4 kHz (*right*) for a group with normal hearing. The data (means with standard deviations) are produced in the following way: after averaging the test-tone levels used ("x") with test-tone and masker frequency being equal, the individual tuning-curve data are normalized to that mean value. Vertical bars represent standard deviations

Simplified psychoacoustical tuning curves have been measured for *normally hearing* subjects using the method described above. Because threshold in quiet varies somewhat for normally hearing subjects and varies even more for hearing-impaired listeners, a meaningful averaging procedure has to be introduced. Such a procedure leads to the masker levels that are plotted as open circles in Fig. 16.14 for $n = 33$ normal hearing subjects. The data connected by straight lines represent the simplified tuning curves. The frequency scaling drawn on the abscissa is chosen so that equal distance on the abscissa corresponds to equal distance along the basilar membrane, i.e., critical-band rate.

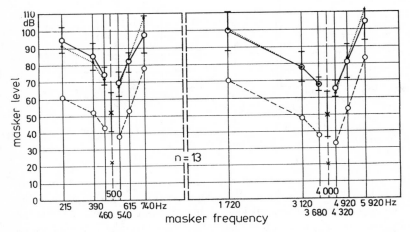

Fig. 16.15. Simplified tuning curves in the frequency regions of 0.5 (*left*) and 4 kHz (*right*) for a group with conductive hearing loss (—) but not otosclerosis. Note that frequency-resolving power remains normal. (- - -) : data for normal hearing; (...): the same data but shifted upwards so that the test-tone levels coincide). Other details as in Fig. 16.14

Pathological ears show tuning curves that may differ from those produced by normally hearing subjects. To make this comparison possible, the tuning curve of normally hearing subjects is also indicated in the data of listeners with pathological hearing. The first example is of a group with conductive hearing loss (Fig. 16.15). It can be clearly seen, by comparing the tuning curves of normally hearing subjects with those produced by listeners with conductive hearing loss, that in the latter case the whole tuning curve is shifted upwards by about 30 dB. These 30 dB correspond to the hearing loss. This means that although there exists a hearing loss of 30 dB, the frequency resolution outlined in the psychoacoustical tuning curve does not change for the group with conductive hearing loss.

The group with the degenerative hearing loss also shows a threshold shift of about 30 dB at 500 Hz (Fig. 16.16), however, the form of the tuning curve is very different. It is very shallow towards lower frequencies of the masker and flatter still towards higher frequencies, although there remains an increment of about 10 dB between the masker frequency of 540 Hz and that of 740 Hz. For the high frequencies of 4 kHz, the hearing loss is even greater, the form

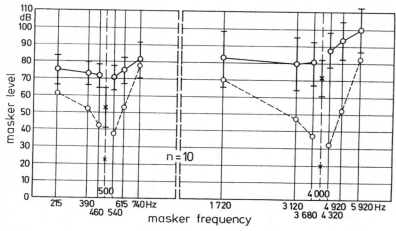

Fig. 16.16. As in Fig. 16.15, but for a group with degenerative hearing loss (——). Note that frequency-resolving power is greatly reduced

of the tuning curve is again very flat towards low frequencies, and the slope towards high frequencies has an angle on these coordinates that is only one quarter that of normal hearing subjects. This means that the frequency selectivity of patients in this group is distinctly worse. Under such conditions, an amplifying hearing aid may bring back the sensitivity as such, but cannot restore the impaired frequency selectivity.

The last example given here is that of noise-induced hearing loss. The tuning curve in the 500-Hz frequency range outlined in Fig. 16.17 shows a normal configuration, while the form of the tuning curves measured at 4 kHz depends on the size of the hearing loss. For this reason two groups have been separated, one with a hearing loss less than 55 dB and another one with a hearing loss greater than 55 dB. This separation makes it clear that frequency resolution becomes worse with larger hearing loss. It should be pointed out, however, that noise-induced hearing loss, which produces mostly a threshold shift only at frequencies above about 1500 or 2000 Hz, produces conditions by which cubic or quadratic difference tones may become audible. Therefore, an additional masking noise should be added in the low frequency range in order to mask this possible difference tone which, if heard, would alter the shape of the tuning curve. The form of the tuning curve of this group indicates clearly that it is not easy to restore normal hearing conditions through a hearing aid if the listener suffers from a noise-induced hearing loss.

Using the procedure of measuring psychoacoustical tuning curves for hearing-impaired listeners, the reduction of the hearing system's frequency selectivity can be estimated. Frequency resolution, as measured for normally hearing subjects, is necessary for good speech discrimination. It is understandable that speech discrimination often fails in conditions in which very

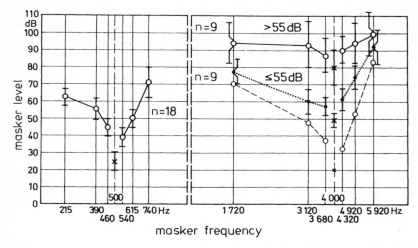

Fig. 16.17. As in Fig. 16.15, but for a group with noise-induced hearing loss. For the 4-kHz frequency range, the group is divided into two subgroups, one (...) with hearing losses of 55 dB or less and the other (—) with hearing losses greater than 55 dB

flat tuning curves are measured, especially in those cases where temporal resolution also fails.

16.2.3 Amplitude Resolution

The discrimination of amplitude or level differences is a prerequisite for the discrimination of speech and music. In audiology, level discrimination is normally evaluated by the use of the SISI test and the Lüscher/Zwislocki test. These tests use modulated tones or direct increases in level without pauses between the presentations. Another test which is effective in the early detection of acoustic neurinoma uses 200-ms tone pulses with a relatively long Gaussian rise/decay time of 20 ms. In contrast with the two tests mentioned, there are silent intervals of 200-ms duration between the tones to be compared. The measurements are performed at 6 frequencies in octave steps between 0.25 and 8 kHz. The data are measured at a level of 30 dB above hearing threshold, and the patient has to judge whether the loudness of the consecutive tone pulses is the same or different. The level difference of the alternating tones can be increased in 0.25 dB steps.

The results show that most of the hearing-impaired patients indicate level discriminations that are the same or almost the same as those of normally hearing subjects. In this case, the median values of the just-perceptible level differencies, ΔL, may be a little different from those of hearing-impaired subjects, but the interquartile ranges overlap so that no clear decision can be made about whether the level discrimination ability is changed or not. Such

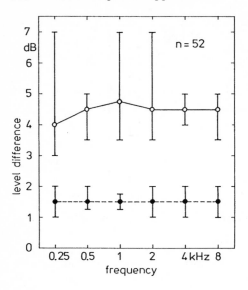

Fig. 16.18. Threshold for level differences for normally hearing subjects (black circles connected with broken lines) and for patients with acoustic neurinoma (open circles connected with solid lines). Note that the interquartile ranges (vertical bars) do not overlap

small differences occur for the impairment of conductive hearing loss, sudden hearing loss, noise-induced hearing loss, infantile or toxic non-progressive sensory neural hearing loss, Menière's disease and presbyacusis. For these hearing impairments no differentiation is possible, because the interquartile ranges of the data and those produced by normally hearing subjects clearly overlap. The hearing impairment resulting from acoustic neurinoma, however, shows a strong difference as indicated in Fig. 16.18. While the normally hearing subjects show a level-difference threshold near 1.5 dB, results for the patients with acoustic neurinoma are about a factor of 3 larger, i.e. 4.5 dB. The medians range from 4 to 4.7 dB and the interquartile ranges do not overlap at all. Because there were 52 patients for the data set, it is clear that there is a distinct increase in the level-difference threshold for patients with acoustic neurinoma. The finding that 49 out of 52 patients with proven acoustic neurinoma showed distinctly increased level-difference threshold emphasizes the attractiveness of this method. Comparable measurements using the SISI test and the Lüscher/Zwislocki test have proven that the method, which measures the level-difference threshold for sounds with intervening intervals, is superior to others in detecting acoustic neurinoma.

16.2.4 Temporal Resolution

Reduced speech discrimination of hearing-impaired listeners may be influenced not only by reduced frequency resolution but also by reduced temporal resolution. The equipment normally used to measure temporal resolution is not simple and the methods used are time consuming. This means that such methods are not useful for clinical routines. Speech was developed over

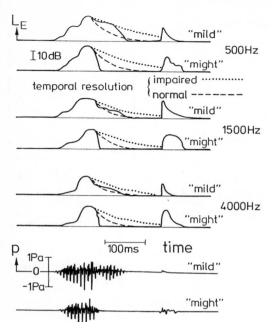

Fig. 16.19. Influence of temporal resolution on the identification of consonants following a loud vowel. The solid lines indicate excitation level as a function of time for normal hearing. The broken lines indicate the decay of normal hearing, while the dotted lines indicate the decay of reduced temporal resolution. The sound pressure versus time function of the two spoken words is shown in the two lower diagrams

long evolutionary time spans so that it just fits the temporal resolution of our hearing system under normal conditions. A prolongation of postmasking can influence speech discrimination strongly. An example is illustrated in Fig. 16.19. The excitation level versus time pattern, and sound pressure versus time function of the two words "mild" and "might" are illustrated. The vowels "i" produce the largest excitation levels. The consonant "l" follows the "i" directly in the word "mild". Although postmasking is effective, the "l" is still audible with normal hearing. For a prolonged decay, however, the "l" may be totally masked. As a consequence, the discrimination between the two words is not possible for someone with reduced temporal resolution.

The effects of postmasking and premasking were described in detail in Chap. 4 (see Figs. 4.17 and 4.25). Temporal masking effects are normally measured by using short test-tone bursts. The results are expressed in temporal masking patterns. In order to reduce the time needed for such measurements, it is necessary to concentrate on the most important part of the masking pattern. Figure 16.20a shows a set of temporal masking patterns produced by an octave-band noise at 4 kHz that is periodically switched on and off for 32 ms. The period of 64 ms corresponds to a repetition frequency of 16 Hz. The temporal masking pattern was measured with short tone bursts with a duration of 5 ms and a frequency of 4 kHz. Temporal resolution is usually given by two values, that of the peak, which is reached during the on-time of the masker, and that of the valley, which is near the end of the off-time of the masker.

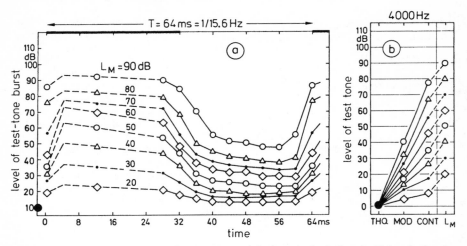

Fig. 16.20a,b. Temporal masking pattern (**a**), i.e. masked threshold level of a 4-kHz test-tone burst of 5-ms duration triggered by a square-wave-modulated noise masker (one octave band at 4 kHz), as a function of the temporal position within the period T of the masker, the level L_M of which is the parameter. The simplified temporal resolution diagram (**b**) shows the level of a just-audible 4-kHz test tone (500 ms on, 500 ms off): at threshold in quiet ('THQ'), masked by square-wave-modulated octave-band noise ('MOD') and masked by the same but continuous masker ('CONT'). The level of the masker is the parameter and is indicated by 'L_M'. Corresponding data points are connected. For normal temporal resolution, the three corresponding points are located on nearly straight lines with 'MOD' in the middle between 'THQ' and 'CONT'

To measure only the most interesting difference between these two values, it may be sufficient to use long test tones that are already familiar to the patient and that are normally interrupted once a second. Such a sequence of 500-ms tones will be heard in the valleys of the temporal masking patterns, following one another with the repetition rate of 16 Hz. The peak value can be measured easily by this test tone if the gap of the masker is eliminated. If threshold in quiet, i.e. "without masker", is also measured, three test-tone thresholds have to be measured by the patient during which the masking noise is either presented continuously, modulated, or not at all.

Figure 16.20b presents the three values for the same normally hearing listener who produced the temporal masking patterns. The masker level, L_M, is the parameter in the diagram and scaled at the right side. Threshold in quiet, which lies at 2.5 dB for this subject, is the point that is included in all measurements. The values obtained for continuous noise depend on the masker level in a similar form as in the temporal masking pattern. The masked threshold reached for continuous masking noise is plotted on the abscissa at the point called "cont" and linked with the L_M values by broken lines. The threshold levels obtained using rectangularly modulated noise as the masker are entered under "mod". These values lie between the levels measured for the "cont" – condition and threshold in quiet. For a masker level L_M about 40 dB above threshold in quiet, an almost straight line is seen connecting the three measured values: threshold in quiet, modulated noise, and continuous noise.

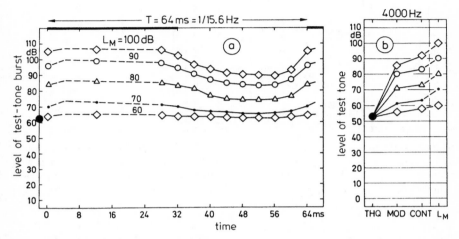

Fig. 16.21a,b. Temporal masking pattern as in Fig. 16.20 but for a listener with a hearing loss of about 50 dB at 4 kHz. Reduced temporal resolution is indicated by the fact that in the right part of the figure the data at 'MOD' lie above the straight line connecting 'THQ' and 'CONT'

Data produced with similar parameters by a listener with noise-induced hearing loss are shown in Fig. 16.21a. The difference is clear because the valleys in the temporal masking patterns are not as deep as for the normally hearing subjects. The temporal-resolution pattern, which is shown in Fig. 16.21b for the same impaired listener, also indicates different behaviour from that in Fig. 16.20b. Not only is the threshold in quiet increased by almost 50 dB, but also elevation of the data points corresponding to the "mod"- condition prevent a straight line from being drawn through the three data points. The straight lines in Fig. 16.20b indicate that the ratio between the level difference of the two upper values ("cont" and "mod" on one hand, and "mod" and "threshold in quiet" on the other) is about one to one. This ratio is called the temporal resolution factor (TRF) and amounts to unity in the case of normally hearing subjects, as outlined in Fig. 16.20b. Similar values near unity have been collected at other frequencies. In contrast, the hearing-impaired listener shows a drastically reduced temporal resolution factor: the difference between the two upper data points ("cont" and "mod") is much smaller than the difference between the two lower data points ("mod" and "threshold in quiet"). For a masker level $L_M = 100$ dB, the ratio is 7.5 dB to

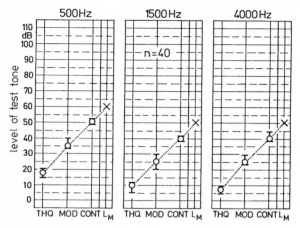

Fig. 16.22. Medians and interquartile ranges of simplified temporal-resolution diagrams for normal hearing at 500, 1500, and 4000 Hz, normalized in such a way that the median of the L_M value is used as a fixed point. The level of the masker is chosen to be about 40 dB above threshold in quiet

30 dB, i.e. only 0.25. This indicates that the temporal resolution factor, TRF, is greatly reduced.

For normally hearing subjects and a setting of the masker level about 40 dB above threshold in quiet, three temporal-resolution patterns for frequencies of 500, 1500, and 4000 Hz are shown in Fig. 16.22. Individual differences for normally hearing subjects are small. The straight lines indicate that a temporal-resolution factor of unity is found for these listeners. Depending on the type of disease, hearing-impaired listeners show either a normal resolution factor near unity or a strongly reduced resolution factor near 0.2, similar to the example outlined in Fig. 16.21b.

Most hearing-impaired listeners have little difficulty in speech communication and speech discrimination if they receive only one voice at high signal-to-noise ratios. However, they have significant problems with background noise, for example at a cocktail party. Therefore, the same procedure was also used to measure the temporal-resolution factor for normally hearing subjects in a condition with background noise. The influence of noise on the temporal-resolution factor is shown in Fig. 16.23. Two background noises were used, one elevating the threshold in quiet to a value near 35 dB and another shifting threshold in quiet up to 55 dB. The results produced under these conditions for normally hearing subjects are compared with the results that are produced without background noise. The results are striking: the straight lines obtained without noise are strongly twisted in such a way that the "mod" value is clearly below the line connecting the masked threshold and the "cont" value, indicating a strongly enlarged temporal-resolution factor, TRF, for normally hearing listeners in background noise. TRF increases

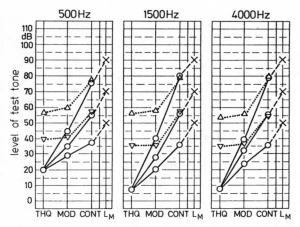

Fig. 16.23. Medians of temporal-resolution data produced by four normally hearing subjects (L_M = 50, 70, and 90 dB). To imitate a threshold shift, an additional continuous masker is presented to raise the threshold in quiet to about 35 dB (∇) and about 55 dB ($\triangle$). Note the upward curvature of the temporal-resolution diagrams (*dotted lines*), contrary to the findings in several pathological ears (downward curvature as shown in Fig. 16.21)

to values up to 5 and sometimes almost 10. This means that a normally hearing subject has a very good temporal-resolution factor in noisy backgrounds, as in the case of a cocktail party, and uses it.

The hearing-impaired listener, however, shows different characteristics. These are outlined in Fig. 16.24. In part (a), a temporal masking pattern is shown for the conditions "in quiet" and "with a background noise". The corresponding temporal-resolution patterns are indicated in part (b). There, the values for "mod" are above the connecting line between "threshold" and "cont". This means that this listener does not gain in temporal resolution when a background noise is added. The temporal-resolution factor remains as reduced as it is without background noise. The data of a normally hearing listener under corresponding conditions are shown in part (c). There, it becomes clear that the normally hearing subject gains temporal resolution in the background noise if the voice level of the communication partner is increased. The hearing-impaired listener, however, does not gain in the background noise even through the voice level of the partner is increased. This means that the difference in TRF between the normally hearing subject and the hearing-impaired listener may reach values up to ten or even more than that by which the factor is reduced in normal listening. This seems to be a dominant reason why hearing-impaired listeners often have such problems with speech discrimination in noisy conditions (cf. Sect. 16.2.7).

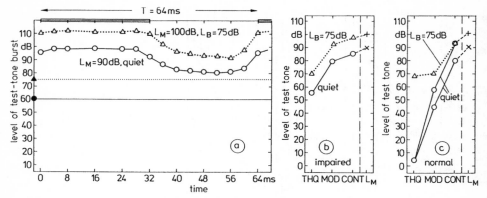

Fig. 16.24a–c. Temporal masking pattern (**a**), i.e. masked threshold level of a 4-kHz test-tone burst (5-ms duration) triggered by a square-wave-modulated octave-band masker (centre frequency 4 kHz), as a function of the temporal position within the period of the masker with level L_M. For additional background noise of level L_B, the upper pattern is created. The simplified temporal-resolution diagram (**b**) shows the level of a just-audible 4-kHz tone (600 ms on, 600 ms off) at threshold in quiet ('THQ'), masked by a modulated masker ('MOD') or by a continuous masker ('CONT'). Data with background noise (*dotted*) are also given. The subject shows noise-induced hearing loss above 2-kHz, leading to reduced temporal resolution as measured in this experiment. Panel (**c**) indicates corresponding data produced by a normally hearing subject

16.2.5 Temporal Integration

When sound bursts are reduced in duration to less than about 200 ms, their level must be raised to remain audible. As described in Sect. 4.4.1, the increment in sound level needed is about 10 dB for a factor of 10 reduction in sound duration. This dependence of threshold in quiet and masked threshold on the sound duration corresponds to a temporal integration of the sound intensity within a time window of 200 ms. This is why the effect is frequently called temporal integration in the hearing system. For hearing-impaired listeners, temporal integration often follows a distinctly different course compared to that of subjects with normal hearing. In extreme cases, the same threshold may be obtained by a hearing-impaired listener for sounds of 2 ms and 200 ms duration, even though over this range a rise in level of 20 dB is necessary to maintain the audibility of the sound burst for subjects with normal hearing.

It is therefore of interest to measure temporal integration in listeners with different kinds of hearing impairment. An example is given in Fig. 16.25 for noise-induced hearing loss. The data were obtained in the frequency range around 4 kHz, and a duration of 300 ms produced the reference threshold value. Tone bursts with durations of 300, 100, 30, and 10 ms were used, and the data compared with those produced by normally hearing subjects. The threshold level difference (relative to the 300-ms condition) is indicated as a function of tone-burst duration in Fig. 16.25. The data for normally hear-

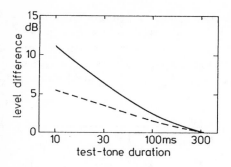

Fig. 16.25. Threshold of 4-kHz tone bursts as a function of their duration, measured in normally hearing subjects (*solid*) and in listeners with noise-induced hearing loss (*broken*), indicating abnormal temporal integration. Note that the threshold data are normalized to the value at the duration of 300 ms

ing subjects correspond closely to those given in Sect. 4.4.1, and indicate an increment of about 10 dB for a factor of 10 reduction from 100 to 10 ms. Hearing-impaired listeners, however, show an increment in level of only about 4 dB over the same range of durations. This increment is much less than with normally hearing subjects, and leads to the fact that the difference between normally hearing and hearing-impaired listeners may even be as much as 10 dB at 3-ms duration. However, this does not mean that hearing-impaired listeners hear better at shorter durations than do normally hearing subjects. Rather, it means that the temporal integration in hearing-impaired listeneres is not developed as much as it is in normally hearing subjects. Very short tone bursts of 3-ms duration produce a relatively broad spectrum. Because many subjects with noise-induced hearing loss show elevated threshold above about 1.5 to 2.5 kHz, but have relatively normal sensitivity for frequencies below that, the spectral splatter of short duration bursts may influence their thresholds.

The spectral splatter can be limited by using relatively large rise times for the 10-ms duration tones. Under such conditions, measurements can be performed with normally hearing subjects, where their thresholds are elevated by special spectrally shaped noise that reproduces the threshold of noise-induced hearing loss as a masked threshold. Under these conditions, measurements of temporal integration have been performed for both categories of subjects. The comparison between the data indicates that the spectral splatter does not influence the measurements either for normally hearing subjects with masked threshold, nor for subjects with the equivalent noise-induced hearing loss measured in quiet. Temporal integration took place for the normally hearing subjects in the same way as for their thresholds in quiet, leading to the conclusion that, unlike the truly hearing-impaired, listeners with impairments simulated by using masking noise show the same amount of temporal integration as they normally show in quiet.

16.2.6 Loudness Summation and Recruitment

As described in Chap. 8, the loudness of a sound is produced by summing the components expressed as specific loudnesses along the critical-band rate axis. Loudness of a band-pass noise increases with increasing bandwidth beyond the critical bandwidth, although the overall intensity is kept constant. The rate by which loudness increases is greatest at moderate levels. Because the increment is based on the fact that loudness components are added up along the critical-band rate to form the total loudness, it is called loudness summation.

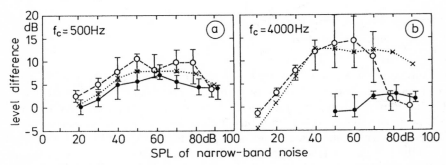

Fig. 16.26a,b. Level difference necessary to produce equal loudness between a narrow- and a broad-band noise as a function of the narrow-band noise level. Data measured for normally hearing subjects (open circles connected with broken lines) agree closely with data produced by the loudness calculation procedure (*dotted*). The data produced by listeners with noise-induced hearing loss (dots connected with solid lines) are similar for a centre frequency of 500 Hz (**a**) but much lower at 4 kHz (**b**), indicating that loudness summation fails in that frequency region

Loudness summation can be measured most easily in normally hearing and hearing-impaired subjects by loudness comparisons. The level difference needed to obtain equal loudness between a narrow-band noise and a broad-band noise can be measured as a function of the sound pressure level of the narrow-band noise. An example of the results of such measurements performed by subjects with noise-induced hearing loss is shown in Fig. 16.26. Part (a) shows the results for 500-Hz, part (b) those for 4-kHz centre frequency. At the 500-Hz centre frequency, noise bandwidths were 85 and 786 Hz, and at 4-kHz centre frequency, bandwidths were 710 and 5900 Hz for the narrow and broad-band noises, respectively. The cut-off frequencies were centred geometrically around the centre frequency. The dots connected with solid lines indicate the data produced by the hearing-impaired subjects. The dotted line indicates results that are produced by the loudness-calculating procedure (DIN 45 631) using third-octave-band analysis. Results produced by normally hearing subjects (broken line) are in reasonably good agreement with the calculated data.

At a centre frequency of 500 Hz, the results of hearing-impaired subjects are also in close agreement with the calculated data. At 4-kHz centre frequency, however, subjects with noise-induced hearing loss produce not only higher thresholds, near 50 dB in this case, but they also do not show any

reasonable loudness summation. Predicted data, and those of normally hearing subjects, show a loudness summation corresponding to values of up to
13 dB. Normally hearing subjects reach such values already 40 to 45 dB above
threshold in quiet, whereas listeners with noise-induced hearing loss show an
increment of only 1 to 2 dB at 40 dB above threshold in quiet, which is actually near 90 dB SPL. This means that in the frequency region of elevated
threshold in quiet, loudness summation is almost negligible for listeners with
noise-induced hearing loss.

This effect may have two reasons: decreased frequency selectivity as discussed in Sect. 16.2.2 for that kind of hearing loss at 4 kHz, or recruitment,
which for such patients appears to be a relatively dominant effect. Recruitment describes the effect in regions in which threshold in quiet is elevated
by an impairment to higher values (normally 30 to 60 dB), whereby loudness
increases much more quickly as a function of increasing level. For example,
40 dB above the elevated threshold, the loudness sensation of the hearing-
impaired listener is almost the same as the loudness sensation of normally
hearing subjects. This means that the loudness function, i.e. loudness as a
function of the level of a tone, is much steeper in such subjects than for normally hearing subjects. Steep loudness functions in normally hearing subjects
are obtained at low levels, 5 dB to 25 dB above threshold in quiet. In these
low-level regions, loudness summation does not take place in normally hearing subjects because of the steepness of the curve relating specific loudness
to excitation level. If the exponent is 1 instead of 0.25, such as at medium
and high levels, then intensity and loudness grow in the same way, i.e. no

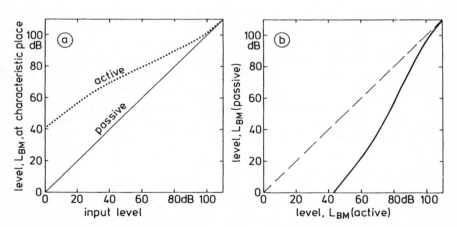

Fig. 16.27a,b. Level of excitation, L_{BM}, at the characteristic place on the basilar membrane shown in (**a**), as a function of input level fed to a model for the
active condition (nonlinear active feedback) and for the passive condition (feedback switched off). The corresponding behaviour using the data produced in the
active condition as the 45°-reference line is plotted in (**b**) indicating the effect of
recruitment

loudness summation takes place. Therefore, the effect of recruitment may also be the reason that loudness summation does not occur in listeners with noise-induced hearing loss.

The effect of recruitment can be illustrated using the nonlinear-active-feedback model of peripheral signal processing. The assumption is that noise-induced hearing loss starts with the outer haircells being impaired, so that they fail in what has been described as active processing. This seems to be a reasonable assumption, although we may also have to consider impairment in the inner haircells at more severe stages of hearing loss. With the outer hair-cells' activity switched off, i.e. the preprocessing model being totally passive, the level at the characteristic place of the tone exciting the model rises in the same way as the level that is put into the model. This effect is shown by the 45°-line (solid) in Fig. 16.27a. With the outer haircells active, this relationship changes drastically so that at lower levels of the input voltage, much higher levels at the characteristic place are found. The difference may reach 40 dB in the case shown. This enhanced relationship, shown as the dotted curve, is normally expected. Hearing impairment has to be characterized in relation to this dotted curve. If we use the dotted curve as the normal situation and plot the effect of impaired outer haircells as the difference between the two curves shown in part (a), we can plot the level at the characteristic place for the active condition as the abscissa and the level at the characteristic place in the passive condition as the ordinate. This is shown in Fig. 16.27b. The dotted curve is now the 45°-line, while the situation of impaired outer haircells, i.e. passive model, is again indicated by the solid line. This function is much steeper than the broken 45°-line; it has an elevated threshold but reaches the broken line at high levels. All these effects are typical of what is called recruitment.

The effects described above assume ideal conditions. Actually, not all outer haircells may be completely destroyed, giving only a small hearing loss. In other cases not only the outer haircells but also the inner haircells may be partially destroyed or impaired, so that effects additional to those described above have to be considered. However, it is interesting to see that the loss of outer haircells alone produces an effect that can describe, at least partly, the effect of recruitment.

16.2.7 Speech in Background-Noise

For persons with hearing deficits, speech intelligibility deteriorates substantially in noisy environments. Therefore, speech-tests in noise are included in the audiological test-battery to detect hearing deficits at an early stage. Frequently a "speech-noise" according to CCITT Rec. G 227 which simulates the average spectral distribution of speech is used as background noise for speech tests. However, an important feature of human speech, the temporal structure is missing.

Therefore, a background noise for speech audiometry was developed simulating the average spectral *and* temporal envelope of speech. The features of this background noise are illustrated in Fig. 16.28. The upper panels show the spectral distribution and the loudness-time function of the CCITT speech noise. The solid curve in the upper left panel shows the spectral weighting with a maximum around 800 Hz in comparison to the average spectral envelope of speech for several different languages as described by Tarnoczy (hatched). The loudness-time function displayed in the upper right panel illustrates that the CCITT speech noise is continuous with almost no temporal fluctuation.

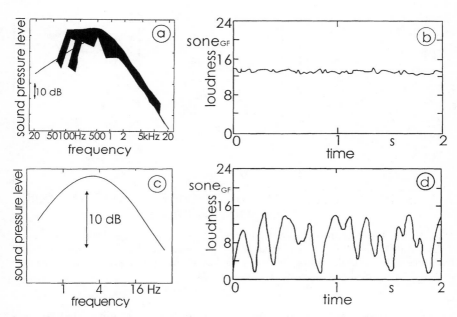

Fig. 16.28a–d. Background noise for speech audiometry. (**a**) Spectral distribution of carrier signal. (**b**) Loudness-time function of stationary background noise according to CCITT Rec. G 227. (**c**) Spectral distribution of modulating signal. (**d**) Loudness-time function of time-varying background noise for speech audiometry (Fastl-noise)

The lower left panel in Fig. 16.28 illustrates that the temporal envelope fluctuation of fluent speech can be characterized by a band-pass with a maximum around 4 Hz. This holds true for many different languages like English, French, German or Hungarian, Polish and even Chinese or Japanese. Generally speaking, at "normal" talking speed about four speech elements like syllables in European languages or mora in Japanese are pronounced per second. As outlined in Chap. 10, the band-pass of fluctuation strength cen-

tred at 4 Hz (Fig. 10.1) and the temporal envelope variation of fluent speech
also centred at 4 Hz illustrate the excellent correlation between speech and
hearing system.

When modulating the CCITT-noise displayed in the upper left panel of
Fig. 16.28 with the average temporal envelope of speech as illustrated in the
lower left panel, a time varying background noise for speech audiometry is
obtained. The loudness-time function of this background noise which was
named by our colleagues Fastl-noise, is shown in the lower right panel of
Fig. 16.28. The noise shows strong temporal variations with about four main
maxima per second.

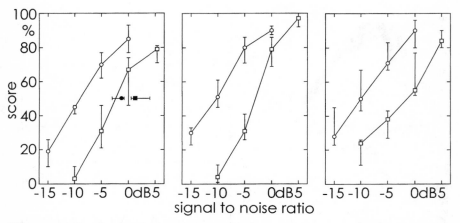

Fig. 16.29. Intelligibility of monosyllables in background noise. Data for German,
Polish, and Hungarian. *Circles*: Fastl-noise. Squares: CCITT-noise. *Open symbols*:
Subjects with normal hearing ability. *Filled symbols*: Subjects with 40 dB hearing
loss at 4 kHz. Level of background noises: 65 dB

With monosyllables as test material, speech intelligibility in noise was
measured for both, the stationary and the time-varying background noise
for different languages. In Fig. 16.29 results are given for German, Polish,
and Hungarian. Data for speech in the Fastl-noise are indicated by circles
and data for the CCITT-noise by squares. Open symbols represent data for
subjects with normal hearing, filled symbols for subjects with 40 dB hearing
loss at 4 kHz.

The results shown in Fig. 16.29 suggest that at a given signal to noise ratio
it is easier to correctly identify a monosyllable in the fluctuating noise than
in the continuous noise. This holds true for all languages investigated. For
50% correct score of subjects with normal hearing, the signal to noise ratio
has to be increased by about 7 dB when proceeding from the fluctuating
Fastl-noise to the continuous CCITT-noise. In comparison to subjects with
normal hearing, for the subjects with 40 dB hearing loss at 4 kHz signal to

noise ratio has to be improved by 7 dB for the Fastl-noise and by 4 dB for the CCITT-noise.

Obviously, the subjects with hearing loss have more difficulty to understand monosyllables presented in "valleys" of the fluctuating noise (cf. Fig. 16.28d). The "benefit" for the fluctuating noise of 7 dB for normal hearing subjects shrinks to 3 dB for the subjects with 40 dB hearing loss at 4 kHz. Since the reduction in performance of subjects with hearing loss versus subjects with normal hearing is larger for the time-varying background noise, it is recommended to use a fluctuating background when testing speech in noise for subjects with hearing loss.

Speech in noise was also tested with 12 patients wearing an 8-channel-cochlea-implant. Sentences were chosen as speech material which eases the test. Again CCITT-noise and Fastl-noise were used as background. Figure 16.30 shows the results: open symbols illustrate data for subjects with normal hearing, filled symbols indicate data for patients with cochlea-implants.

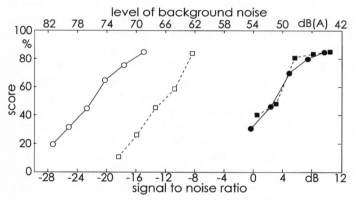

Fig. 16.30. Intelligibility of sentences in background noise. *Open symbols*: subjects with normal hearing. *Filled symbols*: patients with 8-channel-cochlea-implants. *Circles*: Fastl-noise. *Squares*: CCITT-noise. Speech-level 54 dB

For the subjects with normal hearing it is again much easier to understand words of sentences in a time-varying background noise (circles) than in a continuous background noise (squares). At a score of 50%, the advantage in signal to noise ratio amounts to about 10 dB. In contrast, the patients with cochlea-implants reach a 50% score at almost the same signal to noise ratio for both background noises. Even for the continuous background noise, cochlea-implant patients need about 16 dB better signal to noise ratio than subjects with normal hearing. The extremely large difference in signal to noise ratio of about 26 dB between subjects with normal hearing and cochlea-implant patients for speech in fluctuating noise indicates that temporal processing of sound by these patients is not yet sufficiently understood.

16.3 Hearing Aids

Hearing aids are most often used in noisy surroundings. Therefore, it is important to optimize the signal-to-noise ratio by using microphones or an arrangement of microphones that show a strong directional gain. By changing the direction of the head, the signal-to-noise ratio of the sound of interest can then be increased and speech intelligibility frequently improved.

Automatic gain control (AGC) is a technical device often used in hearing aids. There exist different types of AGC-circuits that influence speech intelligibility in different ways. The loudness-time functions that are available in loudness meters are sufficient to assess single-channel hearing aids with AGC. Figure 16.31 indicates total loudness as a function of time for the word "but". The three parts indicate the loudness-time function for (a) the situation without a hearing aid, (b) with a hearing aid that includes AGC with a short recovery time and (c) with a hearing aid with a long recovery time AGC. It becomes clear that the "t" at the end of "but" is almost totally removed in condition (c). For multi-channel hearing aids, the assessment of AGC settings provided for each channel can be undertaken by using a three-dimensional loudness versus critical-band rate versus time pattern.

In cases of hearing impairment that include a loss of frequency selectivity, the hearing aid should include a device that extracts the most important spectral features. Such features are, for example, the formants of vowels that cannot be detected easily if the frequency selectivity of the hearing system is markedly reduced. This can be achieved by using multi-channel devices with additional networks that inhibit activity in channels of lower and unimportant levels, so that only the spectral peaks remain. An example of an 18-tone complex with spectral maxima is shown in Fig. 16.32. Part (a) shows the spectral composition of the input to the system. Part (b) shows the spectral composition remaining as a consequence of the activity of the inhibition matrix. The level differences among the tones lead the inhibition matrix to produce a reduction of the level of the softer tones on both sides of the peak-level tones at 0.7 and at 3 kHz. Speech processed this way sounds very pronounced and sharp to normally hearing subjects. For hard-of-hearing listeners with strongly reduced frequency selectivity, speech preprocessed this way may be more easily understood.

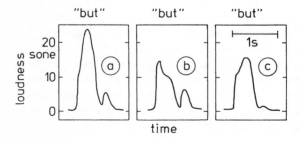

Fig. 16.31a–c. Loudness-time function recorded by a loudness meter and produced by the spoken word "but" without hearing aid in (**a**); using hearing aids that include AGC with short recovery time (**b**) and with long recovery time (**c**)

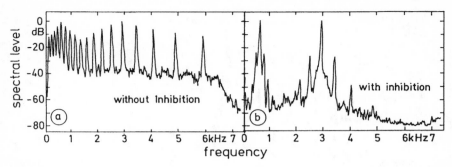

Fig. 16.32a,b. The effect of inhibitory networks, indicated by the processing of the incoming spectrum with two unpronounced peaks (**a**) into a spectrum with two much sharper peaks (**b**)

For deaf people hearing aids can not be used, but hearing protheses such as cochlea implants or devices that stimulate the skin either by electrical current or mechanical vibration have been developed. All these devices have special advantages and disadvantages. The inhibition strategy, however, seems to be useful in the sound preprocessing system used in these devices. The interested reader is referred to the literature for details.

16.4 Broadcasting and Communication Systems

There is a strong tendency in developing broadcasting and communication systems to use digital signal transmission. From this point of view, it becomes important to reduce the information flow in a meaningful way. In this context "meaningful" means that the reduction of information flow should be performed in such a way that listeners are not able to hear a difference between the directly transmitted audio signals and audio signals that are processed for reduced information flow. This holds for music and speech. For speech, however, only the criterion of intelligibility may be important; the reduction of information flow can be much larger if only intelligibility is required.

In order to make the reduction of information flow inaudible, it is best to use the characteristics of the human ear as an information receiver, and reduce the information flow for those events that are already inaudible. The effect of masking is a typical example. It can be introduced into the processing system in such a way that those parts masked by loud sounds are not transmitted to the receiver. Another strategy makes use of those important features that are selected by our hearing system. Spectral pitch, for example, plays an important role in characterizing music and speech. There are many spectral pitches in musical sounds, some of which are very dominant, others that are of less importance and still others that can almost be ignored. The physical equivalent of spectral pitch percepts are "partial tones", which

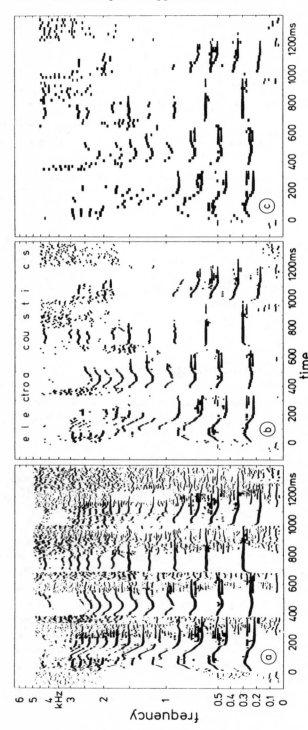

Fig. 16.33a–c. Information-flow reduction established in the partial-tone versus time pattern of the spoken word "electroacoustics": complete information in (**a**); 16 kbit per second in (**b**); 4 kbit per second in (**c**)

can be extracted by a hearing-equivalent spectrum analysis. Such analysis is performed by a modified Fourier transformation which allows adjustment of analysis parameters with regard to the frequency and temporal resolution of the human ear. The second step is temporal smoothing in combination with a threshold criterion during maximum detection that prevents side lobes from being extracted as partial tones. The part-tone time pattern is the pattern that can be processed in such a way that the reduction in information-flow reduction is performed by transmitting all partial-tone events, by ignoring those events with small pitch strength, or by transmitting only the most important events. It seems that speech and music can be well represented by the time-varying frequencies and levels of the partial-tone versus time pattern. The resynthesized signals are nearly indistinguishable from the originals if a rate of 16 kbit per second is used. For speech with a bandwidth of 100 to 5200 Hz and a dynamic range of 60 dB, intelligibility reaches 96 to 99% for a rate of 16 kbit per second, whereas a rate of 4 kbit per second reduces the intelligibility somewhat to values of 92 to 95%, measured by a rhyme-test. Running speech is completely understandable at such percentages of intelligibility.

An example of the word "electroacoustics" is shown for three cases of information-flow reduction in Fig. 16.33. Part (a) shows the complete information, i.e. partial-tone frequencies as a function of time (abscissa). Part (b) shows the effect of the reduction to 16 kbit per second, whereas part (c) indicates the effect of further reduction to 4 kbit per second. The resynthesized signal establishes that the word is clearly understandable, although the reduction is quite dramatic. This example may only be a hint towards the use of psychoacoustics in transmitting audio signals in broadcasting and communication systems at strongly reduced bit rates.

On the other hand, when re-mastering classic program material, e.g. recordings of the famous tenor Caruso, ultimate care has to be taken by the Tonmeister not to "clean" the original too perfectly. Many music enthusiasts miss the "brilliance" of Caruso's voice, if the hiss noise is removed completely. In well controlled psychoacoustic experiments it could be shown that, when adding soft white noise to music low-pass filtered at 8 kHz, experts like sound engineers or professional musicians "hear" that the brilliance of the music increases despite the fact that in both presentations with or without soft background noise, the music contains only frequency components up to 8 kHz. Presumably the experts infer that (non-existent) high-frequency components of the music are masked by the added soft noise. On the other hand, the experts preferred the version without noise although they felt that in this version the music material lost its brilliance.

16.5 Speech Recognition

The human hearing system is still, in every respect, by far the best speech recognition system. Therefore, it is reasonable to simulate this system as much as possible. Such a simulation should be based on available physiological and psychoacoustical knowledge. In addition, linguistic and phonetic rules should be introduced into the recognition procedure. The block diagram of a speech recognition system based on cochlear preprocessing and psychoacoustics is shown in Fig. 16.34. It contains the nonlinear peripheral preprocessing with active feedback, followed by the extraction of basic auditory sensations out of which complex auditory sensations such as virtual pitch or rhythm may be created. All of these sensations are checked for dominant variations. The speech recognition procedure also makes use of nonauditory information such as linguistic and phonetic rules and finally produces a sequence of phonetic items.

We know that our hearing system has the advantage of being able to adapt to the special characteristics of each speaker, or to a certain frequency response of the transmitting system or to some special room acoustics. As yet, we do not know much about these adaptation processes. However, in order to get a more precise recognition system, it is necessary to introduce these capabilities into the recognition system. The dashed lines in Fig. 16.34 show possible feedback loops implementing such adaptation processes. The numbers at the top of the figure indicate the bit rate at certain steps of recognition.

Peripheral preprocessing, auditory sensations and dominant variations have already been discussed in detail. The use of the hearing sensation rhythm, in order to create borders for segmentation, is discussed as an example of the application of psychoacoustics. Figure 13.2 illustrates clearly that the subjectively perceived rhythm correlates strongly with the peak values of

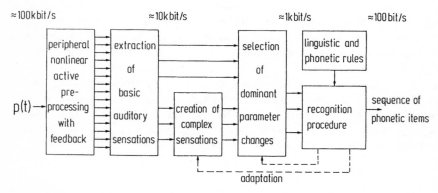

Fig. 16.34. Block diagram of a speech recognition system based on cochlear preprocessing and psychoacoustics

the loudness-time function. This means that the loudness-time function can
be used for setting boundary lines between the single segments. The rhythm
corresponds closely to the syllables. However, a segmentation into half syllables is also possible, because not only the maxima but also the minima in
the loudness-time function can be used.

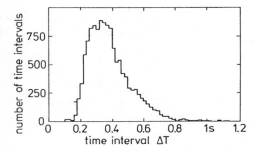

Fig. 16.35. Histogram of subjectively perceived intervals between
neighbouring events in different
languages. The broad maximum
near 300 ms indicates adaptation
of speech to the maximum fluctuation strength near a modulation
frequency of 4 Hz

The temporal intervals between the neighbouring events of the reproduced
rhythm have been measured for German, English, French and Japanese. A
histogram of all reproduced time intervals based on more than 10000 data
points is indicated in Fig. 16.35 with a bin width of 20 ms. All the intervals
are concentrated between 100 and 1000 ms. Almost 90% of the data points
are found between 200 and 600 ms, and the most frequent temporal interval
has a length of about 300 ms. This means that in running speech, about 2 to
5 events are perceived per second. These data correspond nicely to the maximum of the fluctuation strength measured as a function of the modulation
rate. The sequence of processing is evident if the loudness-time function is
considered to be the initial basic information. The maxima of this function
correlate with the perceived rhythm. The segmentation based on this strategy
leads either to syllables or half syllables if the minima in the loudness-time
function are used as well.

16.6 Musical Acoustics

It is clear that psychoacoustics plays an important role in musical acoustics.
There are many basic aspects of musical sounds that are correlated with the
sensations already discussed in psychoacoustics. For example, the preference
of many musicians for natural versus plastic heads of tympani is in line with
the observation that plastic heads produce smaller pitch strength and more
variance in the loudness of tones in different pitch regions. Further examples
may be the different pitch qualities of pure tones and complex sounds, perception of duration, loudness and partially-masked loudness, sharpness as an
aspect of timbre, perception of sound impulses as events within the temporal

patterns leading to rhythm, roughness, and the equivalence of sensational intervals. For this reason, it can be stated that most of this book's contents are also of interest in musical acoustics. At this point we can concentrate on two aspects that have not been discussed so far: musical consonance and the Gestalt principle.

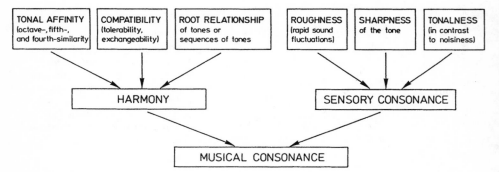

Fig. 16.36. Concept of musical consonance

Although there are different types of musical style, all musicians can distinguish between consonant and non-consonant sounds. It seems that this musical consonance is based on two parts; one which comes from musical acoustics and is related to harmony, the other coming from psychoacoustics, called sensory consonance. Figure 16.36 illustrates the concept of musical consonance in more detail. Harmony is based on three musical components. Tonal affinity refers to relationship that sounds have, for example octave equivalence or fifth- or fourth-similarity. The second contribution from musical acoustics is compatibility. This expression characterizes the effect that in a tonal piece of music, tones can be replaced by other compatible tones without seriously interfering with harmony. Compatibility may also be called the phenomenon of reversibility of chords. Therefore, compatibility may also be called tolerability or interchangeability. The third contribution from musical acoustics to harmony is the "root" – relationship of tones or sequences of tones. This means that key notes are assigned to musical chords. Many of these have the same musical root. Examples for this are the normal bass symbols and the harmony denotation by letters and figures. The perception of root is often ambiguous and the perception of chord roots has a "virtual" character, in the sense that all that has to be produced is the perception of a clearly distinct pitch, which is often a virtual pitch. These three components form harmony, which may be defined more clearly by using this interrelation.

The other contribution to musical consonance is sensory consonance, which is based on psychoacoustic sensations such as roughness, sharpness and tonalness, here defined in contradistinction to noisiness. Sensory consonance represents the general aspect of tonal pleasantness, i.e. a nonmusic-specific aspect. Experience in music shows that roughness should be avoided if pleasantness is to be expected. Sharpness shows a similar effect and should be kept small in order not to reduce pleasantness. Another item from psychoacoustics, the tonalness, describes the hearing-related characteristic of how a sound differs from noise. Consequently, tonalness is the opposite of noisiness. Tonalness of a sound is higher the more audible tonal components are available. This means that noise components should be as small as possible. Tonalness has a close relation to the spectral pitch patterns and can be described quantitatively by them.

Loudness also influences pleasantness. However, this influence is not essential enough as to warrant a detailed discussion here.

The Gestalt principle is very well known from vision. However, it also seems to play an important role in musical acoustics. The concept of hierarchical processing of categories assumes that Gestalt perception is entirely dependent on controlled and categorized sensory representations. It also assumes that perception is organized in hierarchical layers, in each of which one and the same type of processing is established. It seems that the perception of primary contours in vision has its equivalent in the auditory perception of spectral pitches. It is interesting that spectral pitches are processed in a relatively early stage of auditory processing. In vision, the contours lead to what we call Gestalt. In music, it is the pitch that finally leads to Gestalt. Thereby, spectral pitch plays the more important role whereas virtual pitches are equivalent to "illusory" visual contours.

16.7 Room Acoustics

Room acoustics uses mostly physical values. However, room acoustics should also describe the conditions leading to good hearing in a room. Because hearing characteristics are described by psychoacoustical data and values, it seems reasonable to introduce these values into the description of room acoustics. This often means that temporal and spectral effects should be described using total loudness as a function of time, or the three-dimensional distribution of specific loudness versus critical-band rate pattern as a function of time. In addition to that, other psychoacoustical values such as fluctuation strength, partial masking or sharpness, can be used to describe the influence of room acoustics on the characteristics of the sound at the place of a listener. A few examples will illustrate these effects.

It is known that the level produced by a source of constant volume velocity produces a greater level in a room the larger the reverberation time. For the listener, it is not so important that the average level increases; the increment of loudness is much more important. Figure 16.37 indicates the effect of reverberation in rooms when the same speech power produces loudness-time functions under three different conditions: (a) free-field condition, (b)

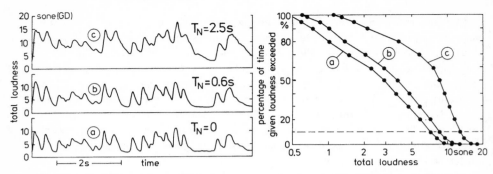

Fig. 16.37. Effect of reverberation time on loudness-time functions (*left*) and on loudness distributions (*right*). Data are obtained in the free-field condition (*a*) and in rooms with reverberation time of 0.6 (*b*) and 2.5 s (*c*), indicating the increase in loudness with increasing reverberation

reverberation time of 0.6 s and (c) reverberation time of 2.5 s. Short periods out of a 10-minute speech are shown in the left part of Fig. 16.37. The right part indicates the cumulative distributions resulting from the loudness-time functions for the three room conditions. Using the loudness exceeded 10% of the time as an indication of the perceived loudness, it can be expected that speech is about 1.21 times louder in the room with 0.6-s reverberation and about 1.78 times louder in the room with 2.5-s reverberation, in comparison with the loudness produced in open-air free-field condition. The increment in loudness is very helpful for the intelligibility of speech as long as the reverberation time does not produce temporal masking.

Speech belongs to the category of sounds that show very strong spectral and temporal variations. Therefore, its total loudness and specific loudness versus critical-band rate distribution depends strongly on time. The variation periodicities lead to the perception of fluctuation. In rooms with long reverberation time, fluctuation strength may be reduced, although the source is the same. Measurements of speech intelligibility and fluctuation strength undertaken in rooms of different characteristics have indicated that speech intelligibility in a room can be predicted using the psychoacoustical model of fluctuation strength. Figure 16.38 compares subjective data with objective data and reverberation time.

Distinct echos should be avoided in room acoustics. In order to indicate these echos a so called "Christmas tree" distribution is often used. It shows the temporal decay produced for a certain location of the listener in the room, if a short noise burst is produced at the location of the speaker or the orchestra. The decay is normally shown as a level-time function that indicates the different echos. Instead of using the level as shown in Fig. 16.39 part (a), it may be more meaningful to use total loudness versus time functions as shown in part (b). Although the decaying curve is more rounded if loudness is used, the stronger echos are seen in the loudness function as well. If the loudness

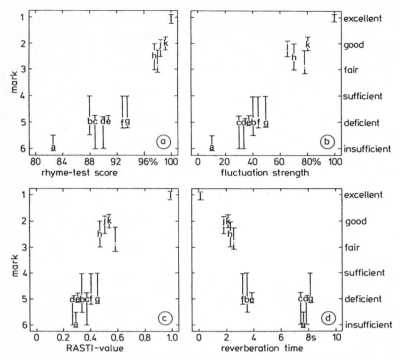

Fig. 16.38a–d. Rhyme-test score, fluctuation strength, RASTI-value and reverberation time in rooms with different characteristics, in relation to subjective data of acceptance expressed in scores between "excellent" and "insufficient". The different letters indicate different rooms

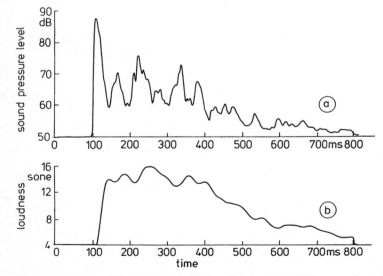

Fig. 16.39a,b. "Christmas-tree" distribution, i.e. SPL as a function of time in response to a short sound burst in a reverberant room (**a**). A more meaningful correlate to subjective acoustical perception may be given with the corresponding loudness-time function (**b**)

of the echos is only a tenth of the initial value, the echos may be ignored in speech perception. The loudness-time function outlined in Fig. 16.39b indicates also that the initial peak in the level-time function does not produce the highest peak in the loudness-time function. The echos and reverberation contribute strongly to perceived loudness.

Literature

Summarizing Books or Handbook Articles

Fastl H.: Beschreibung dynamischer Hörempfindungen anhand von Mithörschwellen-Mustern (Dynamic hearing sensations and masking patterns). (Hochschul, Freiburg 1982)

Feldtkeller R., E. Zwicker: Das Ohr als Nachrichtenempfänger (The ear as a receiver of information), 1. Aufl. (Hirzel, Stuttgart 1956)

Terhardt E.: *Akustische Kommunikation* (Acoustic communication). (Springer, Berlin, Heidelberg, New York 1998)

Zwicker E.: Scaling. In *Handbook of Sensory Physiology*, ed. by W. Keidel, W. Neff, Vol. V, Part 2 (Springer, Berlin, Heidelberg 1975) pp. 401–448

Zwicker E.: Psychoakustik (Psychoacoustics), (Springer, Berlin, Heidelberg 1982)

Zwicker E., R. Feldtkeller: *Das Ohr als Nachrichtenempfänger* (The ear as a receiver of information), 2. erw. Aufl. (Hirzel, Stuttgart 1967)

Zwicker E., M. Zollner: *Elektroakustik* (Electroacoustics), 2. erw. Aufl. (Springer, Berlin, Heidelberg 1987)

Chapter 1
Stimuli and Procedures

Bäuerle R.: Modal-, Zentral- oder Arithmetischer Mittelwert (Modalvalue, median or arithmetic mean). ATM. Blatt JO **21–22**, 33–34 (1974)

Fastl H.: Comparison of DT 48, TDH 49, and TDH 39 earphones. J. Acoust. Soc. Am. **66**, 702–703 (1979)

Fastl H.: Gibt es *den* Frequenzgang von Kopfhörern? (Does the frequency response of earphones exist?) In *NTG-Fachberichie*, Hörrundfunk **7** (VDE, Berlin 1986) pp. 274–281

Fastl H., H. Fleischer: Freifeldübertragungsmaße verschiedener elektrodynamischer und elektrostatischer Kopfhörer (Freefield response of different electrodynamic and electrostatic earphones). Acustica **39**, 182–187 (1978)

Fastl H., S. Namba, S. Kuwano: Freefield response of the headphone Yamaha HP 1000 and its application in the Japanese Round Robin Test on impulsive sound. Acustica **58**, 183–185 (1985)

Fastl H., E. Zwicker: A free-field equalizer for TDH 39 earphones. J. Acoust. Soc. Am. **73**, 312–314 (1983)

Hesse A.: Optimierung einer Abfragemethode zur Bestimmung von Frequenzunterschiedsschwellen (Optimizing a method of constant stimuli for the measurement of frequency differences). Acustica **54**, 181–183 (1984)

Hesse A.: Comparison of several psychophysical procedures with respect to threshold estimates, reproducibility and efficiency. Acustica **59**, 263–273 (1986)

Kaiser W.: Das Békésy-Audiometer der Technischen Hochschule Stuttgart (The Békésy-audiometer of the Technical University Stuttgart). Acustica, Akust. Beihefte AB235–AB238 (1952)

Krump G.: Linienspektren als Testsignale in der Akustik (Line-spectra as test signals in acoustics). In: *Fortschritte der Akustik, DAGA '96*, Verl.: Dt. Gesell. für Akustik e. V., Oldenburg, pp. 288–289 (1996)

Pfeiffer Th.: Filter zur Umformung von Rechteckimpulsen in Gaußimpulse (Filter for transforming of rectangular impulses into Gaussian-shaped impulses). Frequenz **17**, 81–88 (1963)

Port E.: Ein elektrostatischer Lautsprecher zur Erzeugung eines ebenen Schallfeldes (An electrostatic loudspeaker for the production of a plain soundfield). Frequenz **18**, 9–13 (1964)

Schmid. W., G. Jung: Psychoakustische Experimente mit einem elektroakustischen Wiedergabesystem mit variierbarer Sprungantwort (Psychoacoustic experiments with a system of variable step-response). In: *Fortschritte der Akustik, DAGA '95*, Verl.: Dt. Gesell. für Akustik e. V., Oldenburg, pp. 931–934 (1995)

Schorer E.: An active free-field equalizer for TDH 39 earphones [43.66.Yw, 43.88.Si]. J. Acoust. Soc. Am. **80**, 1261–1262 (1986)

Schorer E.: Methodische Einflüsse bei der Bestimmung von Frequenzvariations- und Frequenzunterschiedsschwellen (Influence of different measurement methods on thresholds for frequency variations and frequency differences). In *Fortschritte der Akustik, DAGA '87* (DPG, Bad Honnef 1987) pp. 577–580

Suchowerskyj W.: Beurteilung von Unterschieden zwischen aufeinanderfolgenden Schallen. Acustica **38**, 131–139 (1977)

Terhardt E.: Fourier transformation of time signals: conceptual revision. Acustica **57**, 242–256 (1985)

Terhardt E.: Evaluation of linear-system responses by Laplace-transformation. Critical review and revision. Acustica **64**, 61–72 (1987)

Zollner M.: Einfluß von Stativen und Halterungen auf den Mikrofonfrequenzgang (Influence of stands and clamps on the frequency response of microphones). Acustica **51**, 268–272 (1982)

Zollner M.: Terzanalyse mit Normterzreihe: Eine hinreichende Meßmethode? (1/3 octave-band analyses with standardized filter sets: A sufficient measurement method?) In *Fortschritte der Akustik, DAGA '84* (DPG, Bad Honnef 1984) pp. 267–270

Zwicker E., G. Gässler: Die Eignung des dynamischen Kopfhörers zur Untersuchung frequenzmodulierter Töne (The suitability of a dynamic earphone for the measurement of FM-tones). Acustica, Akust. Beihefte AB134–AB139 (1952)

Zwicker E., E. Hojan: Einfluß von Bedienperson und Geräteform auf den Mikrofon-Frequenzgang von Schallpegelmeßgeräten (Influence of experimenter and shape of the apparatus on the frequency response of sound level meters). Acustica **51**, 263–267 (1982)

Zwicker E., D. Maiwald: Über das Freifeldübertragungsmaß des Kopfhörers DT 48 (On the freefield response of the earphone DT 48). Acustica **13**, 181–182 (1963)

Chapter 2
Threshold of Hearing

Fastl H., H. Baumgartner: Ruhehörschwellen für Sinustöne bzw. Schmalbandrauschen (Threshold in quiet for pure tones vs. narrow noise bands). Acustica **34**, 111–114 (1975)

Zwicker E.: Über die Schwelle des Ohrendruckes für verschiedene Schallereignis-se (On the threshold of the "ear pressure" for different sounds). Frequenz **13**, 238–242 (1959)

Zwicker E., W. Heinz: Zur Häufigkeitsverteilung der menschlichen Hörschwelle (On the statistical distribution of the human hearing threshold). Acustica **5**, 75–80 (1955)

Chapter 3
Processing of Sound in the Auditory System

Harris F.P., E. Zwicker: Characteristics of the $(2f_1 - f_2)$-difference tone in human subjects and in a hardware cochlear nonlinear preprocessing model with active feedback. J. Acoust. Soc. Am. **85**, Suppl.1, S68 (1989)

Kronester-Frei A.: Postnatale Entwicklung des Randfasernetzes der Membrana tectoria beim Kaninchen (Development of the terminal net of the tectorial membrane in the rabbit). Laryngologie, Rhinologie, Otologie **55**, 687–700 (1976)

Kronester-Frei A.: Ultrastructure of the different zones of the tectorial membrane. Cell Tiss. Res. **193**, 11–23 (1978)

Kronester-Frei A.: Localization of the marginal zone of the tectorial membrane in situ, unfixed, and with in vivo-like ionic milieu. Arch. Otorhinolaryngol. **224**, 3–9 (1979)

Kronester-Frei A.: The effect of changes in endolymphatic ion concentrations on the tectorial membrane. Hearing Res. **1**, 81–94 (1979)

Manley G.A., A. Kronester-Frei: Organ of corti: Observation technique in the living animal. Hearing Res. **2**, 87–91 (1980)

Manley G.A., A. Kronester-Frei: The electrophysiological profile of the organ of corti. In *Psychophysical, Physiological and Behavioural Studies in Hearing*, ed. by G. van den Brink, F.A. Bilsen (University Press, Delft 1980) pp. 24–33

Oelmann J.: Die Zeitfunktionen von Cochleapotentialen, hervorgerufen durch gaußförmige Druck- und Sogimpulse am Trommelfell (Time functions of cochlear potentials evoked by Gaussian-shaped condensation and rarefaction impulses at the ear drum). Acustica **46**, 39–50 (1980)

Oelmann J.: Microphonic potentials recorded from the ear canal in man evoked by Gaussian-shaped sound pressure impulses. Scand. Audiol. Suppl. **11**, 59–64 (1980)

Scherer A.: Beschreibung der simultanen Verdeckung mit Effekten aus Mithör-schwellen- und Suppressionsmustern (Description of simultaneous masking on the basis of effects from masking patterns and suppression patterns). In *Fortschritte der Akustik, DAGA '87* (DPG, Bad Honnef 1987) pp. 569–572

Scherer A.: Erklärung der spektralen Verdeckung mit Hilfe von Mithörschwellen-und Suppressionsmustern (Explanation of spectral masking on the basis of masking patterns and suppression patterns). Acustica **67**, 1–18 (1988)

Terhardt E.: Untersuchungen über die Datenreduktion durch das menschliche Gehör (Investigations on the data reduction in the human hearing system). In *Kybernetik 1968*, ed. by H. Marko, G. Färber (Oldenbourg, München 1968) pp. 383–395

Zwicker E.: Die Abmessungen des Innenohrs des Hausschweines (The dimensions of the inner ear of the domestic pig). Acustica **25**, 232–239 (1971)

Zwicker E.: Is the frequency selectivity of the inner ear influenced by the size of the tectorial membrane? Proc. 7th ICA Budapest, Vol.3 (1971) pp. 385–388

Zwicker E.: Investigation of the inner ear of the domestic pig and the squirrel monkey with special regard to the hydromechanics of the cochlear duct. In *Symposium on Hearing Theory* (IPO, Eindhoven 1972) pp. 182–185

Zwicker E.: Vergleich der Struktur des Innenohres bei Hausschwein und Totenkopf-affe (Comparison of the structure of the inner ear from domestic pig and squirrel monkey). In *Akustik und Schwingungstechnik* (VDE, Berlin 1972) pp. 288–291

Zwicker E.: Masking period patterns and cochlear acoustical responses. Hearing Res. **4**, 195–202 (1981)

Zwicker E. On peripheral processing in human hearing. In *Hearing – Physiological Bases and Psychophysics*, ed. by R. Klinke, R. Hartmann (Springer, Berlin, Heidelberg 1983) pp. 104–110

Zwicker E.: Das Innenohr als aktives schallverarbeitendes und schallaussendendes System (The inner ear as an active sound processing and sound emitting system). In *Fortschritte derAkustik, DAGA '85* (DPG, Bad Honnef 1985) pp. 29–44

Zwicker E.: Psychophysics and physiology of peripheral processing in hearing. In *Basic Issues in Hearing*, ed. by H. Duifhuis, J.W. Horst, H.P. Wit (Academic, London 1988) pp. 14–25

Zwicker E.: The inner ear, a sound processing and a sound emitting system. J. Acoust. Soc. Jpn. (E) **9**, 59–74 (1988)

Zwicker E., F.P. Harris: Psychoacoustical and ear canal cancellation of $(2f_1 - f_2)$-distortion products. J. Acoust. Soc. Am. **87**, 2583-2591 (1990)

Zwicker E., A. Hesse: Temporary threshold shifts after onset and offset of moderately loud low frequency maskers. J. Acoust. Soc. Am. **75**, 545–549 (1984)

Zwicker E., G. Manley: The auditory system of mammals and man. In *Biophysics*, ed. by W. Hoppe, W. Lohmann, H. Markl, H. Ziegler (Springer, Berlin, Heidelberg 1983) pp. 671–682

Otoacoustic Emissions

Dallmayr C.: Spontane oto-akustische Emissionen: Statistik und Reaktion auf akustische Störtöne (Spontaneous otoacoustic emissions: statistics and reactions on acoustic masking tones). Acustica **59**, 67–75 (1985)

Dallmayr C.: Suppressions-Periodenmuster von spontanen oto-akustischen Emissionen. (Suppression-period patterns of spontaneous otoacoustic emissions.) In *Fortschritte der Akustik, DAGA '85* (DPG, Bad Honnef 1985) pp. 479–482

Dallmayr C.: Stationary and dynamic properties of simultaneous evoked otoacoustic emissions (SEOAE). Acustica **63**, 243–255 (1987)

Jurzitza D.: Zusammenspiel von Quelle und Last bei der Messung von otoakustischen Emissionen (Influence of probe impedance on measured otoacoustic emissions). In: *Fortschritte der Akustik, DAGA '91*, Verl.: DPG-GmbH, Bad Honnef, pp. 613–616 (1991)

Jurzitza D., W. Hemmert: Quantitative measurements of simultaneous evoked otoacoustic emmissions. Acustica **77**, 93–99 (1992)

Peisl W.: Simulation von zeitverzögerten evozierten oto-akustischen Emissionen mit Hilfe eines digitalen Innenohrmodells (Simulation of delayed evoked otoacoustic emissions in a digital model of the inner ear). In *Fortschritte der Akustik, DAGA '88* (DPG, Bad Honnef 1988) pp. 553–556

Peisl W., E. Zwicker: Simulation der Eigenschaften oto-akustischer Emissionen mit Hilfe eines analogen und eines digitalen Innenohrmodells (Simulation of otoacoustic emission's behaviour by means of an analog and a digital inner-ear model). In *Fortschritte der Akustik, DAGA '89* (DPG, Bad Honnef 1989) pp. 419–422

Scherer A.: Evozierte oto-akustische Emissionen bei Vor- und Nachverdeckung (Evoked otoacoustic emissions with pre- and postmasking). Acustica **56**, 34–40 (1984)

Scherer A.: Zeitverzögerte, evozierte oto-akustische Emissionen bei der Vor-, Simultan- und Nachverdeckung (Delayed evoked otoacoustic emissions and pre-, simultaneous, and postmasking). In *Fortschritte der Akustik, DAGA '84* (DPG, Bad Honnef 1984) pp. 765–768

Scherer A.: Die Amplitude evozierter oto-akustischer Emissionen als Maß für die Verdeckung (The amplitude of evoked otoacoustic emissions as tool for the description of masking). In *Fortschritte der Akustik, DAGA '86* (DPG, Bad Honnef 1986) pp. 413–416

Scherer A.: Beschreibung der simultanen Verdeckung mit Effekten aus Mithörschwellen- und Suppressionsmustern (Description of simultaneous masking on the basis of effects from masking patterns and suppression patterns). In *Fortschritte der Akustik, DAGA '87* (DPG, Bad Honnef 1987) pp. 569–572

Scherer A.: Erklärung der spektralen Verdeckung mit Hilfe von Mithörschwellen- und Suppressionsmustern (Explanation of spectral masking on the basis of masking patterns and suppression patterns). Acustica **67**, 1–18 (1988)

Schloth E.: Amplitudengang der im äußeren Gehörgang gemessenen akustischen Antworten auf Schallreize (Amplitude characteristic of acoustic responses to sound stimuli as measured in the outer ear canal). Acustica **44**, 239–241 (1980)

Schloth E.: Akustisch meßbare stationäre, tonale Signale aus dem Gehör (Acoustically measurable stationary tonal signals out of the ear). In *Fortschritte der Akustik, DAGA '81* (VDE, Berlin 1981) pp. 689–692

Schloth E.: Im äußeren Gehörgang akustisch meßbare Signale aus dem Gehör (Signals out of the ear as measured acoustically in the outer earcanal). In *Fortschritte der Akustik, DAGA '80* (VDE, Berlin 1980) pp. 591–594

Schloth E.: Relation between spectral composition of spontaneous otoacoustic emissions and finestructure of threshold in quiet. Acustica **53**, 250–256 (1983)

Schloth E., E. Zwicker: Mechanical and acoustical influences on spontaneous otoacoustic emissions. Hearing Res. **11**, 285–293 (1983)

Sutton G.J.: Suppression effects in the spectrum of evoked otoacoustic emissions. Acustica **58**, 57–63 (1985)

Zwicker E.: Mithörschwellen-Periodenmuster und Suppressions-Periodenmuster tieffrequenter gaußförmiger Druck- und Sogimpulse (Masking period patterns and suppresion period patterns of low frequency Gaussian-shaped pressure and rarefaction impulses). In *Fortschritte der Akustik, FASE/DAGA '82* (DPG, Bad Honnef 1982) pp. 1239–1242

Zwicker E.: Delayed evoked otoacoustic emissions and their suppression by Gaussian-shaped pressure impulses. Hearing Res. **11**, 359–371 (1983)

Zwicker E.: On peripheral processing in human hearing. In *Hearing – Physiological Bases and Psychophysics*, ed. by R. Klinke, R. Hartmann (Springer, Berlin, Heidelberg 1983) pp. 104–110

Zwicker E.: Das Innenohr als aktives schallverarbeitendes und schallaussendendes System (The inner ear as an active sound processing and sound emitting system). In *Fortschritte derAkustik, DAGA '85* (DPG, Bad Honnef 1985) pp. 29–44

Zwicker E.: Spontaneous otoacoustic emissions, threshold in quiet, and just noticeable amplitude modulation at low levels. In *Auditory Frequency Selectivity*, ed. by B.C.J. Moore, R.D. Patterson (Plenum, New York 1986) pp. 49–56

Zwicker E.: Objective otoacoustic emissions and their uncorrelation to tinitus. Proc. III. Intern. Tinnitus Seminar, Münster 1987, ed, by H. Feldmann (Harsch, Karlsruhe 1987) pp. 75–81

Zwicker E.: Psychophysics and physiology of peripheral processing in hearing. In *Basic Issues in Hearing*, ed. by H. Duifhuis, J.W. Horst, H.P. Wit (Academic, London 1988) pp. 14–25

Zwicker E.: The inner ear, a sound processing and a sound emitting system. J. Acoust. Soc. Jpn. **9** (E), 59–74 (1988)

Zwicker E.: Otoacoustic emissions and cochlear travelling waves. In *Cochlear Mechanisms*, ed. by J.P. Wilson, D.T. Kemp (Plenum, New York 1989) pp. 359–366

Zwicker E.: Otoacoustic emissions in research of inner ear signal processing. In *2nd Intern. Symposium on Cochlear Mechanics and Otoacoustic Emissions*. In *Cochlear Mechanisms and Otoacoustic Emissions*, ed. by F. Grandori et al. (Karger Basel 1990) pp. 63–76

Zwicker E.: On the frequency separation of simultaneously evoked otoacoustic emissions' consecutive extrema and its relation to cochlear travelling waves. J. Acoust. Soc. Am. **88**, 1639–1641 (1990)

Zwicker E.: On the influence of acoustical load on evoked otoacoustic emissions. Hearing Res. **47**, 185–190 (1990)

Zwicker E., F.P. Harris: Psychoacoustical and ear canal cancellation of $(2f_1 - f_2)$-distortion products. J. Acoust. Soc. Am. **87**, 2583–2591 (1990)

Zwicker E., G. Manley: Acoustical responses and suppressions-period patterns in guinea pigs. Hearing Res. **4**, 43–52 (1981)

Zwicker E., A. Scherer: Correlation between time functions of sound pressure, masking, and OAE-suppression. J. Acoust. Soc. Am. **81**, 1043–1049 (1987)

Zwicker E., E. Schloth: Interrelation of different otoacoustic emissions. J. Acoust. Soc. Am. **75**, 1148–1154 (1984)

Zwicker E., M. Stecker, J. Hind: Relations between masking, otoacoustic emissions, and evoked potentials. Acustica **64**, 102–109 (1987)

Zwicker E., J. Wesel: The effect of „addition" in suppression of delayed evoked otoacoustic emissions and in masking. Acustica **70**, 189–196 (1990)

Models of Peripheral Processing

Bauch H.: Die Schwingungsform der Basilarmembran bei Erregung durch Impulse und Geräusche gemessen an einem elektronischen Model des Innenohres (Oscillations of the basilar membrane for stimulation with impulses and noises as measured on an electronic model of the inner ear). Frequenz **10**, 222–234 (1956)

Helle R.: A hydromechanical cochlea model including corti-organ and tectorial membrane. Proc. 8th ICA London, Vol.1 (1974) p.177

Helle R.: Enlarged hydromechanical cochlea model with basilar membrane and tectorial membrane. *In Facts and Models in Hearing*, ed. by E. Zwicker, E. Terhardt (Springer, Berlin, Heidelberg 1974) pp. 77–85

Helle R.: Selektivitätssteigerung in einem hydromechanischen Innenohrmodell mit Basilar- und Deckmembran (Increase in selectivity in a hydromechanic model of the inner ear with basilar membrane and tectorial membrane). Acustica **30**, 301–312 (1974)

Helle R.: Investigation of the vibrational processes in the inner ear with the aid of hydromechanical models. J. Audiol. Techn. **16**, 138–163 (1977)

Lumer G.: Computer model of cochlear preprocessing (steady-state condition) I. Basics and results for one sinusoidal input signal. Acustica **62**, 282–290 (1987)

Lumer G.: Computer model of cochlear preprocessing (steady-state condition) II. Two-tone suppression. Acustica **63**, 17–25 (1987)

Oetinger R.: Die Erregung der Basilarmembran durch sehr kurze Druckimpulse (The excitation of the basilar membrane by very short pressure impulses). Proc. 3rd ICA Stuttgart (1959) pp. 122–125

Oetinger R., H. Hauser: Ein elektrischer Kettenleiter zur Untersuchung der mechanischen Schwingungsvorgänge im Innenohr (An electric transmission line for evaluating the mechanical oscilllations within the inner ear). Acustica **11**, 161–177 (1961)

Richter A.: Ein Modell zur Beschreibung der pegelabhängigen Frequenzselektivität des Gehörs (A model for the description of the level dependent frequency selectivity of the hearing system). Biol. Cybernetics **26**, 225–230 (1977)

Richter A.: Ein Scheibenmodell zur Untersuchung der hydrodynamischen Vorgänge im Spalt zwischen Deckmembran und Cortischem Organ (A slicemodel for research of the hydrodynamic effects in the gap between tectorial membrane and organ of corti). Acustica **38**, 148–150 (1977)

Zwicker E.: Über ein einfaches Funktionsschema des Gehörs (On a simple model of the hearing organ). Acustica **12**, 22–28 (1962)

Zwicker E.: Funktionsmodelle bei der Erforschung des Gehörs (Hardware models and the research of the hearing system). Umschau **14**, 337–346 (1963)

Zwicker E.: A hydrodynamic model of the scala media. Proc. 8th ICA London, Vol.1 (1974) p.173

Zwicker E.: A „second filter" established within the scala media. In *Facts and Models in Hearing*, ed. by E. Zwicker E. Terhardt (Springer, Berlin, Heidelberg 1974) pp. 95–98

Zwicker E.: Ein hydromechanisches Ausschnittmodell des Innenohres zur Erforschung des adäquaten Reizes der Sinneszellen (A section of a hydromechanic model of the inner ear for research of the adequate stimulus for the sensory cells). Acustica **30**, 313–319 (1974)

Zwicker E.: Spaltweite und Spaltströmung in einem Ausschnittmodell des Innenohres (Width of the gap and flow within the gap of a section of a model of the inner ear). Acustica **31**, 47–49 (1974)

Zwicker E.: A model describing nonlinearities in hearing by active processes with saturation at 40 dB. Biol. Cybernetics **35**, 243–250 (1979)

Zwicker E.: A hardware cochlear nonlinear preprocessing model with active feedback. J. Acoust. Soc. Am. **80**, 146–153 (1986)

Zwicker E.: "Otoacoustic" emissions in a nonlinear cochlear hardware model with feedback. J. Acoust. Soc. Am. **80**, 154–162 (1986)

Zwicker E.: Suppression and $(2f_1 - f_2)$-difference tones in a nonlinear cochlear preprocessing model with active feedback. J. Acoust. Soc. Am. **80**, 163–176 (1986)

Zwicker E., G. Lumer: Evaluating travelling wave characteristics in man by an active nonlinear cochlea preprocessing model. In *Peripheral Auditory Mechanisms*, ed. by J.B. Allen et. al. (Springer, Berlin, Heidelberg 1985) pp. 250–257

Zwicker E., W. Peisl: Cochlear preprocessing in analog models, in digital models and in human inner ear. Hearing Res. **44**, 209–216 (1990)

Chapter 4
Masking, Steady-State Effects

Buus S., E. Schorer, M. Florentine, E. Zwicker: Decision rules in detection of simple and complex tones. J. Acoust. Soc. Am. **80**, 1646-1657 (1986)

Fastl H.: Masking patterns of subcritical versus critical band-maskers at 8.5 kHz. Acustica **34**, 167–171 (1976)

Fastl H.: Mithörschwellen-Muster und Hörempfindungen (Masking patterns and hearing sensations). In *Fortschritte der Akustik, DAGA '78* (VDE, Berlin 1978) pp. 103–111

Fastl H.: Masking patterns of maskers with extremely steep spectral skirts. Acustica **48**, 346–347 (1981)

Gralla G.: Wahrnehmungskriterien bei Mithörschwellenmessungen (Perceptual criteria when measuring masking). In: *Fortschritte der Akustik, DAGA '92*, Verl.: DPG-GmbH, Bad Honnef, pp. 861–864 (1992)

Lumer G.: „Addition" von Mithörschwellen ("Addition" of masked thresholds). In *Fortschritte der Akustik, DAGA '84* (DPG, Bad Honnef 1984) pp. 753–756

Lumer G.: Überlagerung von Mithörschwellen an den unteren Flanken schmalbandiger Schalle (Addition of masked thresholds at the lower slopes of narrow-band maskers). Acustica **54**, 154–160 (1984)

Schöne P.: Nichtlinearitäten im Mithörschwellen-Tonheitsmuster von Sinustönen (Nonlinearities in the masking pattern of pure tones). Acustica **37**, 37–44 (1977)

Schöne P.: Mithörschwellen-Tonheitsmuster maskierender Sinustöne (Masking patterns for pure tone maskers). Acustica **43**, 197–204 (1979)

Scholl H.: Über die Bildung der Hörschwellen und Mithörschwellen von Dauerschallen (On the production of absolute thresholds and masked thresholds of continuous sounds). Frequenz **15**, 58–64 (1961)

Scholl H.: Über ein objektives Verfahren zur Ermittlung von Hörschwellen und Mithörschwellen (On an objective procedure for the determination of absolute thresholds and masked thresholds). Frequenz **17**, 125–133 (1963)

Sonntag B.: Zur Abhängigkeit der Mithörschwellen-Tonheitsmuster maskierender Sinustöne von deren Tonheit (On the dependence of masking patterns for pure tones on their critical-band rate). Acustica **52**, 95–97 (1983)

Zwicker E.: On a psychoacoustical equivalent of tuning curves. In *Facts and Models in Hearing*, ed. by E. Zwicker E. Terhardt (Springer, Berlin, Heidelberg 1974) pp. 132–141

Zwicker E.: Reversed behaviour of masking at low levels. Audiology **19**, 330–334 (1980)

Zwicker E.: Masking, a peripheral effect! Proc. 11th ICA Paris, Vol.3 (1983) pp. 71–74

Zwicker E.: Masking in normal ears-psychoacoustical facts and physiological correlates. Proc. III. Intern. Tinnitus Seminar, Münster 1987, ed. by H. Feldmann (Harsch, Karlsruhe 1987) pp. 214–223

Zwicker E., G. Bubel: Einfluß nichtlinearer Effekte auf die Frequenzselektivität des Gehörs (Influence of nonlinear effects on the frequency selectivity of the hearing system). Acustica **38**, 67–71 (1977)

Zwicker E., S. Herla: Über die Addition von Verdeckungseffekten (On the addition of masking effects). Acustica **34**, 89–97 (1975)

Zwicker E., A. Jaroszewski: Inverse frequency dependence of simultaneous tone-on-tone masking patterns at low levels. J. Acoust. Soc. Am. **71**, 1508–1512 (1982)

Temporal Masking Patterns

Bechly M.: Kann die „Suppression" eines Maskierers generell als Dämpfung seines Schallpegels beschrieben werden? (Is it possible to describe the suppression of a masker generally as an attenuation of its sound pressure level?) Acustica **50**, 288–290 (1982)

Bechly M.: Zur Abhängigkeit der "suppression" bei 8 kHz vom Frequenzverhältnis und den Schallpegeln der beiden Maskierer (On the dependence of suppression at 8 kHz on the frequency ratio and the sound pressure levels of the two maskers). Acustica **52**, 113–115 (1983)

Bechly M., H. Fastl: Interaktion der Nachhörschwellen-Tonheits-Zeitmuster zweier maskierender Schalle (Interaction of the transient masking patterns of two masking sounds). Acustica **50**, 70–74 (1982)

Fastl H.: Temporal effects in masking. In *Symposium on Hearing Theory* (IPO, Eindhoven 1972) pp. 35–41

Fastl H.: Mithörschwellen von Ton- und Rauschimpulsen (Masked thresholds for tone bursts and noise bursts). In *Fortschritte der Akustik, DAGA '73* (VDI, Düsseldorf 1973) pp. 455–458

Fastl H.: Temporal masking patterns. Proc. 8th ICA London, Vol. I (1974) p.144

Fastl H.: Transient masking pattern of narrow band maskers. In *Facts and Models in Hearing*, ed. by E. Zwicker, E. Terhardt (Springer, Berlin, Heidelberg 1974) pp. 251–257

Fastl H.: Pulsation patterns of sinusoids vs. critical band noise. Perception & Psychophysics **18**, 95–97 (1975)

Fastl H.: Influence of test-tone duration on auditory masking patterns. Audiology **15**, 63–71 (1976)

Fastl H.: Temporal masking effects: I. Broad band noise masker. Acustica **35**, 287–302 (1976)

Fastl H.: Simulation of a hearing loss at long versus short test tones. Audiology **16**, 102–109 (1977)

Fastl H.: Temporal masking effects: II. Critical band noise masker. Acustica **36**, 317–331 (1977)

Fastl H.: Transient masking patterns and hearing sensations. Proc. 9th ICA Madrid, Contributed papers, Vol. I (1977) p.354

Fastl H.: Mithörschwellen-Muster und Hörempfindungen (Masking patterns and hearing sensations). In *Fortschritte der Akustik, DAGA '78* (VDE, Berlin 1978) pp. 103–111

Fastl H.: Temporal masking effects: III. Pure tone masker. Acustica **43**, 282–294 (1979)

Fastl H.: Vergleich von Mithörschwellenmustern und Pulsationsschwellenmustern bei Frequenzgruppenrauschen und Sinustönen (Comparison of masking patterns and pulsation patterns for critical band wide noise and pure tones). In *Fortschritte derAkustik, FASE/DAGA '82* (DPG, Bad Honnef 1982) pp. 1219–1222

Fastl H.: Folgedrosselung von Sinustönen durch Breitbandrauschen. Messergebnisse und Modellvorstellungen (Temporal partial masking of pure tones by broad-band noise. Experimental results and models). Acustica **54**, 145–153 (1984)

Fastl H.: Nachverdeckung von Schmalbandrauschen bzw. kubischen Differenzrauschen (Postmasking of narrow-band noise vs. cubic difference noise). In *Fortschritte der Akustik, DAGA ' 87* (DPG, Bad Honnef 1987) pp. 565–568

Fastl H., M. Bechly: Post-masking with two maskers: Effects of bandwidth. J. Acoust. Soc. Am. **69**, 1753–1757 (1981)

Fastl H., M. Bechly: „Suppression" bei Simultanverdeckung ("Suppression" in simultaneous masking). Acustica **51**, 242–244 (1982)

Fastl H., M. Bechly: Suppression in simultaneous masking. J. Acoust. Soc. Am. **74**, 754–757 (1983)

Feldtkeller R., R. Oetinger: Die Hörbarkeitsgrenzen von Impulsen verschiedener Dauer (The limits of audibility for impulses of different duration). Acustica **6**, 489–493 (1956)

Gralla G.: Messungen zur Abhängigkeit der Nachhörschwelle von der Dauer des Testschalles (Measurements on the dependences of postmasking on the duration of the test sound). In *Fortschritte der Akustik, DAGA '89* (DPG, Bad Honnef 1989) pp. 371–374

Gralla G.: Ein Modell zur Simulation von Mithör- und AM-Schwellen (A model to simulate masking- and AM-thresholds). In: *Fortschritte der Akustik, DAGA'90*, Verl.: DPG-GmbH, Bad Honnef, pp. 727–730 (1990)

Gralla G.: Nachhörschwellen in Abhängigkeit von der spektralen Zusammensetzung des Maskierers (Post-masking patterns as a function of the spectral distribution of the masker). In: *Fortschritte der Akustik, DAGA'91*, Verl.: DPG-GmbH, Bad Honnef, pp. 501–504 (1991)

Gralla G.: Modelle zur Beschreibung von Wahrnehmungskriterien bei Mithörschwellenmessungen (Models for the description of perceptual magnitude dominant for measurements of masking). Acustica **78**, 233–245 (1993)

Gralla G.: Wahrnehmungskriterien bei Simultan- und Nachhörschwellenmessungen (Perceptual criteria when measuring simultaneous- and post-masking). Acustica **77**, 243–251 (1993)

Schmidt S.: Abhängigkeit des „Overshoot"-Effekts von der spektralen Zusammensetzung des Maskierers (Dependence of the "overshoot"-effect on the spectral composition of the masker). In *Fortschritte der Akustik, DAGA '89* (DPG, Bad Honnef 1989) pp. 387–390

Schmidt S., E. Zwicker: The effect of masker spectral asymmetry on overshoot in simultaneous masking. J. Acoust. Soc. Am. **89**, 1324-1330 (1991)

Scholl H.: Über die Bildung der Hörschwellen und Mithörschwellen von Impulsen (On the production of absolute thresholds and masked thresholds of impulses). Acustica **12**, 91-101 (1962)

Scholl H.: Über ein objektives Verfahren zur Ermittlung von Hörschwellen und Mithörschwellen (On an objective procedure for the determination of absolute thresholds and masked thresholds). Frequenz **17**, 125–133 (1963)

Stein H.J.: Das Absinken der Mithörschwelle nach dem Abschalten von weißem Rauschen (The decay of the masked threshold after switching off white noise). Acustica **10**, 116–119 (1960)

Zwicker E.: "Negative afterimage" in hearing. J. Acoust. Soc. Am. **36**, 2413–2415 (1964)

Zwicker E.: Temporal effects in simultaneous masking and loudness. J. Acoust. Soc. Am. **38**, 132–141 (1965)

Zwicker E.: Temporal effects in simultaneous masking by white-noise bursts. J. Acoust. Soc. Am. **37**, 653–663 (1965)

Zwicker E.: A model describing temporal effects in loudness and threshold. Proc. 6th ICA Tokyo (1968) A-3-4

Zwicker E.: Time constants (characteristic durations) of hearing. J. Audiol. Technique **13**, 82–102 (1974)

Zwicker E.: Einfluß von Zeitstrukturen des Schallreizes auf die Hörempfindungen (Influence of temporal structures of sound stimuli on hearing sensations). In *Kybernetik 1977*, ed. by G. Hauske, E. Butenandt (Oldenbourg, München 1978) pp. 248–262

Zwicker E.: Masking, a peripheral effect! Proc. 11th ICA Paris, Vol.3 (1983) pp. 71–74

Zwicker E.: Dependence of post-masking on masker duration und its relation to temporal effects in loudness. J. Acoust. Soc. Am. **75**, 219–223 (1984)

Zwicker E., H. Fastl: Zur Abhängigkeit der Nachverdeckung von der Störimpulsdauer (On the dependence of post-masking on masker duration). Acustica **26**, 78–82 (1972)

Zwicker E., H. Schütte: On the time pattern of the threshold of tone impulses masked by narrow band noise. Acustica **29**, 343–347 (1973)

Zwicker E., E. Wright: Temporal summation for tones in narrow-band noise. J. Acoust. Soc. Am. **35**, 691–699 (1963)

Masking Period Patterns

Kemp S.: Masking period patterns of frequency modulated tones of different frequency deviations. Acustica **50**, 63–69 (1982)

Scherer A.: Charakteristische Eigenschaften der Mithörschwellen-Periodenmuster (Characteristic features of masking period patterns). In *Fortschritte der Akustik, DAGA '85* (DPG, Bad Honnef 1985) pp. 511–514

Zwicker E.: Der Einfluß der Zeitstruktur verdeckender Klänge auf die Mithörschwelle (On the influence of the temporal structure of masking complex tones on the masked threshold). In *Fortschritte der Akustik, DAGA '75* (Physik, Weinheim 1975) pp. 323–326

Zwicker E.: A model for predicting masking period patterns. Biol. Cybernetics **23**, 49–60 (1976)

Zwicker E.: Die Abbildung der Schalldruckzeitfunktion im Mithörschwellen-Periodenmuster (The representation of the sound pressure-time function in masking period patterns). Acustica **34**, 189–199 (1976)

Zwicker E.: Influence of a complex masker's time structure on masking. Acustica **34**, 138–146 (1976)

Zwicker E.: Masking period patterns of harmonic complex tones. J. Acoust. Soc. Am. **60**, 429–439 (1976)

Zwicker E.: Mithörschwellen-Periodenmuster amplitudenmodulierter Töne (Masking period patterns of amplitude-modulated tones). Acustica **36**, 113–120 (1976)

Zwicker E.: Psychoacoustic equivalent of period histograms. J. Acoust. Soc. Am. **59**, 166–175 (1976)

Zwicker E.: Masking period patterns and hearing theories. In *Psychophysics and Physiology of Hearing*, ed. by E.F. Evans, J.P. Wilson (Academic, London 1977) pp. 393–402

Zwicker E.: Masking period patterns produced by very-low-frequency maskers and their possible relation to basilar-membrane displacement. J. Acoust. Soc. Am. **61**, 1031–1040 (1977)

Zwicker E.: Nonlinear rise of masking period patterns with masker level. Proc. 9th ICA Madrid, Contributed papers, Vol. I (1977) p.352

Chapter 5
Spectral Pitch

Fastl H.: Über Tonhöhenempfindungen bei Rauschen (Pitch of noise). Acustica **25**, 350–354 (1971)

Feldtkeller R.: Lautheit und Tonheit (Loudness and ratio pitch). Frequenz **17**, 207–212 (1963)

Hesse A.: Beschreibung der Pegelabhängigkeit der Spektraltonhöne von Sinustönen anhand von Mithörschwellenmustern (Description of the level dependence of spectral pitch for pure tones on the basis of masking patterns). In *Fortschritte der Akustik, DAGA '86* (DPG, Bad Honnef 1986) pp. 445–448

Hesse A.: Ein Funktionsschema der Spektraltonhöhe von Sinustönen (A model of spectral pitch for pure tones). Acustica **63**, 1–16 (1987)

Schmid W.: Akzentuierende Wirkung von „Zeigertönen" auf Spektraltonhöhen Komplexer Töne (Effects of "pointers" on the spectral pitches within complex-tones). In: *Fortschritte der Akustik DAGA '98*, Verl.: Dt. Gesell. für Akustik e. V., Oldenburg, 468–469 (1998)

Schmid W., W. Auer: Zur Tonhöhenempfindung bei Tiefpaßrauschen (Pitch of low-pass noise). In: *Fortschritte der Akustik, DAGA '96*, Verl.: Dt. Gesell. für Akustik e. V., Oldenburg, 344–345 (1996)

Sonntag B.: Zur Tonhöhenverschiebung gedrosselter Sinustöne (On pitch shifts of partially masked pure tones). In *Fortschritte der Akustik, DAGA '81* (VDE, Berlin 1981) pp. 729–732

Sonntag B.: Tonhöhenverschiebungen von Sinustönen durch Terzrauschen bei unterschiedlichen Frequenzlagen (Pitch shifts of pure tones by 1/3 octave band noise in different frequency regions). Acustica **53**, 218 (1983)

Terhardt E.: Pitch of pure tones: Its relation to intensity. In *Facts and Models in Hearing*, ed. by E. Zwicker, E. Terhardt (Springer, Berlin, Heidelberg 1974) pp. 353–360

Terhardt E.: Pitch shift of monaural pure tones caused by contralateral sounds. Acustica **37**, 56–57 (1977)

Terhardt E.: Absolute and relative pitch revisited on psychoacoustic grounds. Proc. 11th ICA Paris, Vol.4 (1983) pp. 427–430

Terhardt E.: Psychophysics of audio signal processing and the role of pitch in speech. In *The Psychophysics of Speech Perception*, ed. by M.E.H. Schouten (M. Nijhoff, Dordrecht 1987) pp. 271–283

Terhardt E., H. Fastl: Zum Einfluß von Störtönen und Störgeräuschen auf die Tonhöhe von Sinustönen (On the influence of masking tones and noises on the pitch of pure tones). Acustica **25**, 53–61 (1971)

Virtual Pitch

Fastl H.: Pitch and pitch strength of peaked ripple noise. In *Basic Issues in Hearing*, ed. by H. Duifhuis, J.W. Horst, H.P. Wit (Academic, London 1988) pp. 370–379

Grubert A.: Tonhöhen harmonischer und inharmonischer komplexer Töne (Pitch of harmonic vs. inharmonic complex tones) In *Fortschritte der Akustik DAGA '87* (DPG, Bad Honnef 1987) pp. 573–576

Stoll G.: Psychoakustische Messungen der Spektraltonhöhenmuster von Vokalen (Psychoacoustic measurements of the spectral pitch pattern for vowels). In *Fortschritte der Akustik, DAGA '80* (VDE, Berlin 1980) pp. 631–634

Stoll G.: Spectral-pitch pattern: A concept representing the tonal features of sounds. In *Music, Mind and Brain*, ed. by M. Clynes (Plenum, New York 1982) pp. 271–278

Stoll G.: Pitch of vowels: Experimental and theoretical investigation of its dependence on vowel quality. Speech Communication **3**, 137–150 (1984)

Stoll G.: Pitch shift of pure and complex tones induced by masking noise. J. Acoust. Soc. Am. **77**, 188–192 (1985)

Terhardt E.: Frequency analysis und periodicity detection in the sensations of roughness and periodicity pitch. In *Frequency Analysis and Periodicity Detection in Hearing*, ed. by R. Plomp, G.F. Smoorenburg (Sijthoff, Leiden, Netherlands 1970) pp. 278–290

Terhardt E.: Die Tonhöhe harmonischer Klänge und das Oktavintervall (The pitch of harmonic complex tones and the octave interval). Acustica **24**, 126–136 (1971)

Terhardt E.: Pitch shifts of harmonics, an explanation of the octave enlargement phenomenon. Proc. 7th ICA Budapest, Vol.3 (1971) pp. 621–624

Terhardt E.: Frequency and time resolution of the ear in pitch perception of complex tones. In *Symposium on Hearing Theory* (IPO, Eindhoven 1972) pp. 142–153

Terhardt E.: Zur Tonhöhenwahrnehmung von Klängen. I. Psychoakustische Grundlagen (Pitch of complex tones. I. Psychoacoustic facts). Acustica **26**, 173–186 (1972)

Terhardt E.: Zur Tonhöhenwahrnehmung von Klängen. II. Ein Funktionsschema (Pitch of complex tones. 11. A model). Acustica **26**, 187–199 (1972)

Terhardt E.: Die Tonhöhenwahrnehmung: Ergebnis eines komplexen Verarbeitungs- und Lernprozesses (Pitch perception: Result of a complex process of learning and processing). Umschau **73**, 441–442 (1973)

Terhardt E.: Influence of intensity on the pitch of complex tones. Acustica **33**, 334–348 (1975)

Terhardt E.: Frequenz und Tonhöhe (Frequency and pitch). Instrumentenbau **30**, 232–235 (1976)

Terhardt E.: Calculating virtual pitch. Hearing Res. **1**, 155–182 (1979)

Terhardt E.: Evaluation of pitch-related attributes of complex sounds. Proc. 27th Open Seminar on Acoustics, Warschau, Vol.3 (1980) pp. 141–144

Terhardt E.: Toward understanding pitch perception: Problems, concepts and solutions. In *Psychophysical, Physiological and Behavioural Studies in Hearing*, ed. by G. van den Brink, F.A. Bilsen (University Press, Delft 1980) pp. 353–360

Terhardt E.: Comments on "Noise-induced shifts in the pitch of pure and complex tones" (J. Acoust. Soc. Am. **70**, 1661–1668 (1981)). J. Acoust. Soc. Am. **73**, 1069–1070 (1983)

Terhardt E.: Pitch perception and frequency analysis. Proc. 6th FASE, Sopron, Hungary, ed. by T. Tarnóczy (1986) pp. 221–228

Terhardt E.: On the role of ambiguity of perceived pitch in music. Proc. 13th ICA Belgrade, Vol.3 (1989) pp. 35–38

Terhardt E., A. Grubert: Factors affecting pitch judgments as a function of spectral composition. Perception & Psychophysics **42**, 511–514 (1987)

Terhardt E., G. Stoll, M. Seewann: Algorithm for extraction of pitch and pitch salience from complex tonal signals. J. Acoust. Soc. Am. **71**, 679–688 (1982)

Terhardt E., Stoll G., M. Seewann: Pitch of complex signals according to virtual-pitch theory: Tests, examples, and predictions. J. Acoust. Soc. Am. **71**, 671–678 (1982)

Walliser K.: Über die Abhängigkeiten der Tonhöhenempfindung von Sinustönen vom Schallpegel, von überlagertem drosselndem Störschall und von der Darbietungsdauer (On the dependence of the pitch of pure tones on level, partially masking sound and duration). Acustica **21**, 211–221 (1969)

Walliser K.: Über ein Funktionsschema für die Bildung der Periodentonhöhe aus dem Schallreiz (On a model for creating virtual pitch from the sound stimulus). Kybernetik **6**, 65–72 (1969)

Walliser K.: Zusammenhänge zwischen dem Schallreiz und der Periodentonhöhe (Correlations between sound stimuli and virtual pitch). Acustica **21**, 319–329 (1969)

Zwicker Tone

Fastl H.: Auditory after-image produced by complex tones with a spectral gap. Proc. 12th ICA Toronto, Vol.I (1986) B 2–5

Fastl H.: Zum Zwicker-Ton bei Linienspektren mit spektralen Lücken (On the Zwicker-tone for line spectra with spectral gaps). Acustica **67**, 177–186 (1989)

Fastl H., G. Krump: Pitch of the Zwicker-tone and masking patterns. In: *Advances in Hearing Research*, (G.A. Manley et al. eds.), World Scientific, Singapore, 457–466 (1995)

Krump G.: Zum akustischen Nachton bei Linienspektren (On the Zwicker-tone of line spectra). In: *Fortschritte der Akustik, DAGA '90*, Verl.: DPG-GmbH, Bad Honnef, pp. 767–770 (1990)

Krump G.: Zum Zwicker-Ton bei unterschiedlichen Konfigurationen der spektralen Lücke (Zwicker-tones for different configurations of the spectral gap). In: *Fortschritte der Akustik, DAGA '91*, Verl.: DPG-GmbH, Bad Honnef, pp. 513–516 (1991)

Krump G.: Zum Zwicker-Ton bei Linienspektren unterschiedlicher Phasenlagen (Zwicker-tones produced by line-spectra with different phases). In: *Fortschritte der Akustik, DAGA '92*, Verl.: DPG-GmbH, Bad Honnef, pp. 825–828 (1992)

Krump G.: Zum Zwicker-Ton bei zeitlich gepulsten Erzeugerschallen (Zwicker-tones of pulsed sounds). In: *Fortschritte der Akustik, DAGA '92*, Verl.: DPG-GmbH, Bad Honnef, pp. 889–892 (1992)

Krump G.: Zum Zwicker-Ton bei unterschiedlicher Bandbreite der Anregung (Zwicker-tones for exciters of different bandwidth). In: *Fortschritte der Akustik, DAGA '93*, Verl.: DPG-GmbH, Bad Honnef, pp. 808–811 (1993)

Krump G.: Zum Zwicker-Ton bei binauraler Anregung (On the Zwicker-tone for binaural stimulation). In: *Fortschritte der Akustik, DAGA '94*, Verl.: DPG-GmbH, Bad Honnef, pp. 1005–1008 (1994)

Krump G.: Zum Zwicker-Ton bei Linienspektren mit spektraler Überhöhung (Zwicker-tones for line-spectra with spectral enhancement). In: *Fortschritte der Akustik, DAGA '94*, Verl.: DPG-GmbH, Bad Honnef, pp. 1009–1012 (1994)

Krump G.: Ein Funktionsschema zur Bestimmung der Tonhöhe des Zwicker-Tones (A model for estimating the pitch of Zwicker-tones). In: *Fortschritte der Akustik, DAGA '95*, Verl.: Dt. Gesell. für Akustik e. V., Oldenburg, pp. 943–946 (1995)

Zwicker E.: "Negative afterimage" in hearing. J. Acoust. Soc. Am. **36**, 2413–2415 (1964)

Pitch Strength

Chalupper J., W. Schmid: Akzentuierung und Ausgeprägtheit von Spektraltonhöhen bei harmonischen Komplexen Tönen (Enhancement and pitch strength of spectral pitches in harmonic complex tones). In: *Fortschritte der Akustik, DAGA '97*, Verl.: Dt. Gesell. für Akustik e. V., Oldenburg, 357–358 (1997)

Fastl H.: Pitch strength and masking patterns of low-pass noise. In *Psychophysical, Physiological and Behavioural Studies in Hearing*. ed. by G. van den Brink, F.A. Bilsen (University Press, Delft 1980) pp. 334–339

Fastl H.: Ausgeprägtheit der Tonhöhe pulsmodulierter Breitbandrauschen (Pitch strength of gated broad-band noise). In *Fortschritte der Akustik, DAGA '81* (VDE Berlin 1981) pp. 725–728

Fastl H.: Pitch and pitch strength of peaked ripple noise. In *Basic Issues in Hearing*, ed. by H. Duifhuis, J.W. Horst, H.P. Wit (Academic, London 1988) pp. 370–379

Fastl H.: Pitch strength of pure tones. Proc. 13th ICA Belgrade, Vol. 3 (1989) pp. 11–14

Fastl H.: Pitch strength and frequency discrimination for noise bands or complex tones. In *Psychophysical and Physiological Advances in Hearing* (A. Palmer et. al. Eds.) Wharr London (1998)

Fastl H., G. Stoll: Scaling of pitch strength. Hearing Res. **1**, 293–301 (1979)

Hesse A.: Zur Ausgeprägtheit der Tonhöhe gedrosselter Sinustöne (Pitch strength of partially masked pure tones). In *Fortschritte der Akustik, DAGA '85* (DPG, Bad Honnef 1985) pp. 535–538

Huth Ch., W. Schmid: Zur Ausgeprägtheit der Tonhöhe von Tönen mit und ohne Vibrato (Pitch strength of musical tones with and without vibrato). In: *Fortschritte der Akustik DAGA '98*, Verl.: Dt. Gesell. für Akustik e. V., Oldenburg, 448–449 (1998)

Schmid W.: Zur Tonhöhe inharmonischer Komplexer Töne (On the pitch of inharmonic complex tones). In: *Fortschritte der Akustik, DAGA '94*, Verl.: DPG-GmbH, Bad Honnef, pp. 1025–1028 (1994)

Schmid W.: Zur Ausgeprägtheit der Tonhöhe gedrosselter und amplitudenmodulierter Sinustöne (On the pitch strength of partially masked amplitude modulated pure tones). In: *Fortschritte der Akustik, DAGA '97*, Verl.: Dt. Gesell. für Akustik e. V., Oldenburg, pp. 355–356 (1997)

Schmid W.: Zur Ausgeprägtheit der Tonhöhe von Rauschen mit zeitvarianter Bandbegrenzung (On the pitch strength of noise with temporally varying cut-off frequency). In: *Fortschritte der Akustik DAGA '98*, Verl.: Dt. Gesell. für Akustik e. V., Oldenburg, 470–471 (1998)

Schmid W., J. Chalupper: Spektraltonhöhen Komplexer Töne: Psychoakustische Experimente und Berechnung der Ausgeprägtheit der Tonhöhe (Spectral pitches of complex-tones: Psychoacoustic experiments and calculation of pitch strength). In: *Fortschritte der Akustik DAGA '98*, Verl.: Dt. Gesell. für Akustik e. V., Oldenburg, 480–481 (1998)

Schmidt M., H. Fastl, E. Hafter: Detektion und Ausgeprägtheit der Tonhöhe bei Impulsfolgen (Detection and pitch strength of pulses). In: *Fortschritte der Akustik, DAGA '95*, Verl.: Dt. Gesell. für Akustik e. V., Oldenburg, pp. 903–906 (1995)

Wiesmann N., H. Fastl: Ausgeprägtheit der Tonhöhe und Frequenzunterschiedsschwellen von Bandpass-Rauschen (Pitch strength and frequency discrimination for bandpass-noise). In: *Fortschritte der Akustik, DAGA '91*, Verl.: DPG-GmbH, Bad Honnef, pp. 505–508 (1991)

Wiesmann N., H. Fastl: Ausgeprägtheit der virtuellen Tonhöhe und Frequenzunterschiedsschwellen von harmonischen komplexen Tönen (Pitch strength and frequency discrimination of harmonic complex tones). In: *Fortschritte der Akustik, DAGA '92*, Verl.: DPG-GmbH, Bad Honnef, pp. 841–844 (1992)

Chapter 6
Critical Bands and Excitation

Bauch H.: Die Bedeutung der Frequenzgruppe für die Lautheit von Klängen (The relevance of critical bands for the loudness of complex tones). Acustica **6**, 40–45 (1956)

Fastl H., E. Schorer: Critical bandwidth at low frequencies reconsidered. In *Auditory Frequency Selectivity*, ed. by B.C.J. Moore, R.D. Patterson (Plenum, New York 1986) pp. 311–318

Feldtkeller R.: Über die Zerlegung des Schallspektrums in Frequenzgruppen durch das Gehör (Division of the sound spectrum in critical bands). Elektron. Rundschau **9**, 387–389 (1955)

Gässler G.: Über die Hörschwelle für Schallereignisse mit verschieden breitem Frequenzspektrum (On the threshold in quiet of sounds with frequency spectra of different width). Acustica **4**, 408–414 (1954)

374   Literature

Maiwald D.: Beziehungen zwischen Schallspektrum, Mithörschwelle und der Erregung des Gehörs (Relations between sound spectrum, masked threshold and excitation of the hearing system). Acustica **18**, 69–80 (1967)

Oetinger W.: Die Erregung des Gehörs durch Dauergeräusche und durch kurze Impulse (The excitation of the hearing system by continuous sounds and short impulses). NTZ **12**, 391–399 (1959)

Scholl H.: Das dynamische Verhalten des Gehörs bei der Unterteilung des Schallspektrums in Frequenzgruppen (The dynamic performance of the hearing system when separating the sound spectrum into critical bands). Acustica **12**, 101–107 (1962)

Scholl H.: Verdeckung von Tönen durch pulsierende Geräusche (Masking of tones by pulsating noises). Proc. 4th ICA Kopenhagen, (1962) H 56

Schorer E.: Phasengrenzfrequenz und Frequenzgruppenbreite (Critical modulation frequency and bandwidth of the critical band). In *Fortschritte der Akustik, DAGA '85* (DPG, Bad Honnef 1985) pp. 507–510

Schorer E.: Critical modulation frequency based on detection of AM versus FM tones. J. Acoust. Soc. Am. **79**, 1054–1057 (1986)

Schorer E.: Zum Einfluß der Kopfhörerentzerrung bei der Messung der Frequenzgruppenbreite des Gehörs (Influence of headphone equalization on measurements of the width of the critical band). In *Fortschritte der Akustik, DAGA '86* (DPG, Bad Honnef 1986) pp. 437–440

Zwicker E.: Die Verdeckung von Schmalbandgeräuschen durch Sinustöne (Masking of narrow-band noise by pure tones). Acustica **4**, 415–420 (1954)

Zwicker E.: Über die Rolle der Frequenzgruppe beim Hören (On the role of the critical band in hearing). Erg. Biol. **23**, 187–203 (1960)

Zwicker E.: Zur Unterteilung des hörbaren Frequenzbereichs in Frequenzgruppen (Division of the audible frequency range in critical bands). Acustica **10**, 185 (1960)

Zwicker E.: Subdivision of the audible frequency range into critical bands (Frequenzgruppen). J. Acoust. Soc. Am. **33**, 248 (1961)

Zwicker E.: Masking and psychological excitation as consequences of the ear's frequency analysis. In *Frequency Analysis and Periodicity Detection in Hearing*, ed. by R. Plomp, G.F. Smoorenburg (Sijthoff, Leiden, Netherlands 1970) pp. 376–396

Zwicker E.: Introduction to round-table-discussion on "critical bands". Proc. 7th ICA Budapest, Vol. 3 (1971) pp. 189–192

Zwicker E.: Zusammenhänge zwischen neueren Ergebnissen der Psychoakustik (Relations between new results in psychoacoustics). In *Akustik und Schwingungstechnik* (VDI, Düsseldorf 1971) pp. 9–21

Zwicker E.: Temporal effects in psychoacoustical excitation. In *Basic Mechanisms in Hearing*, ed. by A. Möller (Academic, New York 1973) pp. 809–827

Zwicker E.: Loudness and excitation patterns of strongly frequency modulated tones. In *Sensation and Measurement* (papers in honor of S.S. Stevens) (Reidel, Dordrecht 1974) pp. 325–335

Zwicker E.: Mithörschwellen und Erregungsmuster stark frequenzmodulierter Töne (Masking patterns and excitation patterns of strongly frequency modulated tones). Acustica **31**, 243–256 (1974)

Zwicker E.: Über die Phasenbeziehungen zwischen Schalldruck und Erregung (On the phaserelation between sound pressure and excitation). In *Fortschritte der Akustik, DAGA '76* (VDI, Düsseldorf 1976) pp. 605–608

Zwicker E.: Recent developments in psychoacoustics. Proc. 9th ICA Madrid, Invited lectures (1977) pp. 43–53

Zwicker E., H. Fastl: On the development of the critical band. J. Acoust. Soc. Am. **52**, 699–702 (1972)

Zwicker E. G. Flottorp, S.S. Stevens: Critical band width in loudness summation. J. Acoust. Soc. Am. **29**, 548–557 (1957)

Zwicker E., A. Scherer: Zur Verdeckung von Dauertönen durch rechteckförmig moduliertes Breitbandrauschen (Masking of continuous tones by rectangularly gated broad-band noise). Acustica **52**, 115–117 (1983)

Zwicker E., E. Terhardt: Analytical expressions for critical-band rate and critical bandwidth as a function of frequency. J. Acoust. Soc. Am. **68**, 1523–1525 (1980)

Chapter 7
Just-Noticeable Sound Variations

Fastl H.: Frequency discrimination for pulsed versus modulated tones. J. Acoust. Soc. Am. **63**, 275–277 (1978)

Fastl H., A. Hesse: Frequency discrimination for pure tones at short durations. Acustica **56**, 41–47 (1984)

Feldtkeller R.: Welche Tonhöhenunterschiede kann unser Ohr noch wahrnehmen? (Which pitch differencies can be perceived by our hearing system?) Umschau **61**, 518–521 (1961)

Feldtkeller R., E. Zwicker: Die Größe der Elementarstufen der Tonhöhenempfindung und der Lautstärkeempfindung (The magnitude of the steps in pitch sensation and loudness sensation). Acustica **3**, 97–100 (1953)

Maiwald D.: Die Berechnung von Modulationsschwellen mit Hilfe eines Funktionsschemas (The calculation of modulation thresholds with a model). Acustica **18**, 193–207 (1967)

Maiwald D.: Ein Funktionsschema des Gehörs zur Beschreibung der Erkennbarkeit kleiner Frequenz- und Amplitudenänderungen (A model for the description of the perception of small frequency and amplitude variations). Acustica **18**, 81–92 (1967)

Oetinger R.: Die Grenzen der Hörbarkeit von Frequenz- und Tonzahländerungen bei Tonimpulsen (The limits of the audibility of frequency variations for tone bursts). Acustica **9**, 430–434 (1959)

Schorer E.: Frequenz-Diskrimination bei Sinustönen und bandbegrenzten Rauschen (Frequency discrimination for pure tones and band limited noise). In *Fortschritte der Akustik, DAGA '88* (DPG, Bad Honnef 1988) pp. 617–620

Schorer E.: Ein Funktionsschema eben wahrnehmbarer Frequenz- und Amplitudenänderungen (A model of just-noticeable frequency- and amplitude changes). Acustica **68**, 268–287 (1989)

Schorer E.: Vergleich eben erkennbarer Unterschiede und Variationen der Frequenz und Amplitude von Schallen (Comparison of just-noticeable differences and variations in frequency and amplitude of sounds). Acustica **68**, 183–199 (1989)

Walliser K.: Zur Unterschiedsschwelle der Periodentonhöhe (On the JND of virtual pitch). Acustica **21**, 329–336 (1969)

Zwicker E.: Die Grenzen der Hörbarkeit der Amplitudenmodulation und der Frequenzmodulation eines Tones (The limits of audibility for amplitude modulation and frequency modulation of a pure tone). Acustica, Akust. Beihefte AB125–AB133 (1952)

Zwicker E.: Die Veränderung der Modulationsschwellen durch verdeckende Töne und Geräusche (Change of modulation thresholds by masking tones and noises). Acustica **3**, 274–278 (1953)

Zwicker E.: Über die Hörbarkeit nicht sinusförmiger Tonhöhenschwankungen (On the detectibility of nonsinusoidal pitch variations). Funk und Ton **7**, 342–346 (1953)

Zwicker E.: Die elementaren Grundlagen zur Bestimmung der Informationska-
pazität des Gehörs (The foundations for the determination of the information
capacity of the hearing system). Acustica **6**, 365–381 (1956)

Zwicker E.: Direct comparisons between the sensations produced by frequency modulation and amplitude modulation. J. Acoust. Soc. Am. **34**, 1425–1430 (1962)

Zwicker E., L. Graf: Modulationsschwellen bei Verdeckung (Modulation thresholds
for partial masking). Acustica **64**, 148–154 (1987)

Zwicker E., W. Kaiser: Der Verlauf der Modulationsschwellen in der Hörfläche
(Modulation thresholds and hearing area). Acustica, Akust. Beihefte AB239–
AB246 (1952)

Just-Noticeable Sound Differences

Fastl H.: Frequency discrimination for pulsed versus modulated tones. J. Acoust.
Soc. Am. **63**, 275–277 (1978)

Fastl H., A. Hesse: Frequency discrimination for pure tones at short durations.
Acustica **56**, 41–47 (1984)

Fastl H., M. Weinberger: Frequency discrimination for pure and complex tones.
Acustica **49**, 77–78 (1981)

Fleischer H.: Gerade wahrnehmbare Phasenänderungen bei Drei-Ton-Komplexen
(Just audible phase variations for three-tone-complexes). Acustica **32**, 44–50
(1975)

Fleischer H.: Hörbarkeitsgrenzen für Phasenänderungen bei Drei-Ton-Komplexen
(Limits of audibility for phase variations with three-tone-complexes). In
Fortschritte der Akustik, DAGA '75, (Physik, Weinheim 1975) pp. 319–322

Fleischer H.: Hörbarkeit von Phasenunterschieden bei verschiedenen Arten der
Schalldarbietung (Audibility of phase differences for different kinds of sound presentation). Acustica **36**, 90–99 (1976)

Fleischer H.: Schema zur Berechnung der Hörbarkeitsschwellen von Phasenunterschieden (Model for the calculation of the audibility of differences in phase). Biol.
Cybernetics **23**, 161–170 (1976)

Fleischer H.: Subjektive Bewertung von Unterschieden in den Phasenspektren stationärer Klänge (Subjective evaluation of differences in the phase spectra of stationary sounds). In *Fortschritte der Akustik, DAGA '76* (VDI, Düsseldorf 1976)
pp. 581–584

Fleischer H.: Über die Wahrnehmbarkeit von Phasenänderungen (On the audibility
of phase variations). Acustica **35**, 202–209 (1976)

Fleischer H.: Über die Größe der durch Phasenänderungen hervorgerufenen
Empfindungsänderungen (On the magnitude of the variation in hearing sensation
produced by phase variations). Acustica **37**, 83–93 (1977)

Schorer E.: Frequenz-Diskrimination bei Sinustönen und bandbegrenzten Rauschen
(Frequency discrimination for pure tones and band limited noise). In *Fortschritte
derAkustik, DAGA '88* (DPG, Bad Honnef 1988) pp. 617–620

Schorer E.: Ein Funktionsschema eben wahrnehmbarer Frequenz- und Amplitudenänderungen (A model of just-noticeable frequency- and amplitude changes).
Acustica **68**, 268–287 (1989)

Suchowerskyj W.: Zur subjektiven Bewertung zeitlich variabler Schallparameter
(Subjective evaluation of temporally variable sound parameters). In *Fortschritte
der Akustik, DAGA '75* (Physik, Weinheim 1975) pp. 315–318

Suchowerskyj W.: Beurteilung kontinuierlicher Schalländerungen (Evaluation of
continuous sound variations). Acustica **38**, 140–147 (1977)

Suchowerskyj W.: Beurteilung von Unterschieden zwischen aufeinanderfolgenden Schallen (Evaluation of differences for consecutive sounds). Acustica **38**, 131–139 (1977)

Suchowerskyj W.: Funktionsschema zur Beschreibung der subjektiven Bewertung von Schalländerungen (Model for the description of the subjective evaluation of sound variations). Biol. Cybernetics **26**, 169–174 (1977)

Terhardt E.: Über ein Äquivalenzgesetz für Intervalle akustischer Empfindungsgrößen. (On a law of equivalence for intervals of hearing sensations). Kybernetik **5**, 127–133 (1968)

Chapter 8
Loudness, Steady-State Effects

Deuter K.: Gedrosselte Lautheit bei tiefen, mittleren und hohen Frequenzen (Partial masked loudness at low, medium and high frequencies). In *Fortschritte der Akustik, DAGA '88 (DPG*, Bad Honnef 1988) pp. 577–580

Fastl H.: Lautstärkeunterschiede von Schallen mit Bandbreiten innerhalb einer Frequenzgruppe (Loudness differences of sounds with bandwidths within a critical band). In Proc. F.A.S.E. 75, Koll. 1 (1975) pp. 165–173

Fastl H.: Methodenvergleich zur Lautheitsbeurteilung (Comparison of methods for evaluating loudness). In *Akustik zwischen Physik und Psychologie*, ed. by A. Schick (Klett-Cotta, Stuttgart 1981) pp. 103–109

Fastl H., A. Jaroszewski, E. Schorer, E. Zwicker: Equal loudness contours between 100 and 1000 Hz for 30, 50, and 70 phon. Acustica **70**, 197–201 (1990)

Fastl H., S. Namba, S. Kuwano: Cross-cultural investigations of loudness evaluation for noises. In *Contributions to Psychological Acoustics*, ed. by A. Schick et. al. (Kohlrenken, Oldenburg 1986) pp. 354–369

Fastl H., W. Schmid, G. Theile, E. Zwicker: Schallpegel im Gehörgang für gleichlaute Schalle aus Kopfhörern oder Lautsprechern (Sound level in the ear canal for equally loud sounds from headphones versus loudspeakers). In *Fortschritte der Akustik, DAGA '85* (DPG, Bad Honnef 1985) pp. 471–474

Fastl H., E. Zwicker: Lautstärkepegel bei 400 Hz: Psychoakustische Messung und Berechnung nach ISO 532 B (Loudness level at 400 Hz: psychoacoustic measurements and calculations according to ISO 532 B). In *Fortschritte der Akustik, DAGA '87* (DPG, Bad Honnef 1987) pp. 189–192

Feldtkeller R.: Lautheit und Tonheit (Loudness and ratio pitch). Frequenz **17**, 207–212 (1963)

Feldtkeller R., E. Zwicker, E. Port: Lautstärke, Verhältnislautheit und Summenlautheit (Loudness, ratio loudness and summating loudness). Frequenz **13**, 108–117 (1959)

Gleiss N., E. Zwicker: Loudness function in the presence of masking noise. J. Acoust. Soc. Am. **36**, 393–394 (1964)

Hellman R.P., E. Zwicker: Overall loudness of tone-noise complexes: measured and calculated. J. Acoust. Soc. Am. **82**, S 25 (1987)

Hellman R.P., E. Zwicker: Measured and calculated loudness of complex sounds. In Proc. Inter-Noise '87, Vol. II (1987) pp. 973–976

Hellman R.P., E. Zwicker: Loudness of two-tone-noise complexes. In Proc. Inter–Noise '89, Vol.II (1989) pp. 827–832

Scharf B.: Partial masking. Acustica **14**, 16–23 (1964)

Zwicker E.: Über psychologische und methodische Grundlagen der Lautheit (Psychological and methodical basis of loudness). Acustica **8**, 237–258 (1958)

Zwicker E.: Ein graphisches Verfahren zur Bestimmung der Lautstärke und der Lautheit aus dem Terzpegeldiagramm (A graphic procedure for the determination of loudness level and loudness form 1/3 octave band spectra). Frequenz **13**, 234–238 (1959)

Zwicker E.: Lautstärke und Lautheit (Loudness level and loudness). In Proc. 3rd ICA Stuttgart (1959) pp. 63–78

Zwicker E.: Ein Verfahren zur Berechnung der Lautstärke (A procedure for calculating loudness). Acustica **10**, 304–308 (1960)

Zwicker E.: Der gegenwärtige Stand der objektiven Lautstärkemessung, Teil I und Teil II (The present state of the art in objective measurements of loudness, Part I and Part II). ATM, V55-6, V55-7 (1961)

Zwicker E.: Über die Lautheit von ungedrosselten und gedrosselten Schallen (On the loudness of unmasked and partially masked sounds). Acustica **13**, 194–211 (1963)

Zwicker E., R. Feldtkeller: Über die Lautstärke von gleichförmigen Geräuschen (On the loudness of continuous noises). Acustica **5**, 303–316 (1955)

Zwicker E., B. Scharf: A model of loudness summation. Psych. Rev. **72**, 3–26 (1965)

Zwicker E., Y. Yamada: Lautstärke von Tonkomplexen in Abhängigkeit von der Bandbreite, vom Schallpegel und von zugefügtem Breitbandrauschen (Loudness of complex tones as a function of bandwidth, sound level, and added broad band noise). Acustica **53**, 26–30 (1983)

Loudness, Temporal Effects

Bauch H.: Über die Sonderstellung periodischer kurzer Druckimpulse bei der Empfindung der Lautstärke (On the special features of periodic short pressure impulses with respect to the sensation of loudness). Acustica **6**, 494–511 (1956)

Fastl H.: Loudness and masking patterns of narrow noise bands. Acustica **33**, 266–271 (1975)

Fastl H.: Schallpegel und Lautstärke von Sprache (Sound level and loudness of speech). Acustica **35**, 341–345 (1976)

Fastl H.: Loudness of running speech. J.Audiol. Technique **16**, 2–13 (1977)

Fastl H.: Average loudness of road traffic noise. In Proc. Inter-Noise '89, Vol.II (1989) pp. 815–820

Fastl H.: Beurteilung und Messung der wahrgenommenen äquivalenten Dauerlautheit (Evaluation and measurement of perceived average loudness). In *Contributions to Psychological Acoustics V*, ed. by A. Schick et al. (BIS Uni Oldenburg 1990)

Fastl H.: Masking effects and loudness evaluation. In: Recent Trends in Hearing Research (H. Fastl et al. Eds.) Bibliotheks- und Informationssystem der Carl von Ossietzky Universität Oldenburg, Oldenburg, 29–50 (1996)

Feldtkeller R.: Die Kurven konstanter Lautstärke für Dauertöne und für einzelne Druckimpulse (Equal loudness contours for steady-state tones and single pressure impulses). Frequenz **10**, 356–358 (1956)

Henning G.B., E. Zwicker: The effect of low-frequency masker on loudness. Submitted to Hearing Research

Kuwano S., H. Fastl: Loudness evaluation of various kinds of non-steady-state sound using the method of continuous judgement by category. In Proc. 13th ICA Belgrade, Vol.1 (1989) pp. 365–368

Namba S., S. Kuwano, H. Fastl: On the loudness of impulsive noise and road traffic noise – a comparison between Japanese and German subjects. In *Transactions of the Commitee on Noise*, N 85-01-02, The Acoust. Soc. Jpn. (1985)

Namba S., S. Kuwano, H. Fastl: Loudness of road traffic noise using the method of continuous judgement by category. In *Noise as a Public Health Problem* (Swedish Council for Building Res., Stockholm 1988) pp. 241–246

Oetinger R., E. Port: Ein Verfahren zur Berechnung der Lautstärke von Druckimpulsen (A procedure for calculating the loudness of pressure impulses). In Proc. 3rd ICA Stuttgart (1959) pp. 125–128

Port E.: Die Lautstärke von Tonimpulsen verschiedener Dauer (The loudness of tone impulses of different duration). Frequenz **13**, 242–245 (1959)

Port E.: Zur Lautstärke einzelner Rauschimpulse (On the loudness of single noise bursts). In Proc. 4th ICA Kopenhagen, (1962) H 45

Port E.: Über die Lautstärke einzelner kurzer Schallimpulse (On the loudness of single short sound bursts). Acustica **13**, 212–223 (1963)

Port E.: Zur Lautstärkeempfindung und Lautstärkemessung von pulsierenden Geräuschen (On the loudness perception and the loudness measurement of pulsating noises). Acustica **13**, 224–233 (1963)

Port E.: Die Lautstärke pulsierender Geräusche im diffusen Schallfeld eines Hallraumes (The loudness of pulsating noises in the diffuse sound field of a reverberation room). Acustica **14**, 167–173 (1964)

Vogel A.: Ein gemeinsames Funktionsschema zur Beschreibung der Lautheit und der Rauhigkeit (A common model for the description of loudness and roughness). Biol. Cybernetics **18**, 31–40 (1975)

Zwicker E.: Temporal effects in simultaneous masking and loudness. J. Acoust. Soc. Am. **38**, 132–141 (1965)

Zwicker E.: Über die Dynamik der Lautheit (On the dynamics of loudness). In Proc. 5th ICA Liege, (1965) B 24

Zwicker E.: Ein Beitrag zur Lautstärkemessung impulshaltiger Schalle (A contribution to the measurement of loudness of impulsive sounds). Acustica **17**, 11–22 (1966)

Zwicker E.: A model describing temporal effects in loudness and threshold. In Proc. 6th ICA Tokyo, (1968) A-3-4

Zwicker E.: Der Einfluß der zeitlichen Struktur von Tönen auf die Addition von Teillautheiten (The influence of the time structure of sounds on the addition of partial loudnesses). Acustica **21**, 16–25 (1969)

Zwicker E.: Procedure for calculating loudness of temporally variable sounds. J. Acoust. Soc. Am. **62**, 675–682 (1977). Erratum: J. Acoust. Soc. Am. **63**, 283 (1978)

Chapter 9
Sharpness and Sensory Pleasantness

Aures W.: Wohlklangsbeurteilung von Kirchenglocken (Evaluation of sensory pleasantness of church bells). In *Fortschritte der Akustik, DAGA '81* (VDE, Berlin 1981) pp. 733–736

Aures W.: Der Wohlklang: Eine Funktion der Schärfe, Rauhigkeit und Klanghaftigkeit (Sensory pleasantness: a function of sharpness, roughness and tonality). In *Fortschritte der Akustik, DAGA '84* (DPG, Bad Honnef 1984) pp. 735–738

Aures W.: Berechnungsverfahren für den sensorischen Wohlklang beliebiger Schallsignale (A procedure for calculating sensory pleasantness of various sounds). Acustica **59**, 130–141 (1985)

Aures W.: Der sensorische Wohlklang als Funktion psychoakustischer Empfindungsgrößen (Sensory pleasantness as a function of psychoacoustic sensations). Acustica **58**, 282–290 (1985)

Benedini K.: Vokalähnlichkeit schmalbandiger Klänge (Similarity of narrow-band sounds to vowels). In *Fortschritte der Akustik, DAGA '78* (VDE, Berlin 1978) pp. 535–538

Benedini K.: Ein Funktionsschema zur Beschreibung von Klangfarbenunterschieden (A model describing differences in timbre). Biol. Cybernetics **34**, 111–117 (1979)

Benedini K.: Klangfarbenunterschiede zwischen tiefpaßgefilterten harmonischen Klängen (Differences in timbre for low-pass filtered harmonic sounds). Acustica **44**, 129–134 (1980)

Benedini K.: Messung der Klangfarbenunterschiede zwischen schmalbandigen harmonischen Klängen (Evaluation of timbre differences for narrow-band harmonic complex tones). Acustica **44**, 188–193 (1980)

v. Bismarck G.: Psychometrische Untersuchungen der Klangfarbe stationärer Schalle (Psychometric investigations of timbre of steady-state sounds). In *Akustik und Schwingungstechnik* (VDI, Düsseldorf 1971) pp. 371–375

v. Bismarck G.: Timbre of steady sounds: Scaling of sharpness. In Proc. 7th ICA Budapest, Vol.3 (1971) pp. 637–640

v. Bismarck G.: Vorschlag für ein einfaches Verfahren zur Klassifikation stationärer Sprachschalle (A proposal for a simple procedure for classification of stationary speech sounds). Acustica **28**, 186–188 (1973)

v. Bismarck G.: Sharpness as an attribute of the timbre of steady sounds. Acustica **30**, 159–172 (1974)

v. Bismarck G.: Timbre of steady sounds: a factorial investigation of its verbal attributes. Acustica **30**, 146–159 (1974)

Terhardt E., G. Stoll: Bewertung des Wohlklangs verschiedener Schalle (Evaluation of sensory pleasantness of different sounds). In *Fortschritte der Akustik, DAGA '78* (VDE, Berlin 1978) pp. 583–586

Terhardt E., G. Stoll: Skalierung des Wohlklangs (der sensorischen Konsonanz) von 17 Umweltschallen und Untersuchung der beteiligten Hörparameter (Scaling the sensory pleasantness of 17 environmental sounds and investigation of correlated hearing sensations). Acustica **48**, 247–253 (1981)

Chapter 10
Fluctuation Strength

Fastl H.: Fluctuation strength and temporal masking patterns of amplitude-modulated broad-band noise. Hearing Res. **8**, 59–69 (1982)

Fastl H.: Fluctuation strength of FM-tones. In Proc. 11th ICA Paris, Vol.3 (1983) pp. 123–126

Fastl H.: Fluctuation strength of modulated tones and broad-band noise. In *Hearing – Physiological Bases and Psychophysics*, ed. by R. Klinke, R. Hartmann (Springer, Berlin, Heidelberg 1983) pp. 282–288

Fastl H.: Schwankungsstärke und zeitliche Hüllkurve von Sprache und Musik (Fluctuation strength and temporal envelope of speech and music). In *Fortschritte der Akustik, DAGA '84* (DPG, Bad Honnef 1984) pp. 739–742

Fastl H.: Fluctuation strength of narrow band noise. In: *Auditory Physiology and Perception*, Advances in the Biosciences Vol. **83**, Pergamon Press Oxford, pp. 331–336 (1992)

Fastl H., A. Hesse, E. Schorer, J. Urbas, P. Müller-Preuss: Searching for neural correlates of the hearing sensation fluctuation strength in the auditory cortex of squirrel monkeys. Hearing Res. **23**, 199–203 (1986)

Fastl H., U. Widmann, P. Müller-Preuss: Correlations between hearing and vocal activity in squirrel monkey. Acustica **73**, 35–36 (1991)

Schöne P.: Vergleich dreier Funktionsschemata der akustischen Schwankungsstärke (Comparison between three different models of the fluctuation strength). Biol. Cybernetics **29**, 57–62 (1978)
Schöne P.: Messungen zur Schwankungsstärke von amplitudenmodulierten Sinustönen (Measurement of fluctuation strength of amplitude-modulated sinusoidal tones). Acustica **41**, 252–257 (1979)
Terhardt E.: Über akustische Rauhigkeit und Schwankungsstärke (On the acoustic roughness and fluctuation strength). Acustica **20**, 215–224 (1968)
Terhardt E.: Schallfluktuationen und Rauhigkeitsempfinden (Sound fluctuations and the sensation of roughness). In *Akustik und Schwingungstechnik* (VDI, Düsseldorf 1971) pp. 367–370

Chapter 11
Roughness

Aures W.: Psychoakustische Untersuchungen der durch Rauschen verschiedener Bandbreiten hervorgerufenen Rauhigkeitswahrnehmung (Psychoacoustic investigations of the sensation of roughness produced by noises with different bandwidth). In *Fortschritte der Akustik, DAGA '80* (VDE, Berlin 1980) pp. 623–626
Aures W.: Ein Berechnungsverfahren der Rauhigkeit (A procedure for calculating roughness). Acustica **58**, 268–281 (1985)
Fastl H.: Rauhigkeit und Mithörschwellen-Zeitmuster sinusförmig amplitudenmodulierter Breitbandrauschen (Roughness and masked threshold time pattern of sinusoidally amplitude-modulated broad-band noise). In *Fortschritte der Akustik, DAGA '76* (VDI, Düsseldorf 1976) pp. 601–604
Fastl H.: Roughness and temporal masking patterns of sinusoidally amplitude-modulated broad-band noise. In *Psychophysics and Physiology of Hearing*, ed. by E.F. Evans, J.P. Wilson (Academic, London 1977) pp. 403–414
Fastl H.: The hearing sensation roughness and neuronal responses to AM-tones. Hearing Res. **46**, 293–295 (1990)
Kemp S.: Roughness of frequency-modulated tones. Acustica **50**, 126–133 (1982)
Müller-Preuss P., A. Bieser, A. Preuss, H. Fastl: Neural processing of AM-sounds within central auditory pathway. In *Auditory pathway*, ed. by J. Syka, B. Masterton (Plenum, New York 1988) pp. 327–331
Terhardt E.: Über akustische Rauhigkeit und Schwankungsstärke (On the acoustic roughness and fluctuation strength). Acustica **20**, 215–224 (1968)
Terhardt E.: Über die durch amplitudenmodulierte Sinustöne hervorgerufene Hörempfindung (On hearing sensations produced by amplitude-modulated sinusoidal tones). Acustica **20**, 210–214 (1968)
Terhardt E.: Frequency analysis and periodicity detection in the sensations of roughness and periodicity pitch. In *Frequency Analysis and Periodicity Detection in Hearing*, ed. by R. Plomp, G.F. Smoorenburg (Sijthoff, Leiden 1970) pp. 278–290
Terhardt E.: On the perception of periodic sound fluctuations (roughness). Acustica **30**, 201–213 (1974)
Vogel A.: Roughness and its relation to the time-pattern of psychoacoustical excitation. In *Facts and Models in Hearing*, ed. by E. Zwicker, E. Terhardt (Springer, Berlin, Heidelberg 1974) pp. 241–250
Vogel A.: Ein gemeinsames Funktionsschema zur Beschreibung der Lautheit und der Rauhigkeit (A common model for describing loudness and roughness). Biol. Cybernetics **18**, 31–40 (1975)

Vogel A.: Über den Zusammenhang zwischen Rauhigkeit und Modulationsgrad (On the relation between roughness und degree of amplitude modulation). Acustica **32**, 300–306 (1975)

Widmann U., H. Fastl: Calculating roughness using time-varying specific loudness spectra. In: *Proc. Sound Quality Symposium*, Ypsilanti Michigan USA. Ed. by P. Davies, G. Ebbitt, 55–60 (1998)

Chapter 12
Subjective Duration

Burghardt H.: Subjective duration of sinusoidal tones. In Proc. 7th ICA Budapest, Vol.3 (1971) pp. 353–356

Burghardt H.: Einfaches Funktionsschema zur Beschreibung der subjektiven Dauer von Schallimpulsen und Schallpausen (A simple model for describing subjective duration of sound bursts and sound pauses). Kybernetik **12**, 21–29 (1972)

Burghardt H.: Die subjektive Dauer schmalbandiger Schalle bei verschiedenen Frequenzlagen (Subjective duration of narrow-band sounds at different frequency regions). Acustica **28**, 278–284 (1973)

Burghardt H.: Über die subjektive Dauer von Schallimpulsen und Schallpausen (On the subjective duration of sound bursts and sound pauses). Acustica **28**, 284–290 (1973)

Fastl H.: Mithörschwelle und Subjektive Dauer (Masked threshold and subjective duration). Acustica **32**, 288–290 (1975)

Fastl H.: Mithörschwellen-Zeitmuster und Subjektive Dauer bei Sinustönen (Masked threshold time pattern and subjective duration for sinusoidal tones). In *Fortschritte der Akustik, DAGA '75* (Physik, Weinheim 1975) pp. 327–330

Fastl H.: Subjective duration and temporal masking patterns of broad-band noise impulses. J. Acoust. Soc. Am. **61**, 162–168 (1977)

Zwicker E.: Subjektive und objektive Dauer von Schallimpulsen und Schallpausen (Subjective and objective duration of sound bursts and sound pauses). Acustica **22**, 214–218 (1970)

Chapter 13
Rhythm

Heinbach W.: Rhythmus von Sprache: Untersuchung methodischer Einflüsse (Rhythm in speech: Investigation of methodic influences). In *Fortschritte der Akustik, DAGA '85* (DPG, Bad Honnef 1985) pp. 567–570

Köhlmann M.: Rhythmische Segmentierung von Sprach- und Musiksignalen und ihre Nachbildung mit einem Funktionsschema (Rhythmic segmentation of speech and music signals and their simulation in a model). Acustica **56**, 193–204 (1984)

Schütte H.: Wahrnehmung von subjektiv gleichmäßigem Rhythmus bei Impulsfolgen (Perception of subjectively uniform rhythm for sequences of sound bursts). In *Fortschritte der Akustik, DAGA '76* (VDI, Düsseldorf 1976) pp. 597–600

Schütte H.: Ein Funktionsschema für die Wahrnehmung eines gleichmäßigen Rhythmusses in Schallimpulsfolgen (A model for the perception of uniform rhythm procuded by sequences of sound bursts). Biol. Cybernetics **29**, 49–55 (1978)

Schütte H.: Subjektiv gleichmäßiger Rhythmus: Ein Beitrag zur zeitlichen Wahrnehmung von Schallereignissen (Subjectively uniform rhythm: A contribution to temporal perception of sound events). Acustica **41** 197–206 (1978)

Terhardt E., W. Aures: Wahrnehmbarkeit der periodiscben Wiederholung von Rauschsignalen (The audibility of periodic repetitions of noise signals). In *Fortschritte der Akustik, DAGA '84* (DPG, Bad Honnef 1984) pp. 769–772

Terhardt E., H. Schütte: Akustische Rhythmus-Wahrnehmung: Subjektive Gleich-mäßigkeit (Hearing sensation of rhythm: Subjective uniformity). Acustica **35**, 122–126 (1976)

Chapter 14
Nonlinear Distortion in the Ear

Feldtkeller R.: Die Hörbarkeit nichtlinearer Verzerrungen bei der Übertragung musikalischer Zweiklänge (The audibility of nonlinear distortions for transmit-ting musical two-component sounds). Acustica, Akust. Beihefte AB 117–AB 124 (1952)

Helle R.: Amplitude und Phase des im Gehör gebildeten Differenztones dritter Ordnung (Amplitude and phase of the third order difference tone produced in the ear). Acustica **22**, 74–87 (1969)

Humes L.E.: An excitation-pattern algorithm for the estimation of $(2f_2 - f_1)$- and $(f_2 - f_1)$-cancellation level and phase. J. Acoust. Soc. Am. **78**, 1252–1260 (1985)

Humes L.E.: Cancellation level and phase of the $(f_2 - f_1)$-distortion product. J. Acoust. Soc. Am. **78**, 1245–1251 (1985)

Ryffert H.: Die Grenzen der Hörbarkeit nichtlinearer Verzerrungen vierter und fünfter Ordnung für die einfache Quint (The threshold for the audibility of non-linear distortions of fourth and fifth order for the simple fifth). Frequenz **15**, 254–261 (1961)

Zwicker E.: Der ungewöhnliche Amplitudengang der nichtlinearen Verzerrungen des Ohres (The strange dependence of nonlinear distortions of our hearing system on amplitude). Acustica **5**, 67–74 (1955)

Zwicker E.: Der kubische Differenzton und die Erregung des Gehörs (The cubic difference tone and the excitation of our hearing system). Acustica **20**, 206–209 (1968)

Zwicker E.: Different behaviour of quadratic and cubic difference tones. Hearing Res. **1**, 283–292 (1979)

Zwicker E.: Zur Nichlinearität ungerader Ordnung des Gehörs (On the odd-order nonlinearity of our hearing system). Acustica **42**, 149–157 (1979)

Zwicker E.: Cubic difference tone level and phase dependence on frequency dif-ference and level of primaries. In *Psychophysical, Physiological and Behavioural Studies in Hearing*, ed. by G. van den Brink, F.A. Bilsen (Univ. Press, Delft 1980) pp. 268–273

Zwicker E.: Nonmonotic behaviour of $(2f_1 - f_2)$ explained by a saturation feedback model. Hearing Res. **2**, 513–518 (1980)

Zwicker E.: Dependence of level and phase of the $(2f_1 - f_2)$-cancellation tone on frequency range, frequency difference, level of primaries, and subject. J. Acoust. Soc. Am. **70**, 1277–1288 (1981)

Zwicker E.: Formulae for calculating the psychoacoustical excitation level of aural difference tones measured by the cancellation method. J. Acoust. Soc. Am. **69**, 1410–1413 (1981)

Zwicker E.: Level and phase of the $(2f_1 - f_2)$-cancellation tone expressed in vector diagrams. J. Acoust. Soc. Am. **74**, 63–66 (1983)

Zwicker E., H. Fastl: Cubic difference sounds measured by threshold- and compen-sation method. Acustica **29**, 336–343 (1973)

Zwicker E., O.Martner: On the dependence of $(f_2 - f_1)$ difference tones on subject and on additional masker. J. Acoust. Soc. Am. **88**, 1351–1358 (1990)

Chapter 15
Binaural Hearing

Henning G. B., S. Wartini: The effect of signal duration on frequency discrimination at low signal-to-noise ratios in different conditions of interaural phase. Hearing Research **48**, 201–207 (1990)

Henning G.B., E. Zwicker: Effects of the bandwidth and level of noise and of the duration of the signal on binaural masking-level differences. Hearing Res. **14**, 175–178 (1984)

Henning G.B., E. Zwicker: Binaural masking-level differences with tonal maskers. Hearing Res. **16**, 279–290 (1985)

Keller H.: Maskierung binauraler Schwebungen (Masking of binaural beats). In *Fortschritte der Akustik, DAGA '85* (DPG, Bad Honnef 1985) pp. 531–534

Schenkel K.D.: Über die Abhängigkeit der Mithörschwellen von der interauralen Phasenlage des Testschalles (On the dependence of masked thresholds on the interaural phase of the test-sounds). Acustica **14**, 337–346 (1964)

Schenkel K.D.: Die Abhängigkeit der beidohrigen Mithörschwellen von der Frequenz des Testschalls und vom Pegel des verdeckenden Schalles (The dependence of diotic and dichotic masked thresholds as a function of frequency of test sound and as a function of level of masking sound). Acustica **17**, 345–356 (1966)

Schenkel K.D.: Accumulation theory of binaural-masked thresholds. J. Acoust. Soc. Am. **41**, 20–31 (1967)

Schenkel K.D.: Die beidohrigen Mithörschwellen von Impulsen (Diotic and dichotic masked thresholds of sound bursts). Acustica **18**, 38–46 (1967)

Wallerus H.: Richtungsauflösungsvermögen des Gehörs für Sinustöne mit interauralen Pegelunterschieden (Direction resolution of the hearing system for sinusoidal tones with interaural level differences). In *Fortschritte der Akustik, DAGA '76* (VDI, Düsseldorf 1976) pp. 589–592

Zwicker E.: Warum gibt es binaurale Mithörschwellendifferenzen? (Why do binaural masking-level differences exist?) In *Fortschritte der Akustik, DAGA '84* (DPG, Bad Honnef 1984) pp. 691–694

Zwicker E., G.B. Henning: Binaural masking-level differences with tones masked by noises of various bandwidths and level. Hearing Res. **14**, 179–183 (1984)

Zwicker E., G.B. Henning: The four factors leading to binaural masking-level differences. Hearing Res. **19**, 29–47 (1985)

Zwicker E., G.B. Henning: On the effect of interaural phase differences on loudness. Hearing Research **53**, 141–152 (1991)

Zwicker E., U.T. Zwicker: Binaural masking-level differences in non simultaneous masking. Hearing Res. **13**, 221–228 (1984)

Zwicker E., U.T. Zwicker: Dependency of binaural loudness summation on interaural level differences, spectral distribution, and temporal distribution. J. Acoust. Soc. Am. **89**, 756–764 (1991)

Zwicker T.: Diotische und dichotische Wahrnehmung von Schallfluktuationen (Diotic and dichotic perception of sound fluctuations). Acustica **55**, 181–186 (1984)

Zwicker T.: Experimente zur dichotischen Oktavtäuschung (Experiments on dichotic octave-illusions). Acustica **55**, 128–136 (1984)

Zwicker T., E. Zwicker: „Alte" und neue Daten zur binauralen Lautheit ("Old" and new data on binaural loudness). In: *Fortschritte der Akustik, DAGA '90*, Verl.: DPG-GmbH, Bad Honnef, pp. 707–710 (1990)

Zwicker U.T.: Auditory recognition of diotic and dichotic vowel pairs. Speech Commun. **3**, 265–277 (1984)

Zwicker U.T., E. Zwicker: Binaural masking-level difference as a function of masker and testsignal duration. Hearing Res. **13**, 215–219 (1984)

Zwicker U.T., E. Zwicker: Effects of binaural loudness summation and their approximation in objective loudness measurements. In Proc. Inter-Noise '90, Vol. II, pp. 1145–1150 (1990)

Chapter 16
Noise Measurements

Beckenbauer T., I. Stemplinger, A. Seiter: Basics and use of DIN 45681 'Detection of tonal components and determination of a tone adjustment for the noise assessment'. In: Proc. Inter-Noise'96, Vol. 6, pp. 3271–3276 (1996)

Berry B., E. Zwicker: Comparison of subjective evaluations of impulsive noise with objective measurements of the loudness-time function given by loudness meter. In Proc. Inter-Noise '86, Vol. II, pp. 821–824 (1986)

Fastl H.: Noise measurement procedures simulating our hearing system. In *IEICE Techn. Rep.*, Vol.87, No.182 (Inform. Commun. Engs., Tokyo 1987) pp. 53–58

Fastl H.: Gehörbezogene Lärmmessverfahren (Hearing equivalent procedures for noise measurement). In *Fortschritte der Akustik, DAGA '88*, Verl.: DPG-GmbH, Bad Honnef, pp. 111–124 (1988)

Fastl H.: Noise measurement procedures simulating our hearing system. J. Acoust. Soc. Jpn. (E) **9**, 75–80 (1988)

Fastl H.: Loudness of running speech measured by a loudness meter. Acustica **71**, 156–158 (1990)

Fastl H.: Calibration signals for meters of loudness, sharpness, fluctuation strength, and roughness. In: Proc. Inter-Noise'93, Vol.III, (1993) pp. 1257–1260

Fastl H., W. Schmid: Comparison of loudness analysis systems. In: Proc. Inter-Noise'97, Vol.II, pp. 981–986 (1997)

Fastl H., W. Schmid: Vergleich von Lautheits-Zeitmustern verschiedener Lautheits-Analysesysteme (Comparison of loudness-time patterns produced by different loudness-analysis-systems). In: *Fortschritte der Akustik DAGA '98*, Verl.: Dt. Gesell. für Akustik e. V., Oldenburg, 466–467 (1998)

Fastl H., U. Widmann: Subjective and physical evaluation of aircraft noise. Noise Contr. Engng. J.**35**, pp. 61–63 (1990)

Fastl H., U. Widmann: Kalibriersignale für Meßsysteme zur Nachbildung von Lautheit, Schärfe, Schwankungsstärke und Rauhigkeit (Calibration signals for measurement systems to simulate loudness, sharpness, fluctuation strength, and roughness). In: *Fortschritte der Akustik, DAGA '93*, Verl.: DPG-GmbH, Bad Honnef, pp. 640–643 (1993)

Lübcke E., G. Mittag, E. Port: Subjektive und objektive Bewertung von Maschinengeräuschen (Subjective and objective evaluation of industry noises). Acustica **14**, 105–114 (1964)

Nitsche V., H. Fastl: Objective measurements of aircraft noise by sound level meter versus loudness meter. In Proc. Inter-Noise '85, Vol.II (1985) pp. 1247–1250

Nitsche V., H. Fastl: Loudness calculation of aircraft noise compared with loudness meter measurements. In Proc. Inter-Noise '87, Vol.II (1987) pp. 1013–1016

Paulus E., E. Zwicker: Programme zur automatischen Bestimmung der Lautheit aus Terzpegeln oder Frequenzgruppenpegeln (Programs for the automatic determination of loudness using thirdoctave-band or critical-band levels). Acustica **27**, 253–266 (1972)

Pfeiffer Th.: Ein neuer Lautstärkemesser (A new loudness meter). Acustica **14**, 162–167 (1964)

Pfeiffer Th.: Ein Lautstärke-Meßgerät für breitbandige und impulshaltige Schalle (A loudness meter for broad-band noises and impulsive sounds). Acustica **17**, 322–334 (1966)

Stemplinger I., H. Fastl: Accuracy of loudness percentile versus measurement time. In: Proc. Inter-Noise'97, Vol. III, pp. 1347–1350 (1997)

Widmann U., R. Lippold, H. Fastl: Ein Computerprogramm zur Simulation der Nachverdeckung für Anwendungen in akustischen Meßsystemen (A computer program for simulation of post-masking). In: *Fortschritte der Akustik DAGA '98*, Verl.: Dt. Gesell. für Akustik e. V., Oldenburg, 96–97 (1998)

Widmann U., R. Lippold, H. Fastl: Decay of postmasking: Applications in Sound Analysis Systems. In: Proc. NOISE-CON'98, Ypsilanti Michigan USA. Ed. by P. Davies, G. Ebbit, 451–456 (1998)

Zwicker E.: Lautstärkeberechnungsverfahren im Vergleich (A comparison of loudness calculating procedures). Acustica **17**, 278–284 (1966)

Zwicker E.: Funktionsmodelle des Gehörs als Instrumente zur Lautstärkemessung und zur automatischen Spracherkennung (Electronically realized models of the hearing system as instruments for measuring loudness and for automatic recognition of speech). In *Kybernetik, Brücke zwischen den Wissenschaften*, ed. by H. Frank, 7. Aufl. (Umschau, Frankfurt 1970) pp. 165–176

Zwicker E.: Advantages of a precise loudness meter. In Proc. Inter-Noise '79, Vol.I (1979) pp. 223–226

Zwicker E.: Procedure for calculating partially masked loudness based on ISO 532 B. In Proc. Inter-Noise '87, Vol.II (1987) pp. 1021–1024

Zwicker E., W. Daxer: A portable loudness meter based on psychoacoustical models. In Proc. Inter-Noise '81, Vol.II (1981) pp. 869–872

Zwicker E., K. Deuter, W. Peisl: Loudness meters based on ISO 532 B with large dynamic range. In Proc. Inter-Noise '85, Vol.II (1985) pp. 1119–1122

Zwicker E., H. Fastl: Kontrolle von Lärmminderungsmaßnahmen mit dem Lautheitsmesser (Evaluation of noise reduction using a loudness meter). In *Fortschritte der Akustik, FASE/DAGA '82* (DPG, Bad Honnef 1982) pp. 1141–1144

Zwicker E., H. Fastl: A portable loudness meter based on ISO 532 B. In Proc. 11th ICA Paris, Vol.8 (1983) pp. 135–137

Zwicker E., H. Fastl, C. Dallmayr: BASIC-Program for calculating the loudness of sounds from their 1/3-oct. band spectra according to ISO 532 B. Acustica **55**, 63–67 (1984)

Zwicker E., H. Fastl, U. Widmann, K. Kurakata, S. Kuwano, S. Namba: Program for calculating loudness according to DIN 45631 (ISO 532 B). J. Acoust. Soc. Jpn. (E) **12**, 39–42 (1991)

Noise Emissions

Fastl H.: Subjektive Rauschverminderung durch ein DNL-System (Subjective reduction of noise by DNL-systems). In *Fortschritte der Akustik, DAGA '78* (VDE, Berlin 1978) pp. 641–644

Fastl H.: Subjektive Beurteilung eines Dolby-B-Rauschverminderungs-Systems (Subjective evaluation of Dolby-B-noise reduction systems). In *Fortschritte der Akustik, DAGA '80* (VDE, Berlin 1980) pp. 747–750

Fastl H.: Loudness and annoyance of sounds: Subjective evaluation and data from ISO 532 B. In Proc. Inter-Noise '85, Vol.II (1985) pp. 1403–1406

Fastl H.: How loud is a passing vehicle? In Proc. Inter-Noise '87, Vol. II (1987) pp. 993–996

Fast. H.: Loudness versus level of aircraft noise. In: Proc. Inter-Noise '91, Vol. I, 33–36 (1991)

Fastl H.: On the reduction of road traffic noise by "whispering asphalt". In: Proc. Congress, Acoust. Soc. of Japan, Nagano, 681–682 (1991)

Fastl H.: Loudness evaluation by subjects and by a loudness meter. In: Sensory Research, Multimodal Perspectives, (R. T. Verrillo eds.) Lawrence Erlbaum Ass., Hillsdale, New Jersey, 199–210 (1993)

Fastl H.: Psychoacoustics and noise evaluation. In: Contr. to Psychological Acoustics, (A.Schick ed.) Oldenburg: Bibliotheks- und Informationssystem der Carl von Ossietzky Univ., 505–520 (1993)

Fastl H.: Psychoacoustics and noise evaluation. In: NAM'94, (H.S. Olesen, ed.) Aarhus, Denmark, Danish Technol. Institute, 1–12 (1994)

Fastl H.: Applications of psychoacoustics in noise control. Acustica **82**, S. 77 (1996)

Fastl H.: Gehörgerechte Geräuschbeurteilung (Noise evaluation based on features of the human hearing system). In: *Fortschritte der Akustik, DAGA '97*, Verl.: Dt. Gesell. für Akustik e. V., Oldenburg, 57–64 (1997)

Fastl H.: Psychoacoustic noise evaluation. In: Proceedings of the 31st International Acoustical Conference, Acoustics – High Tatras '97, 21–26 (1997)

Fastl H., U. Widmann: Subjective and physical evaluation of aircraft noise. Noise Contr. Engng. J. **35**, 61–63 (1990)

Fastl H., S. Namba, S. Kuwano: Cross-cultural study on loudness evaluation of road traffic noise and impulsive noise: Actual sounds and simulations. In Proc. Inter-Noise '86, Vol.II (1986) pp. 825–830

Fastl H., S. Namba, S. Kuwano: On the reduction of road traffic noise by speed limits using the method of continuous judgment by category. In: Proc. TC Noise, Acoust. Soc. of Japan N-91-45 (1991)

Fastl H., U. Widmann, S. Kuwano, S. Namba: Zur Lärmminderung durch Geschwindigkeitsbeschränkungen (On the noise reduction by speed limits). In: *Fortschritte der Akustik, DAGA '91*, Verl.: DPG-GmbH, Bad Honnef, pp. 449–452 (1991)

Gottschling G., H. Fastl: Akustische Simulation von 6-Sektionen-Fahrzeugen des Transrapid (Acoustic simulation of 6-car-trains of Transrapid). In: *Fortschritte der Akustik, DAGA '97*, Verl.: Dt. Gesell. für Akustik e. V., Oldenburg, pp. 254–255 (1997)

Hellman R., E. Zwicker: Why can a decrease in dB(A) produce an increase in loudness? J. Acoust. Soc. Am. **82**, 1700–1705 (1987)

Jäger K., H. Fastl, G. Gottschling, F. Schöpf, U. Möhler: Wahrnehmung von Pegeldifferenzen bei Vorbeifahrten von Güterzügen (Perceptibility of level differences for passing freight trains). In: *Fortschritte der Akustik, DAGA '97*, Verl.: Dt. Gesell. für Akustik e. V., Oldenburg, pp. 228–229 (1997)

Kuwano S., H. Fastl: Loudness evaluation of various kinds of non-steady state sound using the method of continous judgment by category. In Proc. 13th ICA Belgrade, Vol.I (1989) pp. 365–368

Kuwano S., S. Namba, H. Fastl: Loudness evaluation of various sounds by Japanese and German subjects. In Proc. Inter-Noise '86, Vol.II (1986) pp. 835–840

Kuwano S., S. Namba, H. Fastl, A. Schick: Evaluation of the impression of danger signals - comparison between Japanese and German subjects. In: 7. Oldenburger Symposium (A. Schick, M. Klatte, Eds.), BIS Oldenburg, pp. 115–128 (1997)

Seiter A., I. Stemplinger, T. Beckenbauer: Untersuchungen zur Tonhaltigkeit von Geräuschen (Experiments on the tonality of noises). In: *Fortschritte der Akustik, DAGA '96*, Verl.: Dt. Gesell. für Akustik e. V., Oldenburg, 238–239 (1996)

Spatzl M., U. Widmann, H. Fastl: Subjektive und meßtechnische Beurteilung von Pkw-Emissions- und Immissionsgeräuschen (Subjective and physical evaluation of noise emissions and noise immissions from cars). In: *Fortschritte der Akustik, DAGA '93*, Verl.: DPG-GmbH, Bad Honnef, pp. 604–607 (1993)

Widmann U.: Beschreibung der Geräuschemission von Kraftfahrzeugen anhand der Lautheit (Noise emission of vehicles described by loudness). In: *Fortschritte der Akustik, DAGA '90*, Verl.: DPG-GmbH, Bad Honnef, pp. 401–404 (1990)

Widmann U.: Optimierung von akustisch wirksamen Fahrbahnrandmarkierungen anhand psychoakustischer Kriterien (Improvement of acoustically effective bankets by means of psychoacoustic criteria). In: *Fortschritte der Akustik, DAGA '93*, Verl.: DPG-GmbH, Bad Honnef, pp. 888–891 (1993)

Zwicker E.: Weniger L_A = Größere Lautstärke? (Less A-weighted sound pressure level equal to larger loudness?) In *Fortschritte der Akustik, DAGA '80* (VDE, Berlin 1980) pp. 159–162

Zwicker E.: What is a meaningful value for quantifying noise reduction? In Proc. Inter-Noise '85, Vol.I (1985) pp. 47–56

Zwicker E.: Psychophysics of hearing. In *Noise Pollution*, ed. by A.L. Saénz et. al., Chapt.4, SCOPE (Wiley, New York 1986) pp. 147–167

Zwicker E.: Berechnung partiell maskierter Lautheiten auf der Grundlage von ISO 532 B. (Calculation of partial masked loudness using ISO 532 B). In *Fortschritte der Akustik, DAGA '87* (DPG, Bad Honnef 1987) pp. 181–184

Zwicker E.: Meaningful noise measurement and effective noise reduction. Noise Contr. Eng. J. **29**, 66–76 (1987)

Zwicker E.: Loudness patterns (ISO 532 B), an excellent guide to noise-reduced design and to expected public reaction. In Proc. of NOISE-CON 88, ed. by J.S. Bolton, Noise Contr. Found., New York (1988) pp. 15–26

Zwicker E., H. Fastl: Die Reduzierung von Lärm durch Schallschutzfenster (The reduction of noise by sound isolating windows). Arcus **2**, 80–82 (1984), 1. Nachtrag: Arcus **3**, 100 (1984), 2. Nachtrag: Arcus **4**, 148 (1984)

Zwicker E., H. Fastl: Examples for the use of loudness: Transmission loss and addition of noise sources. In Proc. Inter-Noise '86, Vol.II (1986) pp. 861–866

Zwicker E., H. Fastl: Sinnvolle Lärmmessung und Lärmgrenzwerte (Meaningful noise measurement and noise limits). Z. für Lärmbekämpfung **33**, 61–67 (1986)

Zwicker U.T., E. Zwicker: Effects of binaural loudness summation and their approximation in objective loudness measurements. In Proc. Inter-Noise '90, Vol. II (1990) pp. 1145–1150

Noise Immissions

Fastl H.: Average loudness of road traffic noise. In Proc. Inter-Noise '89, Vol.II (1989) pp. 815–820

Fastl H.: Trading number of operations versus loudness of aircraft. In Proc. Inter-Noise '90, Vol. II (1990) pp. 1133–1136

Fastl H.: Evaluation and measurement of perceived average loudness. In: Contributions to Psychological Acoustics, (A. Schick et al. eds.), Bibliotheks- und Informationssystem der Univ. Oldenburg 205–216 (1991)

Fastl H., G. Gottschling: Beurteilung von Geräuschimmissionen beim TRANS-RAPID (Evaluation of Noise Immissions from Transrapid). In: *Fortschritte der Akustik, DAGA '96*, Verl.: Dt. Gesell. für Akustik e. V., Oldenburg, 216–217 (1996)

Fastl H., G. Gottschling: Subjective evaluation of noise immissions from Transrapid. In: Proc. Inter-Noise'96, Vol. 4, 2109–2114 (1996)

Fastl H., J. Hunecke: Psychoakustische Experimente zum Fluglärmmalus (Psychoacoustic experiments on the aircraft-malus). In: *Fortschritte der Akustik, DAGA '95*, Verl.: Dt. Gesell. für Akustik e. V., Oldenburg, pp. 407–410 (1995)

Fastl H., E. Zwicker: Beurteilung lärmarmer Fahrbahnbeläge mit einem Lautheitsmesser (Evaluation of noise reducing street carpets using a loudness meter). In *Fortschritte der Akustik, DAGA '86* (DPG, Bad Honnef 1986) pp. 223–226

Fastl H., S. Kuwano, S. Namba: Psychoakustische Experimente zum Schienenbonus (Psychoacoustic experiments on the railway-bonus). In: *Fortschritte der Akustik, DAGA '94*, Verl.: DPG-GmbH, Bad Honnef, pp. 1113–1116 (1994)

Fastl H., S. Kuwano, S. Namba: Psychoacoustics and rail bonus. In: Inter-Noise'94. Vol. II., 821–826 (1994)

Fastl H., S. Kuwano, S. Namba: Assessing in the railway bonus in laboratory studies. J. Acoust. Soc. Jpn. (E) **17**, 139–148 (1996)

Fastl H., D. Markus, V. Nitsche: Zur Lautheit und Lästigkeit von Fluglärm (On the loudness and the annoyance of aircraft-noise). In *Fortschritte der Akustik, DAGA '85* (DPG, Bad Honnef 1985) pp. 227–230

Fastl H., W. Schmid, S. Kuwano, S. Namba: Untersuchungen zum Schienenbonus in Gebäuden (Experiments on the "railway-bonus" within buildings). In: *Fortschritte der Akustik, DAGA '96*, Verl.: Dt. Gesell. für Akustik e. V., Oldenburg, pp. 208–209 (1996)

Fastl H., E. Zwicker, S. Kuwano, S. Namba: Beschreibung von Lärmimmissionen anhand der Lautheit (Describing noise immission by loudness). In *Fortschritte der Akustik, DAGA '89* (DPG, Bad Honnef 1989) pp. 751–754

Fastl H., E. Zwicker, S. Kuwano, S. Namba: Mittlere Lautheit von Lärmereignissen unterschiedlicher Anzahl und Art (Average loudness of noise events of different number and type). In: *Fortschritte der Akustik, DAGA '90*, Verl.: DPG-GmbH, Bad Honnef, pp. 393–396 (1990)

Fastl H., Th. Filippou, W. Schmid, S. Kuwano, S. Namba: Psychoakustische Beurteilung der Lautheit von Geräuschimmissionen verschiedener Verkehrsträger (Psychoacoustic evaluation of noise immissions from rail-noise, road-noise, and aircraft-noise). In: *Fortschritte der Akustik DAGA '98*, Verl.: Dt. Gesell. für Akustik e. V., Oldenburg, 70–71 (1998)

Gottschling G., H. Fastl: Prognose der globalen Lautheit von Geräuschimmisionen anhand der Lautheit von Einzelereignissen (Prediction of global loudness of noise immissions on the basis of the loudness of single events). In: *Fortschritte der Akustik DAGA '98*, Verl.: Dt. Gesell. für Akustik e. V., Oldenburg, 478–479 (1998)

Peschel U., H. Fastl: Subjektive Beurteilung der Lärmimmission landender Flugzeuge (Subjective evaluation of noise Immissions from approaching aircraft). In: *Fortschritte der Akustik, DAGA '92*, Verl.: DPG-GmbH, Bad Honnef, pp. 441–444 (1992)

Stemplinger I.: Globale Lautheit von gleichförmigen Industriegeräuschen (Global loudness of continuous industrial noise). In: *Fortschritte der Akustik, DAGA '96*, Verl.: Dt. Gesell. für Akustik e. V., Oldenburg, 240–241 (1996)

Stemplinger I.: Beurteilung der Globalen Lautheit bei Kombination von Verkehrsgeräuschen mit simulierten Industriegeräuschen (Evaluation of global loudness for the combination of traffic noise plus simulated industrial noise). In: *Fortschritte der Akustik, DAGA '97*, Verl.: Dt. Gesell. für Akustik e. V., Oldenburg, 353–354 (1997)

Stemplinger I., Th. Filippou: Psychoakustische Untersuchungen zur Lautheit und zur Lästigkeit von Tennislärm (Psychoacoustic investigations of the loudness and annoyance of noise from tennis courts). In: *Fortschritte der Akustik DAGA '98*, Verl.: Dt. Gesell. für Akustik e. V., Oldenburg, 66–67 (1998)

Stemplinger I., G. Gottschling: Auswirkungen der Bündelung von Verkehrswegen auf die Beurteilung der Globalen Lautheit (Effects of the combination of road and rail tracks on the evaluation of global loudness). In: *Fortschritte der Akustik, DAGA '97*, Verl.: Dt. Gesell. für Akustik e. V., Oldenburg, 401–402 (1997)

Stemplinger I., A. Seiter: Beurteilung von Lärm am Arbeitsplatz (Evaluation of noise on the workplace). In: *Fortschritte der Akustik, DAGA '95*, Verl.: Dt. Gesell. für Akustik e. V., Oldenburg, pp. 867–870 (1995)

Widmann U.: Meßtechnische Beurteilung und Umfrageergebnisse bei Straßenverkehrslärm (Correlations of physical measurements and results from questionnaires for road-traffic noise). In: *Fortschritte der Akustik, DAGA '92*, Verl.: DPG-GmbH, Bad Honnef, pp. 369–372 (1992)

Sound Quality

Fastl H.: Hearing sensations and noise quality evaluation. J. Acoust. Soc. Am. **87**, S 134 (1990)

Fastl H.: Psychoakustik und Geräuschbeurteilung (Psychoacoustics and noise evaluation). In: Soundengineering, (Q.Vo u.a. eds.) expert-verl., Renningen, 10–33 (1994)

Fastl H.: The Psychoacoustics of Sound-Quality Evaluation. In: Proc. EAA-Tutorium, Antwerpen, 1–20 (1996)

Fastl H.: The Psychoacoustics of Sound-Quality Evaluation. Acustica - acta acustica **83**, 754–764 (1997)

Fastl H.: Psychoacoustics and Sound Quality Metrics. In: Proc. Sound Quality Symposium, Ypsilanti Michigan USA. Ed. by P. Davies, G. Ebbit, 3–10 (1998)

Fastl H., Y. Yamada: Cross-cultural study on loudness and annoyance of broadband noise with a tonal component. In *Contributions to Psychological Acoustics*, ed. by A. Schick et. al. (Kohlrenken, Oldenburg 1986) pp. 341–353

Hellman R., E. Zwicker: Magnitude scaling: a meaningful method for measuring loudness and annoyance? In: Fechner Day 90, Proc. of the 6th Annual Meeting of the Intern. Soc. for Psychophysics. (F. Müller, ed.), Inst. f. Psychologie, Univ. Würzburg, pp. 123–128 (1990)

Kuwano S., S. Namba, H. Fastl: On the judgement of loudness, noisiness, and annoyance with actual and artifical noises. J. Sound and Vibration **127**, 457–465 (1988)

Namba S., S. Kuwano, H. Fastl: Cross-cultural study on the loudness, noisiness, and annoyance of various sounds. In Proc. Inter–Noise '87, Vol.II (1987) pp. 1009–1012

Namba S., S. Kuwano, H. Fastl: The definition of loudness, noisiness and annoyance in laboratory situations. In: Proc. 14. ICA Beijing, Vol. 3, H1-1 (1992)

Preis A., E. Terhardt: Annoyance of distortions of speech: Experiments on the influence of interruptions and random-noise impulses. Acustica **68**, 263–267 (1989)

Terhardt E.: Wohlklang und Lärm aus psycho-physikalischer Sicht (Sensory consonance and noise from a psycho-physical point of view). In *Ergebnisse des 3. Oldenburger Symposion zur Psychologischen Akustik*, ed. by A. Schick, K.P. Walcher (Lang, Bern 1984) pp. 403–409

Widmann U.: Minderung der Sprachverständlichkeit als Maß für die Belästigung (Reduction of speech intelligibility as a measure of annoyance). In: *Fortschritte der Akustik, DAGA '91*, Verl.: DPG-GmbH, Bad Honnef, pp. 973–976 (1991)

Widmann U.: Zur Lautheit und Lästigkeit von Breitbandrauschen mit einer tonalen Komponente (On the loudness and annoyance of broadband noise with a tonal component). In: *Fortschritte der Akustik, DAGA '93*, Verl.: DPG-GmbH, Bad Honnef, pp. 632–635 (1993)

Widmann U.: Untersuchungen zur Schärfe und zur Lästigkeit von Rauschen unterschiedlicher Spektralverteilung (Sharpness and annoyance of noises with different spectral envelopes). In: *Fortschritte der Akustik, DAGA '93*, Verl.: DPG-GmbH, Bad Honnef, pp. 644–647 (1993)

Widmann U.: Zur Lästigkeit von amplitudenmodulierten Breitbandrauschen (On the annoyance of amplitude modulated broadband noise). In: *Fortschritte der Akustik, DAGA '94*, Verl.: DPG-GmbH, Bad Honnef, pp. 1121–1124 (1994)

Widmann U.: Subjektive Beurteilung der Lautheit und der Psychoakustischen Lästigkeit von PKW-Geräuschen (Subjective evaluation of loudness and psychoacoustic annoyance of car-sounds). In: *Fortschritte der Akustik, DAGA '95*, Verl.: Dt. Gesell. für Akustik e. V., Oldenburg, pp. 875–878 (1995)

Widmann U., S. Goossens: Zur Lästigkeit tieffrequenter Schalle: Einflüsse von Lautheit und Zeitstruktur (On the annoyance of low-frequency sounds: influence of loudness and time-structure). Acustica **77**, 290–292 (1993)

Zwicker E.: Ein Beitrag zur Unterscheidung von Lautstärke und Lästigkeit (A contribution for differentiating loudness and annoyance). Acustica **17**, 22–25 (1966)

Zwicker E.: On the dependence of unbiased annoyance on loudness. In Proc. Inter-Noise '89, Vol.II (1989) pp. 809–814

Zwicker E.: A proposal for defining and calculating the unbiased annoyance. In: Contributions to Psychological Acoustics, (A. Schick et al. eds.), Bibliotheks- und Informationssystem der Univ. Oldenburg 187–202 (1991)

Zwicker E., E. Terhardt: Über die Störwirkung von Impulsfolgen beim Telefonieren (On the disturbing effect of noise bursts during acoustical communication through phones). NTZ **18**, 80–90 (1965)

Applications in Audiology

Arnold B., U. Baumann, I. Stemplinger, K. Schorn: Bezugskurven für die kategorale Lautstärkeskalierung (Reference curve for Lautstärkeskalierung (Reference curve for loudness scaling in categories). Arch. of Otorhinolaryngology, Suppl. 1996, Teil 2 Sitzungsbericht, Springer

Baumann U., I. Stemplinger, B. Arnold, K. Schorn: Kategoriale Lautstärkeskalierung in der klinischen Anwendung (Category scaling of loudness in clinical applications). In: *Fortschritte der Akustik, DAGA '96*, Verl.: Dt. Gesell. für Akustik e. V., Oldenburg, pp. 128–129 (1996)

Baumann U., I. Stemplinger, B. Arnold, K. Schorn: Kategoriale Lautstärkeskalierung in der klinischen Anwendung (Category scaling of loudness in clinical applications). Laryngo-Rhino-Otologie, **8**, 458–465 (1997)

Chalupper J., K. Spasokukotskij, I. Stemplinger, H. Fastl: Ein Zweisilber-Sprachtest für Ukrainisch (A speech test for Ukrainian language). In: *Fortschritte der Akustik DAGA '98*, Verl.: Dt. Gesell. für Akustik e. V., Oldenburg, 310–311 (1998)

Fastl H.: Measuring hearing thresholds with audiometer headphones. J. Audiol. Techn. **18**, 92–98 (1979)

Fastl H.: An instrument for measuring temporal integration in hearing. Audiol. Acoustics **23**, 164–170 (1984)

Fastl H.: Auditory adaptation, post masking and temporal resolution. Audiol. Acoustics **24**, 144–154; 168–177 (1985)

Fastl H.: A background noise for speech audiometry. Audiol. Acoustics **26**, 2–13 (1987)

Fastl H.: The influence of different measuring methods and background noises on the temporal integration in noise-induced hearing loss. Audiol. Acoustics **26**, 66–82 (1987)

Fastl H.: A masking noise for speech intelligibility tests. In: Proc. TC Hearing, Acoust. Soc. of Japan, H-93-70 (1993)

Fastl H., K. Schorn: Discrimination of level differences by hearing-impaired patients. Audiology **20**, 488–502 (1981)

Fastl H., K. Schorn: On the diagnostic relevance of level discrimination. Audiology **23**, 140–142 (1984)

Fastl H., E. Zwicker: A device for measuring level and frequency difference limens. J. Audiol. Technique **18**, 26–34 (1979)

Florentine M., S. Buus, B. Scharf, E. Zwicker: Frequency selectivity in normally hearing and hearingimpaired observers. J. Speech Hearing Res. **23**, 646–669 (1980)

Florentine M., H. Fastl, S. Buus: Temporal integration in normal hearing, cochlear impairment, and impairment simulated by masking. J. Acoust. Soc. Am. **84**, 195–203 (1988)

Florentine M., E. Zwicker: A model of loudness summation applied to noise induced hearing loss. Hearing Res. **1**, 121–132 (1979)

Gottschling G., W. Schmid, H. Fastl: Vergleich psychoakustischer Methoden zur Skalierung der Lautstärke: I. Grundlagen (Comparison of psychoacoustic methods for the scaling of loudness. I. Basics). In: *Fortschritte der Akustik DAGA '98*, Verl.: Dt. Gesell. für Akustik e. V., Oldenburg, 476–477 (1998)

Hautmann I., H. Fastl: Zur Verständlichkeit von Einsilbern und Dreinsilbern im Störgeräusch (Intelligibility of monosyllables and 3-time repeated monosyllables in background noise). In: *Fortschritte der Akustik, DAGA '93*, Verl.: DPG-GmbH, Bad Honnef, pp. 784–787 (1993)

Hojan E., H. Fastl: Zur Verständlichkeit deutscher Sprache im Fastl-Störgeräusch durch polnische Hörer mit verschiedenen Deutschkenntnissen (Intelligibility of German speech in background-noise by Polish listeners with different degrees of proficiency in German). In: *Fortschritte der Akustik, DAGA '95*, Verl.: Dt. Gesell. für Akustik e. V., Oldenburg, pp. 831–834 (1995)

Hojan E., H. Fastl: Intelligibility of speech in noisy environment. In: VI Sympos. on Tonmeistering and Sound Eng. Warszawa, 62–67 (1995)

Hojan E., H. Fastl: Intelligibility of Polish and German speech for the Polish audience in the presence of noise. Archives of Acoustics **21**, 2, 123–130 (1996)

Schorn K., H. Fastl: The measurement of level difference thresholds and its importance for the early detection of the acoustic neurinoma. Audiol. Acoustics **23**, 22–27; 60–62 (1984)

Schorn K., G. Wurzer, M. Zollner, E. Zwicker: Die Bestimmung des Frequenzselektionsvermögens des funktionsgestörten Gehörs mit Hilfe psychoakustischer Tuningkurven (Evaluation of the impaired ears' frequency selectivity using psychoacoustical tuning curves). Laryngologie, Rhinologie, Otologie **56**, 121–127 (1977)

Schorn K., E. Zwicker: Clinical investigation on temporal resolution capacity of hearing for various types of hearing loss. Audiol. Acoustics **25**, 170–184 (1986)

Schorn K., E. Zwicker: Zusammenhänge zwischen gestörtem Frequenz- und gestörtem Zeitauflösungsvermögen bei Innenohrschwerhörigkeiten (Correlations of impaired frequency- and time-resolution for inner ear hearing loss) Arch. of Laryngologie, Rhinologie, Otologie, Suppl.II, 116–118 (1989)

Schorn K., E. Zwicker: Frequency selectivity and temporal resolution in patients with various inner ear disorders. Audiology **29**, 8–20 (1990)

Stemplinger I., H. Fastl, K. Schorn, F. Bruegel: Zur Verständlichkeit von Einsilbern in unterschiedlichen Störgeräuschen (Intelligibility of monosyllables in different background noises). In: *Fortschritte der Akustik, DAGA '94*, Verl.: DPG-GmbH, Bad Honnef, pp. 1469–1472 (1994)

Stemplinger I., M. Schiele, B. Meglic, H. Fastl: Einsilberverständlichkeit in unterschiedlichen Störgeräuschen für Deutsch, Ungarisch und Slowenisch (Speech intelligibility in different background noises for German, Hungarian, and Slovenian). In: *Fortschritte der Akustik, DAGA '97*, Verl.: Dt. Gesell. für Akustik e. V., Oldenburg, 77–78 (1997)

Terhardt E.: On the perception of spectral information in speech. In *Hearing Mechanisms and Speech*, ed. by O. Creutzfeld, H. Scheich, Chr. Schreiner (Springer, Berlin, Heidelberg 1979) pp. 281–291

Tschopp K., H. Fastl: On the loudness of German speech material used in audiology. Acustica **73**, 33–34 (1991)

Zwicker E.: Klassifizierung von Hörschäden nach dem Frequenzselektionsvermögen (Classification of hearing impairments by the ear's frequency selectivity). In *Kybernetik 1977*, ed. by G. Hauske, E. Butenandt (Oldenbourg, München 1978) pp. 413–415

Zwicker E.: A device for measuring the temporal resolution of the ear. Audiol. Acoustics **19**, 94–108 (1980)

Zwicker E.: Temporal resolution in background noise. Brit. J. of Audiol. **19**, 9–12 (1985)

Zwicker E.: The temporal resolution of hearing - An expedient measuring method for speech intelligibility. Audiol. Acoustics **25**, 156–168 (1986)

Zwicker E.: Otoacoustic emissions in research of inner ear signal processing. 2nd Intern. Symposium on Cochlear Mechanics and Otoacoustic Emissions. In *Cochlear Mechanisms and Otoacoustic Emissions*, ed. by F. Grandori et al. (Karger Basel 1990) pp. 63–76

Zwicker E., K. Schorn: Das Frequenzselektionsvermögen des funktionsgestörten Gehörs (Frequency selectivity of the impaired hearing system). Z. Hörger. Akustik, Sonderheft 1977, 44–47 (1977)

Zwicker E., K. Schorn: Psychoacoustical tuning curves in audiology. Audiology **17**, 120–140 (1978)

Zwicker E., K. Schorn: Temporal resolution in hard-of-hearing patients. Audiology **21**, 474–492 (1982)

Zwicker E., K. Schorn: Delayed evoked otoacoustic emissions – an ideal screening test for excluding hearing impairment in infants. Audiology **29**, 241–251 (1990)

Hearing Aids

Beckenbauer T.: Möglichkeiten zur Verbesserung des Signal/Störverhältnisses durch gerichtete Schallaufnahmen (Possibilities for improving the signal-to-noise ratio by sound recordings with special directivity). In *Fortschritte der Akustik, DAGA '87* (DPG, Bad Honnef 1987) pp. 449–452

Beckenbauer T.: Einfluß vielkanaliger Inhibitionen auf die Sprache (The influence of multichannel inhibition on speech). In *Fortschritte der Akastik, DAGA '88 (DPG, Bad Honnef 1988) pp. 713–716

Beckenbauer T.: Technisch realisierte laterale Inhibition und ihre Wirkung auf störbehaftete Sprache (Effects of lateral inhibition on speech in noise). Acustica **75**, 1–16 (1991)

Fastl H.: On the influence of AGC hearing aids of different types on the loudness-time pattern of speech. Audiol. Acoustics **26**, 42–48 (1987)

Fastl H.: Psychoakustik und Hörgeräteanpassung (Psychoacoustics and fitting of hearing aids). In: Zukunft der Hörgeräte, Schriftenreihe der GEERS-Stiftung, Band 11, 133–146, (1996)

Fastl H., H. Oberdanner, W. Schmid, I. Stemplinger, I. Hochmair-Desoyer, E. Hochmair: Zum Sprachverständnis von Cochlea-Implantat-Patienten bei Störgeräuschen (Speech intelligibility in noise for cochlea-implant patients). In: Fortschritte der Akustik DAGA '98, Verl.: Dt. Gesell. für Akustik e. V., Oldenburg, 358–359 (1998)

Hoffmann C.: Elektrokutane Reize als Träger von Sprachschallinformation (Electrocutaneous stimulation of the skin as carrier for speech information). In Fortschritte der Akustik, DAGA '80 (VDE, Berlin 1980) pp. 771–774

Hoffmann C.: Transmission of speech information for the deaf through electric excitation of the skin. Audiol. Acoustics 23, 4–21 (1984)

Leysieffer H.: Polyvinylidenfluorid als elektromechanische Wandler für taktile Reizgeber (Polyvinylidenfluoride as an electromechanic transducer for tactile stimulation). Acustica 58, 196–206 (1985)

Leysieffer H.: Vibrotaktile Reizgeber mit PVDF (Polyvinylidenfluorid) als elektromechanischem Wandler (Vibrotactile stimulator using PVDF as electromechanic transducer). In Fortschritte der Akustik, DAGA '85 (DPG Bad Honnef 1985) pp. 863–866

Leysieffer H.: A wearable multi-channel auditory prosthesis with vibrotactile skin stimulation. Audiol. Acoustics 25, 230–251 (1986)

Leysieffer H.: Eine mehrkanalige, vibrotaktile Hörprothese (A vibrotactile hearing-prosthesis using several channels). In Fortschritte der Akustik, DAGA '86, (DPG, Bad Honnef 1986) pp. 477–480

Leysieffer H.: Mehrkanalige Sprachübertragung mit einer vibrotaktilen Sinnesprothese für Gehörlose (Speech transmission using a vibrotactile prosthesis with several channels for totally deaf). In Workshop Elektronische Kommunikationshilfen, BIG-Tech Berlin '86, ed. by K.-R. Fellbaum (Weidler, Berlin 1987) pp. 287–298

Leysieffer H.: Sprachverständlichkeitsmessungen mit einer vibrotaktilen Hörprothese (Measurements of speech intelligibility using a vibrotactile hearing prosthesis). In Fortschritte der Akustik, DAGA '87 (DPG, Bad Honnef 1987) pp. 613–616

Naumann H.H., E. Zwicker, H. Scherer, K. Schorn, J. Seifert, M. Stecker, H. Leysieffer, M. Zollner: Erfahrungen mit der Implantation einer Mehrkanalelektrode in den Nervus Acusticus (Experience using a several-channel electrode for implantation in the nervus acousticus). Laryngologie, Rhinologie, Otologie 65, 118–122 (1986)

Schorn K., M. Stecker, M. Zollner: Voruntersuchungen gehörloser Patienten zur Cochlea-Implantation (Preexploration of deaf patients before cochlear implantation). Laryngologie, Rhinologie, Otologie 65, 114–117 (1986)

Simmons F.B., J.M. Epley, R.C. Lummis, N. Guttmann, L.S. Frishkopf, L.S. Harmon, E. Zwicker: Auditory nerve: electrical stimulation in man. Science 148, 104–106 (1965)

Theopold H.M., M. Zollner, K. Schorn, J. Spahmann, H. Scherer: Untersuchungen zur Gewebeverträglichkeit lackisolierter Platin-Iridium-Elektroden (Investigations of the sociability of varnish-isolated platin-iridium-electrodes on tissue). Laryngologie, Rhinologie, Otologie 60, 534–537 (1981)

Zollner M.: Ein implantierbarer Empfänger zur elektrischen Reizung der Hörnerven (A implantible receiver for electrical stimulation of the eighth nerve). In Fortschritte der Akustik, DAGA '80 (VDE, Berlin 1980) pp. 775–778

Zwicker E.: Möglichkeiten zur Spracherkennung über den Tastsinn mit Hilfe eines Funktionsmodells des Gehörs (Possibilities for recognizing speech by the sense of vibration using models of the hearing system). Elektron. Rechenanl. **7**, 239–244 (1963)

Zwicker E., H. Leysieffer, K. Dinter: Ein Implantat zur Reizung des Nervus akustikus mit zwölf Kanälen (An implant for stimulation the nervus acousticus using twelve channels). Laryngologie, Rhinologie, Otologie **65**, 109–113 (1986)

Zwicker E., H. Scherer, H. Leysieffer, K. Dinter: Elektrische Reizung eines sensiblen Nerven mit 70-Mikrometer-Elektroden (Electrical stimulation of a sensoric nerve using 70 micrometer-electrodes). Laryngologie, Rhinologie, Otologie **65**, 105–108 (1986)

Zwicker E., M. Zollner: Criteria for VIIIth nerve implants and for feasible coding. Scand. Audiol. Suppl. **11**, 179–181 (1980)

Broadcasting and Communication Systems

Beckenbauer T.: Ein vielkanaliges Inhibitionsnetzwerk für Sprachübertragung (A multichannel inhibition network for speech transmission). In *Fortschritte der Akustik, DAGA '89* (DPG, Bad Honnef 1989) pp. 191–194

Deuter K.: Zweckmäßige Audiosignalübertragung bei gleichzeitig vorhandenem Störgeräusch (Expedient transmission of audio signals in case of simultaneous noise). Acustica **69**, 133–150 (1989)

Feldtkeller R.: Hörbarkeit nichtlinearer Verzerrungen bei der Übertragung von Instrumentenklängen (Audibility of nonlinear distortions when transmitting sounds of musical instruments). Acustica **4**, 70–72 (1954)

Fastl H.: Loudness of running speech measured by a loudness meter. Acustica **71**, 156–158 (1990)

Fastl H., S. Goossens: Psychoakustische Effekte bei der Rauschbefreiung von Archivmaterial (Psychoacoustic effects to be considered when de-noising archival program material). In: *Fortschritte der Akustik, DAGA '93*, Verl.: DPG-GmbH, Bad Honnef, pp. 868–871 (1993)

Gäßler G.: Die Grenzen der Hörbarkeit nichtlinearer Verzerrungen bei der Übertragung von Instrumentenklängen (The limits of audibility of nonlinear distortions when transmitting sounds of instruments). Frequenz **9**, 15–25 (1955)

Goossens S., H. Fastl: Zur Höhenwahrnehmung beim Hören von Musikaufnahmen (On the perception of "brilliance" in remastered music). In: 17. Tonmeistertagung, Karlsruhe 1992, Verl.: K.G. Saur, München, 394–403 (1993)

Heinbach W.: Untersuchung einer gehörbezogenen Spektralanalyse mittels Resynthese (Investigations of hearing equivalent spectral analysis via resynthesis). In *Fortschritte der Akustik, DAGA '86* (DPG, Bad Honnef 1986) pp. 453–456

Heinbach W.: Datenreduktion von Sprache unter Berücksichtigung von Gehöreigenschaften (Bit-rate reduction of speech in consideration of characteristics of the hearing system). NTZ-Archiv **9**, 327–333 (1987)

Heinbach W.: Verständlichkeitsmessungen mit datenreduzierten natürlichen Einzelvokalen (Intelligibility measurements using bit-rate reduced natural single vowels). In *Fortschritte der Akustik, DAGA '87* (DPG, Bad Honnef 1987) pp. 665–668

Heinbach W.: Aurally adequate signal representation: the Part-Tone-Time-Pattern. Acustica **67**, 113–121 (1988)

Terhardt E.: Verfahren zur gehörbezogenen Frequenzanalyse (Procedures for hearing equivalent frequency analysis). In *Fortschritte der Akustik, DAGA '85* (DPG, Bad Honnef 1985) pp. 811–814

Zwicker E.: The acoustic input impedance of the external ear. NTZ-CJ **1**, 53–60 (1962)

Zwicker E.: Die akustischen Eingangswiderstände neuer künstlicher Ohren (The acoustical input impedance of new artificial ears). NTZ **19**, 368 (1966)

Zwicker E., U.T. Zwicker: Audio engineering and psychoacoustics: Matching signals to the final receiver, the human hearing system. J. Audio Eng. Soc. **39**, 115–126 (1991)

Speech Recognition

Anke D., P. Hoeschele: Einfache Erkennungsgeräte für die gesprochenen Zahlen Null bis Neun (Simple equipment for recognition of spoken numbers between zero and nine). Kybernetik **4**, 228–234 (1968)

Burghardt H., H. Heß: Statistische Untersuchungen der Nulldurchgangs- und Extremwertintervalle zur Unterscheidung von Vokalen (Statistical investigation of the intervals of zero-crossings or extrema for differentiating vowels). NTZ **24**, 389–393 (1971)

Daxer W.: Zweckmäßige Dimensionierung der Vorverarbeitung bei mikroprozessorgesteuerten Erkennungssystemen für isolierte Worte (Meaningful implementation of the preprocessing for recognition systems of isolated words using microprocessors). In *Fortschritte der Akustik DAGA '81* (VDE, Berlin 1981) pp. 641–644

Daxer W., E. Zwicker: On-line isolated word recognition using a microprocessor system. Speech Communication **1**, 21–27 (1982)

Fastl H.: Speech intelligibility tests with a vocoder based on the hearing sensation sharpness. Acustica **51**, 99–102 (1982)

Knebel H.: Extraktion sprachbeschreibender Parameter aus Lautheits-Tonheits-Mustern (Extraction of speech relevant parameters form the loudness critical-band rate patterns). In *Fortschritte der Akustik, DAGA '80* (VDE, Berlin 1980) pp. 671–674

Knebel H.: Ein schärfeorientiertes Vocodersystem (A sharpness-oriented vocoder-system). In *Fortschritte der Akustik, DAGA '81*, (VDE, Berlin 1981) pp. 645–648

Köhlmann M.: Sprachsegmentierung mit Hilfe der Rhythmuswahrnehmung (Segmentation of speech using the perception of rhythm). In *Fortschritte der Akustik, FASE/DAGA '82* (DPG, Bad Honnef 1982) pp. 903–906

Köhlmann M.: Bestimmung der Silbenstruktur von fließender Sprache mit Hilfe der Rhythmuswahrnehmung (Determination of the syllable-structure of running speech using the percept of rhythm). Acustica **56**, 120–125 (1984)

Kunert F.: Messungen zur intra- und interindividuellen Vokalvarianz und ihre Repräsentation in Lautheitsmustern (Representation of the variance of vowels in loudness patterns). In: *Fortschritte der Akustik, DAGA '90*, Verl.: DPG-GmbH, Bad Honnef, pp. 1103–1106 (1990)

Mummert M.: Trennung von tonalen und geräuschhaften Anteilen im Sprachsignal (Dividing tone and noise-quality in speech signals). In *Fortschritte der Akustik, DAGA '90*, Verl.: DPG-GmbH, Bad Honnef, pp. 1047–1050 (1990)

Schlang M.F., M. Mummert: Die Bedeutung der Fensterfunktion für die Fourier-t-Transformation als gehörgerechte Spektralanalyse (Search for an optimal window for the Fourier-t-Transformation). In *Fortschritte der Akustik, DAGA '90* (DPG, Bad Honnef 1990)

Terhardt E.: Beitrag zur automatischen Erkennung gesprochener Ziffern (A contribution to automatic recognition of spoken numbers). Kybernetik **3**, 136–143 (1966)

Terhardt E.: Was ist automatische Spracherkennung? (What is automatic speech recognition?) Elektron. Rechenanl. **20**, 178–186 (1978)

Terhardt E.: Sprachgrundfrequenzextraktion nach Prinzipien der Tonhöhenwahrnehmung (Extraction of fundamental frequency of speech using the principles of pitch perception). In *Fortschritte der Akustik, DAGA '80* (VDE, Berlin 1980) pp. 667–670

Terhardt E.: Sprachparameter in der Hörwahrnehmung (Parameters of speech sounds in hearing percepts). In *Interaktion zwischen Artikulation und akustischer Perzeption,* ed. by M. Spreng (Thieme, Stuttgart 1980) pp. 79–86

Terhardt E.: Aspekte und Möglichkeiten der gehörbezogenen Schallanalyse und -bewertung (Aspects and possibilities of the hearing equivalent sound analysis und evaluation). In *Fortschritte der Akustik, DAGA '81* (VDE, Berlin 1981) pp. 99–110

Zollner M.: Verständlichkeit der Sprache eines einfachen Vocoders (Intelligibility of speech produced by a simple vocoder). Acustica **43**, 271–272 (1979)

Zwicker E.: Funktionsmodelle des Gehörs als Instrumente zur Lautstärkemessung und zur automatischen Spracherkennung (Models of the hearing system as instruments for measurement of loudness and for automatic speech recognition). In *Kybernetik, Brücke zwischen den Wissenschaften,* ed. by H. Frank, 7. Aufl. (Umschau, Frankfurt 1970) pp. 165–176

Zwicker E.: A program for automatic speech recognition. In *Pattern Recognition in Biological and Technical Systems* (Springer, Berlin, Heidelberg 1971) pp. 350–356

Zwicker E.: Nachbildung des Gehörs – Nützliches Hilfsmittel bei der automatischen Spracherkennung (Modelling the hearing system - useful aid for automatic speech recognition). In *Kybernetik und Bionik,* (Oldenbourg, München 1974) pp. 316–323

Zwicker E.: Peripheral preprocessing in hearing and psychoacoustics as guidelines for speech recognition. In *Units and their Representation in Speech Recognition,* Proc. Montreal Symposium on Speech Recognition, Canadian Acoustical Association (1986)

Zwicker E., W. Daxer: Automatische Echtzeit-Erkennung von 14 isoliert gesprochenen Worten in einem kompakten Gerät mit Mikroprozessor (Automatic online-recognition of 14 isolated spoken words in a compact unit using microprocessors). In *Fortschritte der Akustik, DAGA '80,* (VDE, Berlin, Heidelberg 1980) pp. 731–734

Zwicker E., W. Hess, E. Terhardt: Erkennung gesprochener Zahlworte mit Funktionsmodell und elektronischer Rechenanlage (Recognition of spoken numbers using a model of hearing and an electronic computer). Kybernetik **3**, 267–272 (1967)

Zwicker E., W. Hess, E. Terhardt: Automatische Erkennung gesprochener Zahlwörter (Automatic recognition of spoken numbers). Umschau **68**, 182 (1968)

Zwicker E., E. Terhardt, E. Paulus: Automatic speech recognition using psychoacoustic models. J. Acoust. Soc. Am. **65**, 487–498 (1979)

Applications in Musical Acoustics

Baumann U.: Akustische Untersuchungen an einer Kirchenorgel (Acoustic investigations of a pipe organ). In *Fortschritte der Akustik, DAGA '90* (DPG, Bad Honnef 1990), pp. 541–544 (1990)

Fastl H.: Schwankungsstärke, Rhythmus und zeitliche Hüllkurve von Musikausschnitten (Fluctuation strength, rhythm, and temporal envelope of pieces of music). In *13. Tonmeistertagung München '84* (Bildungswerk Verband Deutscher Tonmeister, Berlin 1984) pp. 337–345

Fastl H.: Gehörbezogene Lautstärkemeßverfahren in der Musik (Hearing equivalent procedures of loudness measurements in music). Das Orchester **38**, 1–6 (1990)

Fastl H., H. Fleischer: Über die Ausgeprägtheit der Tonhöhe von Paukenklängen (On the pitch strength of timpani). In: *Fortschritte der Akustik, DAGA '92*, Verl.: DPG-GmbH, Bad Honnef, pp. 237–240 (1992)

Fleischer H., H. Fastl: Untersuchungen an Konzertpauken (Investigations of timpani). In: *Fortschritte der Akustik, DAGA '91*, Verl.: DPG-GmbH, Bad Honnef, pp. 885–888 (1991)

Pfaffelhuber K.: Messung des dynamischen Verhaltens einer Saite (Measurement of the dynamic behaviour of a string). In *Fortschritte der Akustik, DAGA '90* (DPG, Bad Honnef 1990) pp. 563–566 (1990)

Seewann M., E. Terhardt: Messungen der wahrgenommenen Tonhöhe von Glocken (Measurements of the perceived pitch of bells). In *Fortschritte der Akustik, DAGA '80* (VDE, Berlin 1980) pp. 635–638

Stoll G., R. Parncutt: Harmonic relationship in similarity judgements of nonsimultaneous complex tones. Acustica **63**, 111–119 (1987)

Terhardt E.: Oktavspreizung und Tonhöhenverschiebung bei Sinustönen (Octave spread and pitch shift using sinusoidal tones). Acustica **22**, 345–351 (1970)

Terhardt E.: Tonhöhenwahrnehmung und harmonisches Empfinden (Pitch perception and the sensation of harmony). In *Akustik und Schwingungstechnik* (VDE Berlin 1972) pp. 59–68

Terhardt E.: Pitch, consonance, and harmony. J. Acoust. Soc. Am. **55**, 1061–1069 (1974)

Terhardt E.: Die Stimmung von Tasteninstrumenten (On the tuning of keyboard instruments). Instrumentenbau **29**, 361–362 (1975)

Terhardt E.: Die Helmholtz'sche Theorie der musikalischen Konsonanz: Mißverständnisse, Ergänzungen, Korrekturen (Helmholtz' theory of musical consonance: misunderstandings, supplements, and corrections). In *Fortschritte der Akustik, DAGA '76* (VDI, Düsseldorf 1976) pp. 593–596

Terhardt E.: Ein psychoakustisch begründetes Konzept der musikalischen Konsonanz (A psycho acoustically based concept of musical consonance). Acustica **36**, 121–137 (1976)

Terhardt E.: The two-component theory of musical consonance. In *Psychophysics and Physiology of Hearing*, ed. by E.F. Evans, J.P. Wilson (Academic, London 1977) pp. 381–390

Terhardt E.: Psychoacoustic evaluation of musical sounds. Perception & Psychophysics **23**, 483–492 (1978)

Terhardt E.: Conceptual aspects of musical tones. The Human. Assoc. Review **30**, 46–57 (1979)

Terhardt E.: Die psychoakustischen Grundlagen der musikalischen Akkordgrundtöne und deren algorithmische Bestimmung (The psychoacoustical fundaments of the musical root of chords and its determination by algorithms). Tiefenstruktur der Musik, ed. by C. Dahlhaus, M. Krause (TU, Berlin 1982) pp. 23–50

Terhardt E.: Music perception and elementary auditory sensations. Audiol. Acoustics **22**, 52–56; 86–96 (1983)

Terhardt E.: The concept of musical consonance: A link between music and psychoacoustics. Music Perception **1**, 276–295 (1984)

Terhardt E.: Some psycho-physical analogies between speech and music. In *Music in Medicine*, ed. by R. Spintge, R. Droh (Mayr, Miesbach 1985) pp. 89–102

Terhardt E.: Gestalt principles and music perception. In *Auditory Processing of Complex Sounds*, ed. by W.A. Yost et. al. (Erlbaum, Hillsdale 1986) pp. 157–166

Terhardt E.: Methodische Grundlagen der Musiktheorie (Methodic basis of music theory). Musicologica Austriaca **6**, 107–126 (1986)

Terhardt E.: Psychophysikalische Grundlagen der Beurteilung musikalischer Klänge (Psychophysical basis of the judgements of musical sounds). In *Qualitätsaspekte bei Musikinstrumenten*, ed. by J. Meyer (Ed. Moeck, Celle, No. 4044, 1988) pp. 9–22

Terhardt E.: Physiophysical principles of musical sound evaluation. In *28th Acoust. Conference*, Strbske Pleso-High Tatras, ed. by C. Goralikova (Dom techniky, Bratislava 1989) pp. 42–50

Terhardt E.: Characteristics of musical tones in relation to auditory acquisition of information. In Proc. of 1st Int'l Conf. on Music Perception and Cognition, ed. by T. Umemoto (Dept. of Music, Kyoto Univ., Kyoto 1989) pp. 191–196

Terhardt E.: A system theory approach to musical stringed instruments: Dynamic behaviour of a string at point of excitation. Acustica **70**, 179–188 (1990)

Terhardt E., H. Fastl, H.P. Haller, J. Meyer, K. Wogram: Stand und Entwicklung der musikalischen Akustik (Review and development of musical acoustics). Umschau **81**, 71–76 (1981)

Terhardt E., A. Grubert: Zur Erklärung des "Tritonus-Paradoxons" (An explanation of the "Tritonusparadoxon"). In *Fortschritte der Akustik, DAGA '88* (DPG, Bad Honnef 1988) pp. 717–720

Terhardt E., T. Horn, K. Pfaffelhuber: Untersuchungen zum Anstreichvorgang von Saiten (Investigations on bowing of strings). In *Fortschritte der Akustik, DAGA '90* (DPG, Bad Honnef 1990) pp. 567–570 (1990)

Terhardt E., M. Seewann: Tonartenidentifikation kurzer Musikdarbietungen (Determination of musical key of short pieces of music). In *Fortschritte der Akustik, FASE/DAGA '82* (DPG, Bad Honnef 1982) pp. 879–882

Terhardt E., M. Seewann: Aural key identification and its relationship to absolute pitch. Music Perception **1**, 63–83 (1983)

Terhardt E., M. Seewann: Auditive und objektive Bestimmung der Schlagtonhöhe von historischen Kirchenglocken (Auditive and objective determination of the pitch of historic church bells). Acustica **54**, 129–144 (1984)

Terhardt E., M. Seewann: Der „akustische Bass" von Orgeln (The "acoustic bass" of organs). In *Fortschritte der Akustik, DAGA '84* (DPG, Bad Honnef 1984) pp. 885–888

Terhardt E., G. Stoll, R. Schermbach, R. Parncutt: Tonhöhenmehrdeutigkeit, Tonverwandtschaft und Identifikation von Sukzessivintervallen (Pitchambiguity, pitch relationship, and identification of successive intervals). Acustica **61**, 57–66 (1986)

Terhardt E., W.D. Ward: Recognition of musical key: Exploratory study. J. Acoust. Soc. Am. **72**, 26–33 (1982)

Terhardt E., M. Zick: Evaluation of the tempered tone scale in normal, stretched, and contracted intonation. Acustica **32**, 268–274 (1975)

Walliser K: Über die Spreizung von empfundenen Intervallen gegenüber mathematisch harmonischen Intervallen bei Sinustönen (On the spread of perceived intervals compared with the mathematically harmonic intervals using sinusoidal tones). Frequenz **23**, 139–143 (1969)

Zwicker E.: Unmasked and partially masked loudness in musical dynamics. J. Acoust. Soc. Am. **85**, Suppl. 1, 140 (1989)

Zwicker E., W. Spindler: Über den Einfluß nichtlinearer Verzerrungen auf die Hörbarkeit des Frequenzvibrators (On the influence of nonlinear distortions on the audibility of frequency modulation). Acustica **3**, 100–104 (1953)

Applications in Room Acoustics

Fastl H.: Reverberation and post-masking. In Proc. FASE '78, Vol.III (1978) pp. 37–40

Fastl H., H. Frisch, E. Zwicker: Schwankungsstärke und Sprachverständlichkeit in Räumen (Fluctuation strength and speech intelligibility in rooms). In *Fortschritte der Akustik, DAGA '88* (DPG, Bad Honnef 1988) pp. 693–696

Fastl H., E. Zwicker, R. Fleischer: Beurteilung der Verbesserung der Sprachverständlichkeit in einem Hörsaal (Evaluation of the improvement in speech intelligibility in a lecture hall). Acustica **71**, 287–292 (1990)

Terhardt E.: Physiologische Aspekte des Hörens in Räumen (Physiological aspects of listening in rooms). In *Räume zum Hören*, arcus **6**, ed. by R. Müller (R. Müller, Köln 1989) pp. 16–23

Subject Index

Springer Series in Information Sciences

Editors: Thomas S. Huang   Teuvo Kohonen   Manfred R. Schroeder

Managing Editor: H. K. V. Lotsch

Printing: Mercedes-Druck, Berlin
Binding: Stürtz AG, Würzburg